UNITEXT for Physics

UNITEXT for Physics series, formerly UNITEXT Collana di Fisica e Astronomia, publishes textbooks and monographs in Physics and Astronomy, mainly in English language, characterized of a didactic style and comprehensiveness. The books published in UNITEXT for Physics series are addressed to upper undergraduate and graduate students, but also to scientists and researchers as important resources for their education, knowledge and teaching.

More information about this series at http://www.springer.com/series/13351

Nicola Cufaro Petroni

Probability and Stochastic Processes for Physicists

Springer

Nicola Cufaro Petroni
Department of Physics
University of Bari
Bari, Italy

ISSN 2198-7882 ISSN 2198-7890 (electronic)
UNITEXT for Physics
ISBN 978-3-030-48410-1 ISBN 978-3-030-48408-8 (eBook)
https://doi.org/10.1007/978-3-030-48408-8

This Springer imprint is published by the registered company Springer Nature Switzerland AG
The registered company address is: Gewerbestrasse 11, 6330 Cham, Switzerland

... Iuvat integros accedere fontis

atque haurire, iuvatque novos decerpere flores[1]

Lucretius, I 927-8

[1]Translation by A. E. Stallings (Penguin, London, 2007):

I thrill to come upon untasted springs and slake my thirst.

I joy to pluck strange flowers ...

Preface

This book originates from the full-semester lessons held yearly within the master's degree course in Physics at the University of Bari. Part I about Probability essentially starts from scratch in Chap. 1 as if nothing had been known before about this topic, and it quickly goes on to conceptually relevant, but less familiar, issues. The key feature of this part no doubt is the notion of random variables (and then also of stochastic processes) which is introduced in Chap. 3 as a paramount concept that differs from that of distribution previously elucidated in Chap. 2 . The interplay between the random variables and their distributions is a main staple of the present volume. Chapter 4 finally deals with a panoply of limit theorems that in any case represent the main gate to the world of the subsequent sections.

Part II is then devoted to Stochastic Processes that are presented in Chap. 5 from a rather general standpoint with particular attention to stationarity and ergodicity. In Chap. 6, the first two examples of classical processes (Poisson and Wiener) are heuristically introduced with a discussion of the familiar procedures used to manufacture their random trajectories; the first presentation of both the white noise and the Brownian motion is also added in. The vast topic of Markovianity is tackled in Chap. 7 with a discussion of the equations ruling the distributions of the jump-diffusion processes: this gives also the opportunity of introducing the Cauchy and Ornstein-Uhlenbeck processes, and of including a quick reference to the general topic of the Lévy processes. The equations ruling instead of the evolution of the trajectories of the diffusion processes, namely the Itō stochastic differential equations, are presented in Chap. 8 in the framework of the stochastic calculus: several examples of solutions of stochastic differential equations are also produced.

The two chapters of Part III deal finally with the modeling of particular physical problems by means of the tools acquired in the previous sections. Chapter 9 gives a broad presentation of a dynamical theory of the Brownian motion modeled mainly by the Ornstein-Uhlenbeck process and is completed with a discussion of the Smoluchowski approximation and of the Boltzmann equilibrium distributions. In Chap. 10 an account is given of the stochastic mechanics, a model leading to a deep connection between the time-reversal invariant diffusion processes and the Schrödinger equation: a topic apparently relevant for physicists and mathematicians alike. On the one hand indeed it shows, for instance, that the statistical Born rule of quantum mechanics could not be just a postulate added by hand to its formal apparatus, and on the other it also hints to wider than usual possible developments of the stochastic analysis.

A number of appendices conclude the book: a few of them are simple complements, lengthy proofs and useful reminders, but others are short discussions about particular relevant topics moved over there to avoid interrupting the main text. For instance, Appendix A deals with the possible incompatibility of legitimate joint distributions; Appendix C discusses the celebrated Bertrand paradox; Appendix H shows the baffling outcomes of a naive application of calculus to stochastic processes; and Appendix I presents the case of a process satisfying the Chapman-Kolmogorov conditions despite it being non-Markovian.

Besides its traditional purposes, one of the aims of this book is to bridge a gap often existing between physicists and mathematicians in the subject area of probability and stochastic processes: a difference usually ranging from notions to notations, from techniques to names. This, however, would arguably be an outcome hard to be achieved within the scope either of one course or even of one book, and therefore we will be satisfied if we only approached the goal of—at least partially—restoring an intellectual osmosis that has always been fruitful along the centuries.

While even this limited prize would not come in any case without a price—it usually requires some effort from the physics students to go beyond the most familiar ways of thinking—it is important to remark that the focus of the following chapters will always be on concepts and models, rather than on methods and mathematical rigor. Most important theorems, for instance, will not be skipped away, but they will be presented and clarified in their role without proofs, and the interested reader will be referred for further details to the existing literature quoted in Bibliography.

Bari, Italy
April 2020

Nicola Cufaro Petroni

Contents

Part I
Probability

Chapter 1
Probability Spaces

1.1 Samples

The first ideas of modern probability came (around the XVII century) from gambling, and we too will start from there. The simplest example is that of a flipped coin with two possible outcomes: *head* (H) and *tail* (T). When we say that the coin is *fair* we just mean that there is no reason to surmise a bias in favor of one of these two results. As a consequence H and T are *equiprobabile*, and to make quantitative this statement it is customary to assign a ***probability*** as a fraction of the unit so that in our example

$$p = \boldsymbol{P}\{H\} = \frac{1}{2} \qquad q = \boldsymbol{P}\{T\} = \frac{1}{2}$$

Remark that $p + q = 1$, meaning that *with certainty* (namely with probability 1) either H or T shows up, and that there are no other possible outcomes. In a similar way for a *fair* dice with six sides labeled as I, II, ..., VI we have

$$p_1 = \boldsymbol{P}\{I\} = \frac{1}{6}\,; \quad \dots \quad ; \; p_6 = \boldsymbol{P}\{VI\} = \frac{1}{6}$$

Of course we still find $p_1 + \cdots + p_6 = 1$.

From these examples a first idea comes to the fore: at least in the elementary cases, we allot probabilities by simple *counting*, a protocol known as ***classical definition*** (see more later) providing the probability of some statement A about a random experiment. For instance in a dice throw let A be "*a side with an even number comes out*": in this case we instinctively add up the probabilities of the outcomes corresponding to A; in other words we count both the *possible* equiprobable results, and the results *favorable* to the event A, and we assign the probability

$$\boldsymbol{P}\{A\} = \frac{\text{number of favorable results}}{\text{number of possible results}}$$

N. Cufaro Petroni, *Probability and Stochastic Processes for Physicists*, UNITEXT for Physics, https://doi.org/10.1007/978-3-030-48408-8_1

Remark that, as before, this probability turns out to be a number between 0 and 1. In short, if in a fair dice throw we take A = *"a side with an even number comes out"*, B = *"a side with a multiple of 3 comes out"*, and C = *"a side different from VI comes out"*, a simple counting entails that, with 6 equiprobable results, and 3, 2 and 5 favorable results respectively for A, B and C, we get

$$\boldsymbol{P}\{A\} = \frac{1}{2}\,, \qquad \boldsymbol{P}\{B\} = \frac{1}{3}\,, \qquad \boldsymbol{P}\{C\} = \frac{5}{6}$$

When instead we throw *two* fair dices the possible result are 36, namely the *ordered* pairs (n, m) with n and m taking the 6 values I, ..., VI. The fairness hypothesis means now that the 36 elementary events (I, I) ; (I, II) ; ... ; (VI, VI) are again equiprobable so that for every pair

$$\boldsymbol{P}\{I, I\} = \frac{1}{36}\,, \quad \boldsymbol{P}\{I, II\} = \frac{1}{36}\,, \quad \dots \quad ; \ \boldsymbol{P}\{VI, VI\} = \frac{1}{36}$$

We then find by counting that A = *"the pair (VI, VI) fails to appear"* comes out with the probability

$$\boldsymbol{P}\{A\} = \frac{35}{36}$$

From the previous discussion it follows that the probability of a random event will be a number between 0 and 1: 1 meaning the certainty of its occurrence and 0 its impossibility while the intermediate values represent all the other cases. These assignments also allow (at least in the simplest cases) to calculate the probabilities of more complicated events by counting equiprobable results. It is apparent then the relevance of preliminarily determining *the set of all the possible results of the experiment*, but it is also clear that this direct calculation method becomes quickly impractical when the number of such results grows beyond a reasonable limit. For example, the possible sequences (without repetition) of the 52 cards of a French card deck are

$$52 \cdot 51 \cdot 50 \cdot \ldots \cdot 2 \cdot 1 = 52! \simeq 8 \cdot 10^{67}$$

a huge number making vain every hope of solving problems by direct counting

Definition 1.1 A **sample space** Ω is the set (either finite or infinite) of all the possible results ω of an experiment.

Remark that Ω is not necessarily a set of numbers: its elements *can be* numbers, but in general they are of an arbitrary nature. In our previous examples the sample space $\Omega = \{\omega_1, \omega_2, \ldots, \omega_N\}$ was *finite* with cardinality N: for a coin it has just two elements

$$\Omega = \{H, T\}\,; \qquad N = 2$$

while for a dice

$$\Omega = \{I, II, \ldots, VI\}\,; \qquad N = 6$$

When instead our experiment consists in two coin flips (or equivalently in one flip of two coins) we have

$$\Omega = \{HH,\ HT,\ TH,\ TT\}\,; \qquad N = 4$$

while for n flips

$$\Omega = \{\omega = (a_1\,,\ldots,\,a_n)\ :\ a_i = \text{either } H \text{ or } T\}\,; \qquad N = 2^n$$

The most relevant instances of *infinite* spaces on the other hand are the sets of the integer numbers N, of the real numbers $\boldsymbol{R}$, of the n-tuples of real numbers $\boldsymbol{R}^n$, of the sequences of real numbers $\boldsymbol{R}^\infty$, and finally $\boldsymbol{R}^T$ the set of the real functions from T to $\boldsymbol{R}$. Of course in the case of finite sample spaces Ω—where we can think of adopting the *classical definition*—it would be paramount to know first its cardinality N as in the following examples.

Example 1.2 Take a box containing M numbered (distinguishable) balls and sequentially draw n balls by replacing them in the box after every extraction: we call it a **sampling with replacement**. By recording the extracted numbers we get that the possible results of the experiment are $\omega = (a_1,\ a_2, \ldots a_n)$ with $a_i = 1,\ 2, \ldots M$ and $i = 1, 2, \ldots n$, and possibly **with repetitions**. The sample spaces—the set of our n-tuples—can now be of two kinds:

1. **ordered samples** $(a_1\,,\ldots,\,a_n)$: the samples are deemed different even just for the order of the extracted labels and are called **dispositions**; for example with $n = 4$ extractions, the sample $(4\,,1\,,2\,,1)$ is considered different from $(1\,,1\,,2\,,4)$; it is easy to find then that the cardinality of Ω is in this case

$$N_d = M^n$$

2. **non-ordered samples** $[a_1\,,\ldots,\,a_n]$: in this case the samples $(4\,,1\,,2\,,1)$ and $(1\,,1\,,2\,,4)$ coincide so that the number of the elements of Ω, called **partitions**, is now smaller than the previous one, and it is possible to show (see [1] p. 6) that

$$N_r = \binom{M+n-1}{n} = \frac{(M+n-1)!}{n!\,(M-1)!}$$

When instead we draw the balls without replacing them in the box we will have a **sampling without replacement**. Apparently in this case the samples $(a_1, \ldots, a_n)$ will exhibit only different labels (**without repetitions**), and $n \le M$ because we can not draw a number of balls larger than the initial box content. Here too we must itemize two sample spaces:

1. **ordered samples** $(a_1\,,\ldots,\,a_n)$: the so called **permutations** of M objects on n places; their number is now

$$N_p = (M)_n = \frac{M!}{(M-n)!} = M(M-1)\dots(M-n+1)$$

because with every draw we leave a chance less for the subsequent extractions. Remark that if $n = M$, we have $N = M!$ namely the number of the permutations of M objects on M places.

2. **non-ordered samples** $[a_1, \dots, a_n]$: we have now the **combinations** of M objects on n places, and their number is

$$N_c = \binom{M}{n} = \frac{M!}{n!(M-n)!} \tag{1.1}$$

because every non-ordered sample $[a_1, \dots, a_n]$ allows $n!$ permutations of its labels (see the previous remark), and then $N_c \cdot n! = N_p$ leading to the required result.

We started this section from the equiprobable elements ω of a sample space Ω, and went on to calculate the probabilities of more complicated instances by sums and counting (classical definition). We also remarked however that this course of action is rather impractical for large Ω, and is utterly inapplicable for uncountable spaces. We will need then further ideas in order to be able to move around these obstacles.

1.2 Events

We already remarked that a subset $A \subseteq \Omega$ represents a statement about the results of a random experiment. For instance, in the case of three coin flips the sample space consists of $N = 2^3 = 8$ elements

$$\Omega = \{HHH,\ HHT, \dots,\ TTT\}$$

and the subset

$$A = \{HHH,\ HHT,\ HTH,\ THH\} \subseteq \Omega$$

stands for the statement "*H comes out at least twice on three flips*". In the following we will call ***events*** the subsets $A \subseteq \Omega$, and we will say that the event A *happens* if the result ω of the experiment belongs to A, namely if $\omega \in A$.

In short the events are a family of *propositions* and the operations among events (as set operations) are a model for the *logical connectives* among propositions. For instance the connectives *OR* and *AND* correspond to the operations *union* and *intersection*:

$$A \cup B = \{\omega : \omega \in A, \text{ OR } \omega \in B\}$$
$$A \cap B = A\,B = \{\omega : \omega \in A, \text{ AND } \omega \in B\}\,.$$

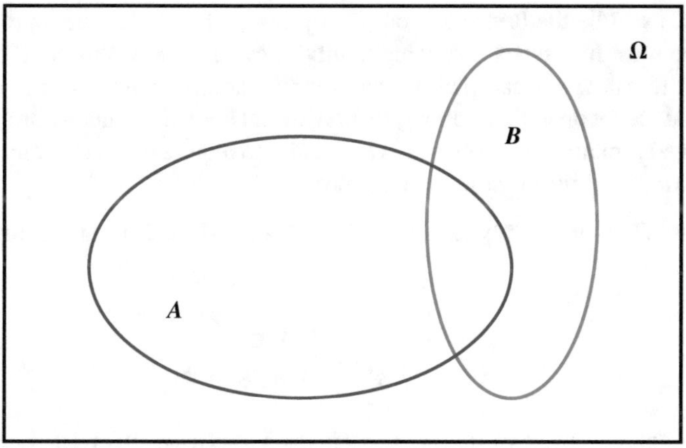

Fig. 1.1 Venn diagrams

The logical meaning of the following operators on the other hand is illustrated by the *Venn diagrams* in the Fig. 1.1:

$$\overline{A} = \{\omega\, :\omega \notin A\}\,;$$
$$A\backslash B = A \cap \overline{B} = \{\omega\, :\omega \in A\,,\ \text{but}\ \omega \notin B\}\,;$$
$$A\triangle B = (A\backslash B)\cup(B\backslash A) \qquad \text{(symmetric difference)}$$

In this context of course Ω is the *sure (certain)* event, since every result ω belongs to Ω, while $\emptyset$ is the *impossible* event since no result belongs to it. We will also say that A e B are *disjoint* or *incompatible* when$A \cap B = \emptyset$. It is apparent that the properties of the set operations replicate the properties of the logical operations as for instance in the identities

$$\overline{A \cap B} = \overline{A} \cup \overline{B}\,; \quad \overline{A \cup B} = \overline{A} \cap \overline{B}$$

known as *de Morgan laws*. As a consequence for two coin flips, from the events

$$A = \{HH,\ HT,\ TH\} = H\ \textit{comes out at least once}$$
$$B = \{HT,\ TH,\ TT\} = T\ \textit{comes out at least once}$$

we can produce other events as for example

$$A \cup B = \{HH,\ HT,\ TH,\ TT\} = \Omega\,, \qquad A \cap B = \{HT,\ TH\}\,, \quad A\backslash B = \{HH\}$$

We should remark at once, however, that our family of events in Ω will *not necessarily coincide* with the collection $\wp(\Omega)$ of all the subsets of Ω. Typically we choose a particular sub-collection of parts of Ω, as for instance when $\Omega = \boldsymbol{R}$ since $\wp(\boldsymbol{R})$ would include also pathological sets that are practically irrelevant. Only when Ω is finite $\wp(\Omega)$ will be the best selection. In any case, when we define a probabilistic model, me must first select a suitable family of events, and we must stop here for an instant to ask if we can pick up an arbitrary family or not. In fact, since we require that the set operations among events (logical operations among propositions) produce again acknowledged events, we should also require that our family of events be closed under all the possible set operations.

Definition 1.3 A non empty family $\mathcal{F} \subseteq \wp(\Omega)$ of parts of Ω is an **algebra** when

$$\begin{aligned} &\Omega \in \mathcal{F} \\ &\overline{A} \in \mathcal{F}, \qquad \forall A \in \mathcal{F} \\ &A \cap B \in \mathcal{F}, \qquad \forall A, B \in \mathcal{F} \end{aligned}$$

Moreover $\mathcal{F}$ is a σ-**algebra** if it is an algebra and also meets the further condition

$$\bigcap_n A_n \in \mathcal{F} \qquad \forall (A_n)_{n\in N} \text{ of elements of } \mathcal{F}$$

Proposition 1.4 *If $\mathcal{F}$ is a σ-algebra, then*

$$\begin{aligned} \emptyset &\in \mathcal{F} \\ A \cup B &\in \mathcal{F} \qquad \forall A, B \in \mathcal{F} \\ A \backslash B &\in \mathcal{F} \qquad \forall A, B \in \mathcal{F} \\ \bigcup_n A_n &\in \mathcal{F} \qquad \forall (A_n)_{n\in N} \textit{ of elements of } \mathcal{F} \end{aligned}$$

Proof Omitted: see [1] p. 132–3. ■

In short a σ-algebra is a family of parts of Ω closed under finite or countable set operations, but not necessarily under uncountable operations on event collections $(A_t)_{t\in T}$ where T is uncountable. From now on we will always suppose that our events constitute a σ-algebra $\mathcal{F}$, and we will also call ***probabilizable space*** every pair $(\Omega, \mathcal{F})$ where $\mathcal{F}$ is a σ-algebra of events of Ω. For a given Ω the simplest examples of σ-algebras are

$$\begin{aligned} &\mathcal{F}_* = \{\emptyset, \Omega\}, \\ &\mathcal{F}_A = \{A, \overline{A}, \emptyset, \Omega\}, \qquad (A \subseteq \Omega), \\ &\mathcal{F}^* = \wp(\Omega). \end{aligned}$$

In particular the σ-algebra $\mathcal{F}_A$ is called σ-***algebra generated*** by A and can be generalized as follows: for a given family $\mathcal{E} \subseteq \wp(\Omega)$ of parts of Ω, we will call σ-algebra generated by $\mathcal{E}$ the smallest σ-algebra $\sigma(\mathcal{E})$ containing $\mathcal{E}$.

Proposition 1.5 *Given a family $\mathcal{E} \subseteq \wp(\Omega)$ of parts Ω, the σ-algebra $\sigma(\mathcal{E})$ generated by $\mathcal{E}$ always exists*

Proof Omitted: see [1] p. 140. ■

Definition 1.6 A (finite or countable) family of subsets $\mathcal{D} = \{D_1, D_2 \dots\}$ is called a **decomposition** of Ω in the **atoms** D_k, if the D_k are non empty, disjoint parts of Ω such that $\bigcup_k D_k = \Omega$.

The decompositions are indeed families of events such that always one, and only one of them occurs, and hence are the models of mutually exclusive events that exhaust all possibilities. Apparently, however, a decomposition is neither a σ-algebra, nor an algebra: it does not contain, for one thing, the unions of atoms. However, based on the Proposition 1.5, from a decomposition $\mathcal{D}$ we will always be able to provide the generated σ-algebra $\mathcal{F} = \sigma(\mathcal{D})$, the simplest example being the decomposition

$$\mathcal{D}_A = \{A, \overline{A}\}$$

supplying the generated σ-algebra $\mathcal{F}_A$.

Example 1.7 We will now quickly discuss a few examples of relevant σ-algebras (for details see [1] p. 143–151):

1. When Ω coincides with the set $\boldsymbol{R}$ of real numbers take first the family $\mathcal{I}$ of (both bounded and unbounded) right-closed intervals

$$I = (a, b], \qquad -\infty \le a < b \le +\infty$$

namely intervals of the type

$$(a, b], \quad (-\infty, b], \quad (a, +\infty), \quad (-\infty, +\infty)$$

with $a, b \in \boldsymbol{R}$ (right-unbounded intervals will conventionally considered right-closed). Since interval unions are not in general intervals, $\mathcal{I}$ is neither a σ-algebra, nor an algebra. Dropping the analytical details, take then the σ-algebra generated by $\mathcal{I}$ denoted as $\mathcal{B}(\boldsymbol{R})$, and called **Borel σ-algebra** of $\boldsymbol{R}$, while its elements will be called **Borel sets** of $\boldsymbol{R}$. The σ-algebra $\mathcal{B}(\boldsymbol{R})$ contains all the $\boldsymbol{R}$ subsets of the type

$$\emptyset, \quad \{a\}, \quad [a, b], \quad [a, b), \quad (a, b], \quad (a, b), \quad \boldsymbol{R}$$

along with their (both countable and uncountable) unions and intersections. As a matter of fact the same σ-algebra can be generated by different families of subsets, notably by that of the open sets of $\boldsymbol{R}$. The corresponding probabilizable space will denoted as

$$(\boldsymbol{R}, \mathcal{B}(\boldsymbol{R}))$$

2. Consider now the case $\Omega = \boldsymbol{R}^n$ of the Cartesian product of n real lines: its elements will now be the n-tuples of real numbers $\omega = \boldsymbol{x} = \{x_1, x_2, \dots, x_n\}$. As in the previous example, here too there are several equivalent procedures

to produce a suitable σ-algebra that can consequently be seen as generated by the open sets of $\boldsymbol{R}^n$: this too will be called **Borel σ-algebra** of $\boldsymbol{R}^n$ and will be denoted as $\mathcal{B}(\boldsymbol{R}^n)$, so that the probabilizable space will be

$$\big(\boldsymbol{R}^n,\ \mathcal{B}(\boldsymbol{R}^n)\big)$$

3. If $(\boldsymbol{R}_n)_{n\in N}$ is a sequence of real lines, the Cartesian product $\boldsymbol{R}^\infty = \boldsymbol{R}_1 \times \ldots \times \boldsymbol{R}_n \times \ldots$ will be the set of the sequences of real numbers with $\omega = x = (x_n)_{n\in N}$. In this case we start from the subsets of $\boldsymbol{R}^\infty$ called **cylinders** and consisting of the sequences $(x_n)_{n\in N}$ such that a finite number m of their components—say $(x_{n_1}, x_{n_2}, \ldots, x_{n_m})$—belong to a Borel set $B \in \mathcal{B}(\boldsymbol{R}^m)$ called **cylinder base**. Since these bases are finite-dimensional, from the results of the previous examples it is possible to produce cylinder families that generate a σ-algebra denoted as $\mathcal{B}(\boldsymbol{R}^\infty)$ and again called **Borel σ-algebra** of $\boldsymbol{R}^\infty$. The corresponding probabilizable space then is

$$\big(\boldsymbol{R}^\infty,\ \mathcal{B}(\boldsymbol{R}^\infty)\big)$$

and it is possible to show that the all following subsets belong to $\mathcal{B}(\boldsymbol{R}^\infty)$

$$\begin{aligned}
&\{x \in \boldsymbol{R}^\infty \ :\ \sup_n x_n > a\}\,,\ \ \{x \in \boldsymbol{R}^\infty \ :\ \inf_n x_n < a\}\,,\\
&\{x \in \boldsymbol{R}^\infty \ :\ \underline{\lim}_n\, x_n \le a\}\,,\ \ \{x \in \boldsymbol{R}^\infty \ :\ \overline{\lim}_n\, x_n > a\}\,,\\
&\{x \in \boldsymbol{R}^\infty \ :\ x \text{ converges}\}\,,\ \ \{x \in \boldsymbol{R}^\infty \ :\ \lim_n x_n > a\}\,,
\end{aligned}$$

4. Take finally the set $\boldsymbol{R}^T$ of the functions defined on a (generally uncountable) subset T of $\boldsymbol{R}$, and denote its elements ω in one of the following ways: $x,\ x(\,\cdot\,),\ x(t),\ (x_t)_{t\in T}$. Following the previous procedures, consider first the cylinders with (finite- or at most countably infinite-dimensional) base B: these consists of the functions that in a (at most countable) set of points t_j take values belonging B. Build then the σ-algebra generated by such cylinders, denoted $\mathcal{B}(\boldsymbol{R}^T)$, and take

$$\big(\boldsymbol{R}^T,\ \mathcal{B}(\boldsymbol{R}^T)\big)$$

as the probabilizable space. It is possible to prove, however, that $\mathcal{B}(\boldsymbol{R}^T)$ exactly coincides with the set of cylinders with finite- or at most countably infinite-dimensional bases, namely with the family of parts of $\boldsymbol{R}^T$ singled out through restrictions on $(x_t)_{t\in T}$ on an at most countable set of points t_j. As a consequence several $\boldsymbol{R}^T$ subsets—looking at the behavior of $(x_t)_{t\in T}$ in an uncountable set of points t—do not belong to $\mathcal{B}(\boldsymbol{R}^T)$: for example, with $T = [0, 1]$, the sets (all relevant for our purposes)

$$\begin{aligned}
A_1 &= \Big\{x \in \boldsymbol{R}^{[0,1]} \ :\ \sup_{t\in[0,1]} x_t < a,\ a \in \boldsymbol{R}\Big\}\\
A_2 &= \Big\{x \in \boldsymbol{R}^{[0,1]} \ :\ \exists t \in [0, 1] \ni' x_t = 0\Big\}\\
A_3 &= \Big\{x \in \boldsymbol{R}^{[0,1]} \ :\ x_t \text{ is continuous in } t_0 \in [0, 1]\Big\}
\end{aligned}$$

do not belong to $\mathcal{B}(\boldsymbol{R}^{[0,1]})$. In order to circumvent this hurdle it is customary to restrict our starting set $\boldsymbol{R}^T$. For instance a suitable σ-algebra $\mathcal{B}(C)$ can be assembled beginning with the set C of the **continuous functions** $x(t)$: in so doing the previous subsets A_1, A_2 ed A_3 all will turn out to belong to $\mathcal{B}(C)$. We will neglect however the details of this approach (see [1] p. 150).

1.3 Probability

For finite probabilizable spaces $(\Omega, \mathcal{F})$, with Ω of cardinality N, a *probability* can be defined by attributing a number $p(\omega_k)$ at every $\omega_k \in \Omega$ so that

$$0 \le p(\omega_k) \le 1\,, \quad k = 1, \dots, N\,; \qquad \sum_{k=1}^{N} p(\omega_k) = 1$$

The probability of an event $A \in \mathcal{F}$ then is

$$\boldsymbol{P}\{A\} = \sum_{\omega_k \in A} p(\omega_k)$$

In this case the triple $(\Omega,\ \mathcal{F},\ \boldsymbol{P})$ is called a ***finite probability space***. We will delay to the general setting a discussion of the usual properties of such a probability, for instance

$$\begin{aligned}
\boldsymbol{P}\{\emptyset\} &= 0 \\
\boldsymbol{P}\{\Omega\} &= 1 \\
\boldsymbol{P}\{A \cup B\} &= \boldsymbol{P}\{A\} + \boldsymbol{P}\{B\} \qquad \text{if } A \cap B = \emptyset \qquad (additivity) \\
\boldsymbol{P}\{\overline{A}\} &= 1 - \boldsymbol{P}\{A\} \\
\boldsymbol{P}\{A\} &\le \boldsymbol{P}\{B\} \qquad \text{if } A \subseteq B
\end{aligned}$$

This definition can be extended (with some care about the convergence) to a countable Ω, but we must also remark that in every case the choice of the numbers $p(\omega_k)$ is not always a straightforward deal. The procedures to deduce the $p(\omega_k)$ from the empirical data would constitute the mission of the ***statistics*** that we will touch here only in passing, while for our examples we will adopt the ***classical definition*** already mentioned in the Sect. 1.1: we first reduce ourselves to some finite sample space of N equiprobable elements so that at every ω_k a probability $p(\omega_k) = 1/N$ is allotted, and then, if $N(A)$ is the number of samples belonging to $A \in \mathcal{F}$, we take

$$\boldsymbol{P}\{A\} = \frac{N(A)}{N}$$

Example 1.8 Coincidence problem: From a box containing M numbered balls draw with replacement an ordered sequence of n balls and record their numbers. We showed in the Sect. 1.1 that the sample space Ω consists of $N = M^n$ equiprobable elements $\omega = (a_1, \ldots, a_n)$. If then

$$A = \{\omega \ : \text{the } \ a_k \text{ are all different}\}$$

and since by simple enumeration it is

$$N(A) = M\,(M-1)\,\cdots\,(M-n+1) = (M)_n = \frac{M!}{(M-n)!}$$

from the classical definition we apparently have

$$\boldsymbol{P}\{A\} = \frac{(M)_n}{M^n} = \left(1 - \frac{1}{M}\right)\left(1 - \frac{2}{M}\right)\cdots\left(1 - \frac{n-1}{M}\right).$$

This result can be interpreted as a **birthdays problem**: for n given persons what is the probability P_n that at least two birthdays coincide? We take the ordered samples—because every different arrangement of the birthdays on n distinguishable persons is a different result—and an answer can be found from the previous discussion with $M = 365$. If indeed $\boldsymbol{P}\{A\}$ is the probability that all the birthdays differ, it is

$$P_n = 1 - \boldsymbol{P}\{A\} = 1 - \frac{(365)_n}{365^n}$$

giving rise to the rather striking results

n	4	16	22	23	40	64
P_n	0.016	0.284	0.467	0.507	0.891	0.997

It is unexpected indeed that already for $n = 23$ the coincidence probability exceeds $1/2$, and that for just 64 people this event is almost sure. Remark in particular that when $n \geq 366$ apparently we have $\boldsymbol{P}\{A\} = 0$ (namely $P_n = 1$) because a zero factor appears in the product: this agrees with the fact that for more than 365 people we surely have coincidences. On the other hand these results are less striking if we compare them with that of a slightly different question: "if I am one of the n persons of the previous problem, what is the probability P'_n that at least one birthday coincides with my birthday?" In this case $N(A) = 365 \cdot 364^{n-1}$ and hence

$$P'_n = 1 - \left(\frac{364}{365}\right)^{n-1}$$

so that now

n	4	16	22	23	40	64	101
P'_n	0.011	0.040	0.056	0.059	0.101	0.159	0.240

while P'_n never coincides with 1 (even for $n \geq 366$) since, irrespective of the number of people, it is always possible that no birthday coincides with my birthday.

Finite or countable probability models soon become inadequate because sample spaces often are ***uncountable***: it is easy to check for instance that even the well known set of all the *T-C infinite sequences* of coin flips is uncountable. In these situations a probability can not be defined by preliminarily attributing numerical weights to the individual elements of Ω. If indeed we would allot non zero weights $p(\omega) > 0$ to the elements of an uncountable set, the condition $\sum_{\omega\in\Omega} p(\omega) = 1 < +\infty$ could never be satisfied, and no coherent probability could be defined on this basis. In short, for the general case, a definition of $\boldsymbol{P}\{A\}$ can not be given by simple enumeration as in the finite (or countable) examples, but it needs the new concept of *set measure* that we will now introduce.

Definition 1.9 Given a σ-algebra $\mathcal{F}$ of parts of a set Ω, we call **measure** on $\mathcal{F}$ every **σ-additive map** $\mu : \mathcal{F} \to [0, +\infty]$, namely a map such that for every sequence $(A_n)_{n\in N}$ of disjoint elements of $\mathcal{F}$ it is

$$\mu\Big\{\bigcup_n A_n\Big\} = \sum_n \mu\{A_n\}$$

We say moreover that μ is a **finite measure** se $\mu\{\Omega\} < +\infty$, and that it is a **σ-finite measure** if Ω is decomposable in the union $\Omega = \bigcup_n A_n,\ A_n \in \mathcal{F}$ of disjoint sets with $\mu\{A_n\} < +\infty,\ \forall n \in N$. A finite measure $\boldsymbol{P}$ with $\boldsymbol{P}\{\Omega\} = 1$ is called a **probability measure**.

Definition 1.10 We say that a statement holds ***P*-almost surely** ($\boldsymbol{P}$-a.s.) if it holds for every $\omega \in \Omega$, but for a set of $\boldsymbol{P}$-measure zero. Sets of $\boldsymbol{P}$-measure zero are also called **negligible**.

Of course for every $A \in \mathcal{F}$ we always have $\mu\{A\} \leq \mu\{\Omega\}$ because μ is additive and positive, and $\Omega = A \cup \overline{A}$. This entails in particular that if μ is finite we will always have $\mu\{A\} < +\infty$. Remark that a finite measure is always σ-finite, but the converse does not hold: for example the usual ***Lebesgue measure*** on the real line, which attributes the measure $|b - a|$ to every interval $[a, b]$, apparently is σ-finite, but not finite.

Proposition 1.11 *Given a probability measure* $\boldsymbol{P} : \mathcal{F} \to [0, 1]$, *the following properties hold:*

1. $\boldsymbol{P}\{\emptyset\} = 0$
2. $\boldsymbol{P}\{A\backslash B\} = \boldsymbol{P}\{A\} - \boldsymbol{P}\{AB\}\ ,\quad \forall A, B \in \mathcal{F}$

3. $\boldsymbol{P}\{A \cup B\} = \boldsymbol{P}\{A\} + \boldsymbol{P}\{B\} - \boldsymbol{P}\{AB\}\ , \quad \forall A, B \in \mathcal{F}$
4. $\boldsymbol{P}\{A \triangle B\} = \boldsymbol{P}\{A\} + \boldsymbol{P}\{B\} - 2\boldsymbol{P}\{AB\}\ , \quad \forall A, B \in \mathcal{F}$
5. $\boldsymbol{P}\{B\} \leq \boldsymbol{P}\{A\}$ *if* $B \subseteq A$, *with* $A, B \in \mathcal{F}$
6. $\boldsymbol{P}\big\{\bigcup_n A_n\big\} \leq \sum_n \boldsymbol{P}\{A_n\}$, *for every sequence of events* $(A_n)_{n \in N}$

The last property is also known as ***subadditivity***.

Proof Omitted: see [1] p. 134. ■

Definition 1.12 Kolmogorov axioms: We call **probability space** every ordered triple $(\Omega, \mathcal{F}, \boldsymbol{P})$ where Ω is a set of elements ω also said **sample space**, $\mathcal{F}$ is a σ-**algebra** of events of Ω, and $\boldsymbol{P}$ is a **probability measure** on $\mathcal{F}$.

Remark that an *event of probability* 0 is not necessarily the empty set $\emptyset$, while an *event of probability* 1 not necessarily coincides with Ω. This is relevant—as we will see later—foremost for *uncountable spaces*, but even for finite spaces it can be useful give zero probability to some sample ω, instead of making Ω less symmetric by eliminating such samples. For instance it is often important to change the probability $\boldsymbol{P}$ on the same Ω, and in so doing the probability of some ω could vanish: it would be preposterous, however, to change Ω by eliminating these ω, and we choose in general to keep them, albeit with 0 probability. An important case of change of probability is discussed in the next section.

1.4 Conditional Probability

Definition 1.13 Given a probability space $(\Omega, \mathcal{F}, \boldsymbol{P})$ ad two events $A, B \in \mathcal{F}$ with $\boldsymbol{P}\{B\} \neq 0$, we will call **conditional probability** of A w.r.t. B

$$\boldsymbol{P}\{A|B\} \equiv \frac{\boldsymbol{P}\{A \cap B\}}{\boldsymbol{P}\{B\}} = \frac{\boldsymbol{P}\{AB\}}{\boldsymbol{P}\{B\}}$$

while $\boldsymbol{P}\{AB\}$ takes the name of **joint probability** of A and B.

Remark that for the time being the requirement $\boldsymbol{P}\{B\} \neq 0$ is crucial to have a coherent definition: we postpone to the Sect. 3.4 its extension to the case $\boldsymbol{P}\{B\} = 0$. Anyhow the new map $\boldsymbol{P}\{\,\cdot\,|B\} : \mathcal{F} \to [0, 1]$ is again a probability with

$$\boldsymbol{P}\{\emptyset|B\} = 0 \qquad \boldsymbol{P}\{\Omega|B\} = \boldsymbol{P}\{B|B\} = 1 \qquad \boldsymbol{P}\big\{\overline{A}|B\big\} = 1 - \boldsymbol{P}\{A|B\}$$
$$\boldsymbol{P}\{A_1 \cup A_2|B\} = \boldsymbol{P}\{A_1|B\} + \boldsymbol{P}\{A_2|B\} \qquad \text{if} \quad A_1 \cap A_2 = \emptyset$$

and so on, so that in fact $(\Omega, \mathcal{F}, \boldsymbol{P}\{\,\cdot\,|B\})$ is a new probability space

Proposition 1.14 Total probability formula: *Given* $(\Omega, \mathcal{F}, \boldsymbol{P})$, *an event* $A \in \mathcal{F}$ *and a decomposition* $\mathcal{D} = \{D_1, \ldots, D_n\}$ *with* $\boldsymbol{P}\big\{D_j\big\} \neq 0$, $j = 1, \ldots, n$ *we get*

$$P\{A\} = \sum_{j=1}^{n} P\{A|D_j\}\ P\{D_j\}$$

Proof Since

$$A = A \cap \Omega = A \cap \left(\bigcup_{j=1}^{n} D_j \right) = \bigcup_{j=1}^{n} (A \cap D_j)$$

it is enough to remark that the events $A \cap D_j$ are disjoint to have

$$P\{A\} = P\left\{ \bigcup_{j=1}^{n} (A \cap D_j) \right\} = \sum_{j=1}^{n} P\{A \cap D_j\} = \sum_{j=1}^{n} P\{A|D_j\}\ P\{D_j\}$$

because P is additive ∎

In particular, when $\mathcal{D} = \{B, \overline{B}\}$ the total probability formula just becomes

$$P\{A\} = P\{A|B\}\ P\{B\} + P\{A|\overline{B}\}\ P\{\overline{B}\} \tag{1.2}$$

a form that will be used in the next example.

Example 1.15 Consecutive draws: Take a box with M balls: m are white and $M - m$ black. Draw now sequentially two balls: neglecting the details of a suitable probability space, consider the two events

$$B = \text{"the first ball is white"}$$
$$A = \text{"the second ball is white"}$$

We will suppose our balls all equiprobable in order to make use of the classical definition. If then the first draw is *with replacement*, it is apparent that $P\{A\} = P\{B\} = \frac{m}{M}$. If on the other hand the first draw is *without replacement*, and if we find the first ball white (namely: if B happens), the probability of A would be $\frac{m-1}{M-1}$; while if the first ball is black (namely: if $\overline{B}$ happens) we would have $\frac{m}{M-1}$. In a third experiment let us draw now consecutively, and without replacement, two balls, and without looking at the first let us calculate the probability that the second is white namely the probability of A. From our opening remarks we know that

$$P\{B\} = \frac{m}{M} \qquad P\{\overline{B}\} = \frac{M-m}{M}$$
$$P\{A|B\} = \frac{m-1}{M-1} \qquad P\{A|\overline{B}\} = \frac{m}{M-1}$$

so that from the Total probability formula (1.2) we get

$$\boldsymbol{P}\{A\} = \frac{m-1}{M-1}\frac{m}{M} + \frac{m}{M-1}\frac{M-m}{M} = \frac{m}{M} = \boldsymbol{P}\{B\}$$

In short the probability of A depends on the available information: if we draw without replacement the first outcome affects the probability of the second, and hence $\boldsymbol{P}\{A|B\}$ differs from $\boldsymbol{P}\{A|\overline{B}\}$, and both differ from $\boldsymbol{P}\{B\}$. If instead the first outcome is unknown, we again get $\boldsymbol{P}\{A\} = \boldsymbol{P}\{B\} = \frac{m}{M}$ as if we had replaced the first ball

Proposition 1.16 Multiplication formula: *Given* $(\Omega, \mathcal{F}, \boldsymbol{P})$ *and the events* $A_1, \dots, A_n$ *with* $\boldsymbol{P}\{A_1 \dots A_{n-1}\} \neq 0$, *it is*

$$\boldsymbol{P}\{A_1 \dots A_n\} = \boldsymbol{P}\{A_n|A_{n-1} \dots A_1\}\, \boldsymbol{P}\{A_{n-1}|A_{n-2} \dots A_1\} \dots \boldsymbol{P}\{A_2|A_1\}\, \boldsymbol{P}\{A_1\}$$

Proof From the definition of conditional probability we have indeed

$$\begin{aligned}
&\boldsymbol{P}\{A_n|A_{n-1} \dots A_1\}\ \boldsymbol{P}\{A_{n-1}|A_{n-2} \dots A_1\}\ \dots\ \boldsymbol{P}\{A_2|A_1\}\ \boldsymbol{P}\{A_1\} \\
&= \frac{\boldsymbol{P}\{A_1 \dots A_n\}}{\boldsymbol{P}\{A_1 \dots A_{n-1}\}} \frac{\boldsymbol{P}\{A_1 \dots A_{n-1}\}}{\boldsymbol{P}\{A_1 \dots A_{n-2}\}} \dots \frac{\boldsymbol{P}\{A_1 A_2\}}{\boldsymbol{P}\{A_1\}} \boldsymbol{P}\{A_1\} = \boldsymbol{P}\{A_1 \dots A_n\}
\end{aligned}$$ ■

This Multiplication formula is very general and will play a role in the discussion of the Markov property in the second part of these lectures

Proposition 1.17 Bayes theorem: *Given* $(\Omega, \mathcal{F}, \boldsymbol{P})$ *and two events* A, B *with* $\boldsymbol{P}\{A\} \neq 0$ *and* $\boldsymbol{P}\{B\} \neq 0$, *it is*

$$\boldsymbol{P}\{A|B\} = \frac{\boldsymbol{P}\{B|A\}\, \boldsymbol{P}\{A\}}{\boldsymbol{P}\{B\}}$$

If moreover $\mathcal{D} = \{D_1, \dots, D_n\}$ *is a decomposition with* $\boldsymbol{P}\{D_j\} \neq 0,\ j = 1, \dots, n$, *we also have*

$$\boldsymbol{P}\{D_j|B\} = \frac{\boldsymbol{P}\{B|D_j\}\, \boldsymbol{P}\{D_j\}}{\sum_{k=1}^{n} \boldsymbol{P}\{B|D_k\}\, \boldsymbol{P}\{D_k\}}$$

Proof The first statement (also called ***Bayes formula***) again follows from the definition of conditional probability because

$$\boldsymbol{P}\{B|A\}\, \boldsymbol{P}\{A\} = \boldsymbol{P}\{AB\} = \boldsymbol{P}\{A|B\}\, \boldsymbol{P}\{B\}$$

The second statement then follows from the first and from the theorem of Total probability. ■

In the statistical applications the events D_j are called (mutually exclusive and exhaustive) *hypotheses* and $\boldsymbol{P}\{D_j\}$ their *a priori probability*, while the conditional probabilities $\boldsymbol{P}\{D_j|B\}$ take the name of *a posteriori probabilities*. As we will see in a forthcoming example of the Sect. 2.1.2, these names originate from the fact that the occurrence of the event B alters the probabilities initially given to the hypotheses D_j.

1.5 Independent Events

Two events are independent when the occurrence of one of them does not affect the probability of the other. By taking advantage, then, of our definition of conditional probability we could say that A is independent from B if $\boldsymbol{P}\{A|B\} = \boldsymbol{P}\{A\}$, and hence if $\boldsymbol{P}\{AB\} = \boldsymbol{P}\{A\}\,\boldsymbol{P}\{B\}$. The plus of this second statement w.r.t. that based on conditioning is that it holds even when $\boldsymbol{P}\{B\} = 0$. From the symmetry of these equations, moreover, it is easy to see that if A is independent from B, even the converse holds.

Definition 1.18 Given $(\Omega, \mathcal{F}, \boldsymbol{P})$, we say that A and B are **independent events** when

$$\boldsymbol{P}\{AB\} = \boldsymbol{P}\{A\}\,\boldsymbol{P}\{B\}$$

We also say that two σ-algebras $\mathcal{F}_1$ e $\mathcal{F}_2$ of events (more precisely: two sub-σ-algebras of $\mathcal{F}$) are **independent σ-algebras** if every event of $\mathcal{F}_1$ is independent from every event of $\mathcal{F}_2$.

This notion of independence can be also extended to more than two events, but we must pay attention first to the fact that for an arbitrary number of events $A, B, C, \dots$ we can speak of *pairwise independence*, namely $\boldsymbol{P}\{AB\} = \boldsymbol{P}\{A\}\,\boldsymbol{P}\{B\}$, but also of *three by three independence*, namely $\boldsymbol{P}\{ABC\} = \boldsymbol{P}\{A\}\,\boldsymbol{P}\{B\}\,\boldsymbol{P}\{C\}$, and so on, and then, and above all, to the circumstance that such independence levels do not imply each other: for instance three events can be pairwise independent without being so three by three, and also the converse holds. We are then obliged to extend our definition in the following way.

Definition 1.19 Given $(\Omega, \mathcal{F}, \boldsymbol{P})$ we say that $A_1, \dots, A_n \in \mathcal{F}$ are **independent events** if however taken k indices $j_1, \dots, j_k$ (with $k = 2, \dots, n$) we have

$$\boldsymbol{P}\{A_{j_1} \dots A_{j_k}\} = \boldsymbol{P}\{A_{j_1}\} \dots \boldsymbol{P}\{A_{j_k}\}$$

namely when they are independent pairwise, three by three, ..., n by n in every possible way.

The notion of independence is contingent on the probability $\boldsymbol{P}\{\,\cdot\,\}$: the same events can be either dependent or independent according to the chosen $\boldsymbol{P}\{\,\cdot\,\}$. This is apparent in particular when we introduce also the idea of *conditional independence* that allows to compare the independence under the two different probabilities $\boldsymbol{P}\{\,\cdot\,\}$ and $\boldsymbol{P}\{\,\cdot\,|D\}$

Definition 1.20 Given $(\Omega, \mathcal{F}, \boldsymbol{P})$, we say that two events A and B are **conditionally independent** w.r.t. D when

$$\boldsymbol{P}\{AB|D\} = \boldsymbol{P}\{A|D\}\, \boldsymbol{P}\{B|D\}$$

if $D \in \mathcal{F}$ is such that $\boldsymbol{P}\{D\} \neq 0$

It would be possible to show with a few examples—that we neglect—that A and B, dependent under a probability $\boldsymbol{P}$, could be made conditionally independent w.r.t. some other event D. Even the notion of conditional independence will be instrumental in the discussion of the Markov property in the second part of these lectures.

Reference

1. Shiryaev, A.N.: Probability. Springer, New York (1996)

Chapter 2
Distributions

2.1 Distributions on N

2.1.1 Finite and Countable Spaces

We will explore now the protocols used to define on $(\Omega, \mathcal{F})$ a probability $\boldsymbol{P}$ also called either ***law*** or ***distribution***, and we will start with finite or countable spaces so that $\boldsymbol{P}$ will be defined in an elementary way.

Example 2.1 Binomial distributions: An example of finite sample space is the set of the first $n+1$ integer numbers $\Omega_n = \{0, 1, \dots, n\}$ (with the σ-algebra $\wp(\Omega_n)$ of all its subsets): given then a number $p \in [0, 1]$, with $q = 1 - p$, we can define a $\boldsymbol{P}$ by first attributing to every $\omega = k$ the probability

$$p_n(k) = \binom{n}{k} p^k q^{n-k} \qquad k = 0, 1, \dots, n \tag{2.1}$$

and then by taking

$$\boldsymbol{P}\{B\} = \sum_{k \in B} p_n(k) \tag{2.2}$$

as the probability of $B \subseteq \Omega_n$. It would be easy to check that such a $\boldsymbol{P}$ is σ-additive, that its values lie in $[0, 1]$, and finally that

$$\boldsymbol{P}\{\Omega_n\} = \sum_{k=0}^{n} p_n(k) = \sum_{k=0}^{n} \binom{n}{k} p^k q^{n-k} = (p+q)^n = 1 \tag{2.3}$$

The numbers $p_n(k)$ ($n = 1, 2, \dots$ and $p \in [0, 1]$) are called **binomial distribution**, and we will denote them as $\mathfrak{B}(n; p)$. The case $\mathfrak{B}(1; p)$ on $\Omega_1 = \{0, 1\}$ with

N. Cufaro Petroni, *Probability and Stochastic Processes for Physicists*,
UNITEXT for Physics, https://doi.org/10.1007/978-3-030-48408-8_2

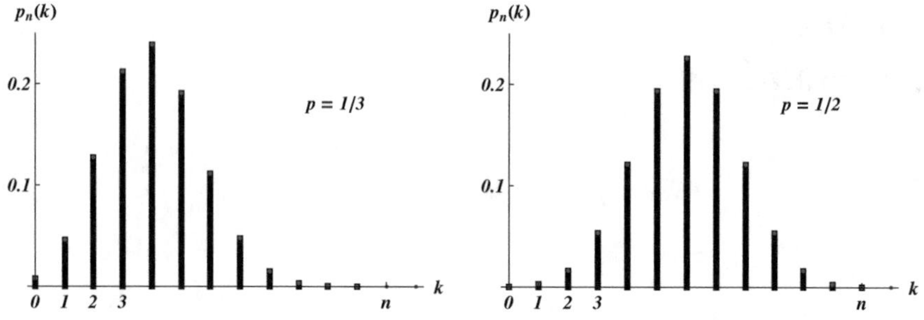

Fig. 2.1 Bar diagrams of binomial distributions $\mathfrak{B}\,(n;\,p)$

$$p_1(1) = p \qquad p_1(0) = q = 1 - p$$

is also called **Bernoulli distribution**. The bar diagram of a typical binomial distribution is displayed in the Fig. 2.1 for two different values of p. The meaning of these laws, and their link with experiments of ball drawing from urns will be discussed in the Sect. 2.1.2.

Example 2.2 Poisson distributions: By going now to the countable set of the integers $\Omega = N = \{0, 1, 2, \ldots\}$, with $\mathcal{F} = \wp(N)$, we again start by allotting to every $\omega = k \in N$ the probability

$$\boldsymbol{P}\{\omega\} = p_\alpha(k) = e^{-\alpha}\frac{\alpha^k}{k!} \qquad \alpha > 0 \tag{2.4}$$

and then we define the probability of $A \in \mathcal{F}$ as

$$\boldsymbol{P}\{A\} = \sum_{k\in A} p_\alpha(k)$$

Positivity and additivity are readily checked, while the normalization follows from

$$\boldsymbol{P}\{\Omega\} = \sum_{k\in N} e^{-\alpha}\frac{\alpha^k}{k!} = e^{-\alpha}\sum_{k\in N}\frac{\alpha^k}{k!} = e^{-\alpha}e^{\alpha} = 1$$

The probabilities (2.4) (which are non-zero for every $k = 0, 1, 2, \ldots$) are called **Poisson distribution** and are globally denoted as $\mathfrak{P}(\alpha)$. For the time being the parameter $\alpha > 0$ and the formula (2.4) itself are arbitrarily taken: their meaning will be made clear in the Sect. 4.5, where it will be also shown that the results of these probability spaces are typically obtained by counting, for instance, the number of particles emitted by a radioactive sample in 5 min, or the number of phone calls at a

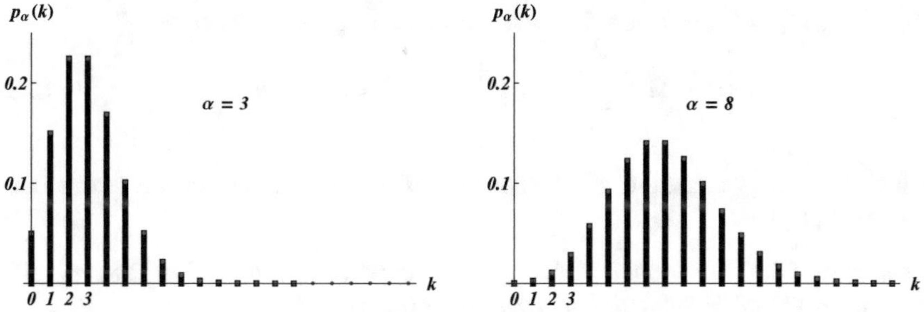

Fig. 2.2 Bar diagrams of Poisson distributions $\mathfrak{P}(\alpha)$

call center in one hour, and so on. Examples of Poisson distributions for different α values are on display in the Fig. 2.2.

2.1.2 Bernoulli Trials

The binomial distribution in Example 2.1 is defined without a reference to some factual problem, so that in particular the allotment of the probabilities $p_n(k)$ looks rather unmotivated, albeit coherent. To find an empirical model for $\mathfrak{B}\,(n; p)$ take ***n* drawings with replacement** from a box containing black and white balls, and the sample space Ω consisting of all the possible ordered n-tuples of results. It is customary to encode the outcomes by numbers—1 for white, and 0 for black—so that our samples will be ordered n-tuples of 0–1 symbols

$$\omega = (a_1, \dots a_n) \qquad a_i = 0, 1; \qquad i = 1, \dots, n \tag{2.5}$$

with the family of all the subsets $\wp\,(\Omega)$ as σ-algebra of the events. Give now to every $\omega = (a_1, \dots a_n)$ the probability

$$\boldsymbol{P}\{\omega\} = p^k q^{n-k} \tag{2.6}$$

where $k = \sum_i a_i$ is the number of white balls in ω, $p \in [0, 1]$ is arbitrary, $q = 1 - p$, and finally define the probability of the events $A \in \mathcal{F}$ as

$$\boldsymbol{P}\{A\} = \sum_{\omega \in A} \boldsymbol{P}\{\omega\} \tag{2.7}$$

The definition (2.6) is again unmotivated, and we will devote the following remarks to make clear its meaning. First it is easy to see that $\boldsymbol{P}$ as defined in (2.7) is positive and additive. To check then its normalization $\boldsymbol{P}\{\Omega\} = 1$ it is expedient to consider the $n + 1$ events

$$D_k = \text{“}\textit{there are } k \textit{ white balls among the } n \textit{ outcomes}\text{”}$$
$$= \left\{ \omega \in \Omega : \sum_{i=1}^{n} a_i = k \right\} \qquad k = 0, \dots, n \tag{2.8}$$

which apparently constitute a decomposition $\mathcal{D}$ of Ω, and to calculate their probabilities $\boldsymbol{P}\{D_k\}$.

Proposition 2.3 *The probabilities $\boldsymbol{P}\{D_k\}$ for the decomposition $\mathcal{D}$ in* (2.8) *coincide with the $p_n(k)$ of the binomial distribution $\mathfrak{B}(n;p)$.*

Proof Since the k symbols 1 in a sample $\omega \in D_k$ can be placed in several different ways on the n available positions without changing the probability, every D_k will be constituted of a certain number—say n_k—of equiprobable samples each with probability $\boldsymbol{P}\{\omega\} = p^k q^{n-k}$, so that

$$\boldsymbol{P}\{D_k\} = \sum_{\omega \in D_k} \boldsymbol{P}\{\omega\} = n_k \, p^k \, q^{n-k}$$

We are left then with the problem of finding n_k: for a given $k = \sum_i a_i$ every sample $\omega = (a_1 \dots a_n)$ is uniquely identified by a set of *occupation numbers* $[b_1, \dots b_k]$ labeling the *positions* of the k symbols 1 on the n places of ω; for example, with $n = 7$

$$\omega = (0, 1, 1, 0, 0, 1, 0) \leftrightarrow [2, 3, 6]$$

Apparently the ordering in $[b_1, \dots b_k]$ is immaterial (in our example both [2,3,6] and [3,6,2] denote the same 7-tuple with a symbol 1 at the 2nd, 3rd e 6th place); moreover the b_j values are all different, and hence n_k will be the number of all the possible non ordered k-tuples, without repetitions $[b_1, \dots b_k]$ where every b_j takes the values $1, 2, \dots n$. From (1.1) we then have that

$$n_k = \binom{n}{k} \qquad \boldsymbol{P}\{D_k\} = p_n(k) = \binom{n}{k} p^k q^{n-k}$$

As a consequence the $p_n(k)$ of a binomial distribution $\mathfrak{B}(n;p)$ are the probabilities $\boldsymbol{P}\{D_k\}$ of the events D_k in the sample space Ω of n drawings, with a $\boldsymbol{P}$ defined as in (2.6) and (2.7). ■

We are now also able to check the coherence of the definition (2.6) because, from the additivity of $\boldsymbol{P}$ and from (2.3), we have

$$\boldsymbol{P}\{\Omega\} = \boldsymbol{P}\left\{ \bigcup_{k=0}^{n} D_k \right\} = \sum_{k=0}^{n} \boldsymbol{P}\{D_k\} = \sum_{k=0}^{n} p_n(k) = 1$$

Finally, to make the meaning of $p \in [0, 1]$ and of (2.6) more apparent, take, for $j = 1, \dots, n$, the events

$$A_j = \text{“}a \text{ white ball comes out at the } j^{\text{th}} \text{ draw”} = \{\omega \in \Omega : a_j = 1\}$$

while $\overline{A}_j$ corresponds to a black ball at the jth draw. At variance with the D_k, however, the events A_j are not disjoint (we can find white balls in different draws) so that they are not a decomposition of Ω.

Proposition 2.4 *The numbers $p \in [0, 1]$ and $q = 1 - p$ respectively are the probabilities of finding a white and a black ball in every single draw, namely*

$$\boldsymbol{P}\{A_j\} = p\,, \qquad \boldsymbol{P}\{\overline{A}_j\} = q = 1 - p$$

Regardless of the value of p, moreover, the events A_j are all mutually independent w.r.t. the $\boldsymbol{P}$ defined in (2.6), *and this elucidates the meaning of that definition.*

Proof For the sake of brevity we will neglect a complete discussion (see [1] pp. 29–31) and we will confine ourselves to a few remarks. For $n = 1$ (just one draw) we have $\Omega = \Omega_1 = \{0, 1\}$ and from (2.6) we get

$$\boldsymbol{P}\{A_1\} = \boldsymbol{P}\{1\} = p \qquad \boldsymbol{P}\{\overline{A}_1\} = \boldsymbol{P}\{0\} = q = 1 - p$$

so that p comes out to be the probability of finding a white ball in one single draw, and we will neglect to show that this is so even for every single draw in a sequence. On the other hand a little algebra, omitted again, would show that for $j \neq \ell$ it is

$$\boldsymbol{P}\{A_j A_\ell\} = p^2 \qquad \boldsymbol{P}\{A_j \overline{A}_\ell\} = pq \qquad \boldsymbol{P}\{\overline{A}_j \overline{A}_\ell\} = q^2$$

so that the events A_j, A_k, together with their complements, are independent w.r.t. $\boldsymbol{P}$ defined in (2.7). This remark can also be extended to three or more events. Finally, since every $\omega \in \Omega$ is the intersection of k events A_j with $n - k$ events $\overline{A}_\ell$, apparently also the choice (2.6) for the probability of ω has been made exactly in view of their independence. ■

In short our space $(\Omega, \mathcal{F}, \boldsymbol{P})$, with $\boldsymbol{P}$ defined as in (2.6), is a model for *n independent verification trials of the event: “a white ball comes out”*, while the $p_n(k)$ of a binomial distribution $\mathfrak{B}\,(n;\, p)$ are the *probabilities of finding k white balls among n independent draws with replacement*. Of course drawing balls from an urn is just an example, and the same model also fits n independent verification trials of an arbitrary event A which occurs with probability p in every trial. The 0 -1 random experiments of this model are also known as ***Bernoulli trials*** and their corresponding probability space is an example of ***direct product***: given n replicas of the space describing a single draw with $\Omega_1 = \{0, 1\}$, $\mathcal{F}_1 = \{1, 0, \Omega_1, \emptyset\}$ and $\boldsymbol{P}_1$ a Bernoulli distribution $\mathfrak{B}\,(1;\, p)$

$$\boldsymbol{P}_1\{1\} = p\,, \qquad \boldsymbol{P}_1\{0\} = 1 - p$$

the direct product has the Cartesian product $\Omega = \Omega_1 \times \ldots \times \Omega_1$ of the n-tuples of 0–1 symbols as sample space, the family of all its parts as σ-algebra $\mathcal{F}$ of the events, and a probability $\boldsymbol{P}$ defined by (2.7) taking for every sample the product

$$\boldsymbol{P}\{\omega\} = \boldsymbol{P}_1\{a_1\} \cdot \ldots \cdot \boldsymbol{P}_1\{a_n\} = p^k q^{n-k} \qquad k = \sum_{j=1}^{n} a_j$$

This *product probability* is not a compulsory choice, but uniquely corresponds to the independence of the trials.

Example 2.5 An application of the Bayes theorem: As foretold at the end of the Sect. 1.4 we are able now to discuss a statistical application of the Bayes theorem (Proposition 10.10). Within the notations of the Sect. 1.4, take two externally identical boxes D_1 e D_2 with black and white balls in different proportions: the fraction of white balls in D_1 is $^1/_2$, while that in D_2 is $^2/_3$. We can not look into the boxes, but it is allowed to sample their content with replacement. Choose then a box and ask which one has been taken. Apparently $\mathcal{D} = \{D_1, D_2\}$ is a decomposition and, lacking further information, the two events must be deemed equiprobable namely

$$\boldsymbol{P}\{D_1\} = \boldsymbol{P}\{D_2\} = \frac{1}{2}$$

To know better, however, we can draw a few balls: a large number of white balls, for example, would hint toward D_2, and vice versa in the opposite case. The Bayes theorem provides now the means to make quantitative these so far qualitative remarks. Suppose for instance to perform $n = 10$ drawings with replacement from the chosen box, finding $k = 4$ white, and $n - k = 6$ black balls, namely that the event

$$B = \text{“}\textit{among the } n = 10 \textit{ drawn out balls } k = 4 \textit{ are white}\text{”}$$

occurs. According to the two possible urns D_1 e D_2, the probabilities of B are respectively the binomial distributions $\mathfrak{B}\left(10; \frac{1}{2}\right)$ and $\mathfrak{B}\left(10; \frac{2}{3}\right)$, namely

$$\boldsymbol{P}\{B|D_1\} = \binom{10}{4} \left(^1/_2\right)^4 \left(^1/_2\right)^{10-4} = \binom{10}{4} \frac{1}{2^{10}}$$

$$\boldsymbol{P}\{B|D_2\} = \binom{10}{4} \left(^2/_3\right)^4 \left(^2/_3\right)^{10-4} = \binom{10}{4} \frac{2^4}{3^{10}}$$

and hence from the Bayes theorem we get

$$\boldsymbol{P}\{D_1|B\} = \frac{\boldsymbol{P}\{B|D_1\}\ \boldsymbol{P}\{D_1\}}{\boldsymbol{P}\{B|D_1\}\ \boldsymbol{P}\{D_1\} + \boldsymbol{P}\{B|D_2\}\ \boldsymbol{P}\{D_2\}} = \frac{\frac{1}{2^{10}}}{\frac{1}{2^{10}} + \frac{2^4}{3^{10}}}$$

$$= \frac{3^{10}}{3^{10} + 2^{14}} = 0.783$$

$$\boldsymbol{P}\{D_2|B\} = \frac{2^{14}}{3^{10} + 2^{14}} = 0.217$$

Predictably the relatively small number of white balls hints toward D_1, but now we have a precise quantitative estimate of its probability. Of course further drawings would change this result, but intuitively these oscillations should stabilize for a large number of trials.

Example 2.6 Multinomial distribution: The Binomial distribution discussed in the Example 2.1 can be generalized by supposing a sample space Ω still made of ordered n-tuples $\omega = (a_1, \dots, a_n)$, but for the fact that now the symbols a_j can take $r + 1$ (con $r \geq 1$) values $b_0, b_1, \dots, b_r$ instead of just two. For instance we can think of drawing with replacement n balls from a box containing balls of $r + 1$ different colors, but even here it is expedient to label the $r + 1$ colors with the numbers $0, 1, 2, \dots, r$. Suppose now that $k_i,\ i = 0, 1, \dots, r$ is the number of balls that in a given sample ω take the color b_i, and start by attributing to ω the probability

$$\boldsymbol{P}\{\omega\} = p_0^{k_0} p_1^{k_1} \cdot \dots \cdot p_r^{k_r}$$

where $k_0 + k_1 + \dots + k_r = n$, while $p_0, p_1, \dots, p_r$ are $r + 1$ arbitrary, non negative numbers such that $p_0 + p_1 + \dots + p_r = 1$. Given then the events

$$D_{k_1 \dots k_r} = \text{“}among\ the\ n\ balls\ we\ find\ k_0\ times\ b_0,\ k_1\ times\ b_1,\ \dots\ , k_r\ times\ b_r\text{”}$$

it is possible to prove that they are a decomposition of Ω, and that each contains

$$\binom{n}{k_1,\ \dots,\ k_r} = \frac{n!}{k_0!\, k_1! \dots k_r!}$$

equiprobable samples ω, so that finally

$$\boldsymbol{P}\{D_{k_1 \dots k_r}\} = p_n(k_1, \dots, k_r) = \binom{n}{k_1,\ \dots,\ k_r} p_0^{k_0} p_1^{k_1} \cdot \dots \cdot p_r^{k_r} \qquad (2.9)$$

The set of these probabilities takes the name of **multinomial distribution** and is denoted with the symbol $\mathfrak{B}\,(n; p_1, \dots, p_r)$. This is a new family of distributions classified by the number n of draws, and by the non negative numbers $p_i \in [0, 1]$, with $p_0 + p_1 + \dots + p_r = 1$, which are the probabilities of finding b_i in every single drawing. Remark that the binomial distribution $\mathfrak{B}\,(n; p)$ is the particular case with $r = 1$: in this instance p_1 and p_0 are usually labeled p e q, while $k_1 = k$ and $k_0 = n - k$.

2.2 Distributions on $\boldsymbol{R}$

To analyze how to define a probability on uncountable spaces we will start with $(\boldsymbol{R}, \mathcal{B}(\boldsymbol{R}))$, by remarking at once that the distributions studied in the previous Sect. 2.1 will constitute the particular case of the *discrete distributions*.

2.2.1 Cumulative Distribution Functions

Suppose first that somehow a probability $\boldsymbol{P}$ is defined on $(\boldsymbol{R}, \mathcal{B}(\boldsymbol{R}))$ and take

$$F(x) = \boldsymbol{P}\{(-\infty, x]\} \qquad \forall x \in \boldsymbol{R} \tag{2.10}$$

Proposition 2.7 *The function $F(x)$ defined in* (2.10) *has the following properties*

1. *$F(x)$ is non decreasing*
2. $F(+\infty) = 1\,, \quad F(-\infty) = 0$
3. *$F(x)$ is right continuous with left limits $\forall x \in \boldsymbol{R}$ (cadlag); moreover it is outright continuous if and only if (iff) $\boldsymbol{P}\{x\} = 0$.*

Proof The properties 1 and 2 easily result from (2.10). As for 3, remark that a monotone and bounded $F(x)$ always admits the right and left limits $F(x^{+})$ for every $x \in \boldsymbol{R}$. Take now a monotone sequence $(x_n)_{n \in N}$ such that $x_n \downarrow x$ from right: since $(-\infty, x_n] \to (-\infty, x]$ we will then have that (see [1] p. 134 for the continuity of a probability)

$$F(x^{+}) = \lim_n F(x_n) = \lim_n \boldsymbol{P}\{(-\infty, x_n]\} = \boldsymbol{P}\{(-\infty, x]\} = F(x)$$

so that $F(x)$ is right continuous. The same can not be said, instead, if $x_n \uparrow x$ from left, because now $(-\infty, x_n] \to (-\infty, x)$ and hence

$$F(x^{-}) = \lim_n F(x_n) = \lim_n \boldsymbol{P}\{(-\infty, x_n]\} = \boldsymbol{P}\{(-\infty, x)\} \neq F(x)$$

Being however $(-\infty, x] = (-\infty, x) \cup \{x\}$, we in general get

$$F(x) = \boldsymbol{P}\{(-\infty, x]\} = \boldsymbol{P}\{(-\infty, x)\} + \boldsymbol{P}\{x\} = F(x^{-}) + \boldsymbol{P}\{x\}$$

namely $F(x^{-}) = F(x) - \boldsymbol{P}\{x\}$, so that $F(x)$ would be also left continuous, and hence outright continuous, *iff* $\boldsymbol{P}\{x\} = 0$. ■

The previous result entail in particular that $\boldsymbol{P}\{x\}$ can be non zero *iff* $F(x)$ is discontinuous in x, and in this case

$$\boldsymbol{P}\{x\} = F(x) - F(x^-) = F(x^+) - F(x^-) \tag{2.11}$$

Moreover, since $(-\infty, b] = (-\infty, a] \cup (a, b]$, from the additivity of $\boldsymbol{P}$ we also have

$$\boldsymbol{P}\{(-\infty, b]\} = \boldsymbol{P}\{(-\infty, a]\} + \boldsymbol{P}\{(a, b]\}$$

and hence

$$\boldsymbol{P}\{(a, b]\} = F(b) - F(a) \tag{2.12}$$

for every $-\infty \leq a < b \leq +\infty$.

Definition 2.8 We call **(cumulative) distribution function** (*cdf*) on $\boldsymbol{R}$ every $F(x)$ satisfying 1, 2 and 3.

The previous discussion shows that at every $\boldsymbol{P}$ on $(\boldsymbol{R}, \mathcal{B}(\boldsymbol{R}))$ it is always joined a *cdf* $F(x)$. The subsequent theorem then points out that the reverse is also true: every *cdf* on $\boldsymbol{R}$ always defines a probability $\boldsymbol{P}$ on $(\boldsymbol{R}, \mathcal{B}(\boldsymbol{R}))$ such that (2.12) holds

Theorem 2.9 *Given a cdf $F(x)$ on $\boldsymbol{R}$, there is always one and only one probability $\boldsymbol{P}$ on $(\boldsymbol{R}, \mathcal{B}(\boldsymbol{R}))$ such that*

$$\boldsymbol{P}\{(a, b]\} = F(b) - F(a)$$

for every $-\infty \leq a < b \leq +\infty$.

Proof Omitted: see [1] p. 152. ∎

There is then a one-to-one correspondence between the laws $\boldsymbol{P}$ on $(\boldsymbol{R}, \mathcal{B}(\boldsymbol{R}))$ and the *cdf* $F(x)$ on $\boldsymbol{R}$, so that a probability on $(\boldsymbol{R}, \mathcal{B}(\boldsymbol{R}))$ is well defined *iff* we know its *cdf* $F(x)$. Since, however, in the following we will make use of measures on $(\boldsymbol{R}, \mathcal{B}(\boldsymbol{R}))$ that are not finite (for instance the Lebesgue measure) it will be expedient to slightly generalize our framework.

Definition 2.10 We say that μ is a **Lebesgue-Stieltjes** (L-S) **measure** on $(\boldsymbol{R}, \mathcal{B}(\boldsymbol{R}))$ if it is σ-additive, and $\mu\{B\} < +\infty$ for every bounded B. We also call **generalized distribution function** on $\boldsymbol{R}$ (*gcdf*) every $G(x)$ on $\boldsymbol{R}$ satisfying the properties 1 and 3, but not in general 2.

It is possible to show that, if μ is a L-S measure on $(\boldsymbol{R}, \mathcal{B}(\boldsymbol{R}))$, the function $G(x)$ defined, but for an additive constant, by

$$G(y) - G(x) = \mu\{(x, y]\}, \qquad x < y$$

is a *gcdf*, while the subsequent theorem encodes the reverse statement that to every *gcdf* $G(x)$ we can always associate a unique L-S measure.

Theorem 2.11 *Given a gcdf $G(x)$ on $\boldsymbol{R}$, there is always one and only one L-S measure μ on $(\boldsymbol{R}, \mathcal{B}(\boldsymbol{R}))$ such that*

$$\mu\{(a, b\,]\} = G(b) - G(a)$$

for every $-\infty \leq a < b \leq +\infty$.

Proof Omitted: see [1] p. 158. ■

It is apparent that a *gcdf* $G(x)$ has the same properties of a *cdf* but for 2, so that $G(x)$ can take both negative and larger than 1 values, while its asymptotic behavior for $x \to \pm\infty$ is not bounded. A well known example of these measures is the ***Lebesgue measure*** on $\boldsymbol{R}$, namely the σ-finite measure λ that to every interval $(a, b] \in \mathcal{B}(\boldsymbol{R})$ assign the measure $\lambda\{(a, b\,]\} = b - a$: in this case the *gcdf* simply is

$$G(x) = x.$$

2.2.2 *Discrete Distributions*

Definition 2.12 We say that a probability $\boldsymbol{P}$ is a **discrete distribution** on $(\boldsymbol{R}, \mathcal{B}(\boldsymbol{R}))$ if its *cdf* $F(x)$ is piecewise constant, and discontinuously changes its value in a (finite or countable) set of points $x_1, x_2, \dots$ where it is $F(x_i) - F(x_i^-) > 0$.

The *cdf* of a discrete law apparently is a typical step function (see for instance the Fig. 2.3) so that $\boldsymbol{P}\{(a, b\,]\} = 0$ if within $(a, b]$ we find no discontinuities x_i, while in general it is

$$\boldsymbol{P}\{(a, b\,]\} = \sum_{x_i \in (a,b]} \left[F(x_i) - F(x_i^-)\right] = F(b) - F(a)$$

As already remarked we find $\boldsymbol{P}\{x\} = 0$ wherever $F(x)$ is continuous, and $p_i = \boldsymbol{P}\{x_i\} = F(x_i) - F(x_i^-)$ wherever $F(x)$ makes jumps. As a consequence the probability $\boldsymbol{P}$ happens to be concentrated in the (at most) countably many points $x_1, x_2, \dots$ and is well defined by giving these points and the numbers $p_1, p_2, \dots$ which also

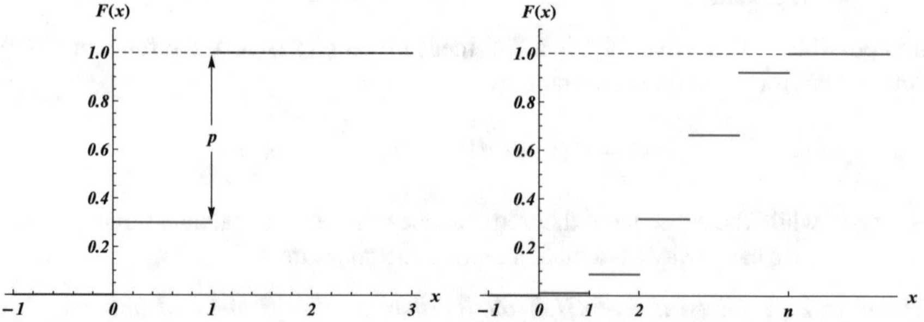

Fig. 2.3 *cdf* of a Bernoulli $\mathfrak{B}(1, p)$ and of a binomial distribution $\mathfrak{B}(n, p)$

are named ***discrete distribution***. The examples of finite and countable probability spaces discussed in the Sect. 2.1 are particular discrete distributions where $x_k = k$ are integer numbers. The main difference with the present approach is that in the Sect. 2.1 the sample space Ω was restricted *just* to the set of points x_k, while here Ω is extended to $\boldsymbol{R}$ and x_k are the points with non-zero probability. This entails, among other, that—beyond the bar diagrams of the Figs. 2.1 and 2.2—we can now represent a discrete distribution by means of its *cdf* $F(x)$ with a continuous variable $x \in \boldsymbol{R}$.

Example 2.13 Notable discrete distributions: Consider first the case where just one value $b \in \boldsymbol{R}$ occurs with probability 1, namely $\boldsymbol{P}$-a.s.: the family of these distributions, called **degenerate distributions**, is denoted by the symbol $\boldsymbol{\delta}_b$, and its *cdf* $F(x)$ show just one unit step in $x = b$, namely is a **Heaviside function**

$$\vartheta(x) = \begin{cases} 1, \text{ if } x \geq 0 \\ 0, \text{ if } x < 0 \end{cases} \tag{2.13}$$

Of course its bar diagram will have just one unit bar located at $x = b$. On the other hand in the family $\mathfrak{B}\,(1; p)$ of the **Bernoulli distributions** two values 1 and 0 occur respectively with probability p and $q = 1 - p$, while in the **binomial distributions** $\mathfrak{B}\,(n; p)$ the values $k = 0, \ldots, n$ occur with the probabilities

$$p_n(k) = \binom{n}{k} p^k q^{n-k}, \quad q = 1 - p, \qquad 0 \leq p \leq 1$$

The corresponding Bernoulli and binomial *cdf*'s are displayed in the Fig. 2.3. Finally in the family $\mathfrak{P}(\alpha)$ of the **Poisson distributions** all the integer numbers $k \in N$ occur with the probabilities

$$p_k = \frac{\alpha^k e^{-\alpha}}{k!}, \qquad \alpha > 0$$

and their *cdf* is shown in the Fig. 2.4.

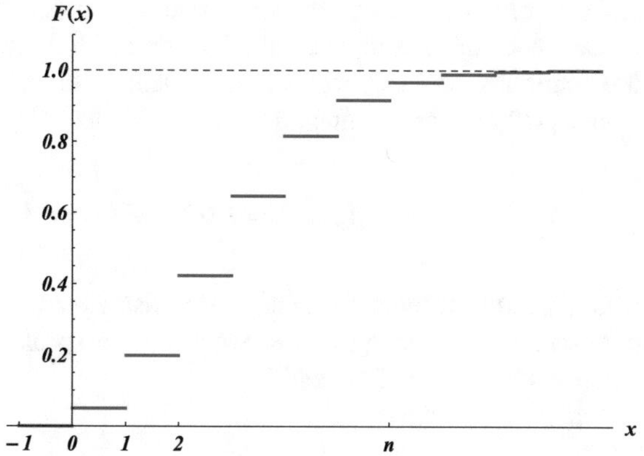

Fig. 2.4 *cdf* of a Poisson distribution $\mathfrak{P}(\alpha)$

2.2.3 Absolutely Continuous Distributions: Density

Definition 2.14 Take two measures μ and ν on the same $(\Omega, \mathcal{F})$: we say that ν is **absolutely continuous** (*ac*) w.r.t. μ (and we write $\nu \ll \mu$) when $\mu(A) = 0$ for $A \in \mathcal{F}$ also entails $\nu(A) = 0$. If in particular $\Omega = \boldsymbol{R}$, when a probability $\boldsymbol{P}$ on $(\boldsymbol{R}, \mathcal{B}(\boldsymbol{R}))$ is *ac* w.r.t. the Lebesgue measure we also say for short that its *cdf* $F(x)$ is *ac*.

Theorem 2.15 Radon-Nikodym theorem on R: *A cdf $F(x)$ on $\boldsymbol{R}$ is ac iff it exists a non negative function $f(x)$ defined on $\boldsymbol{R}$ such that*

$$\int_{-\infty}^{+\infty} f(x)\,dx = 1 \qquad F(x) = \int_{-\infty}^{x} f(z)\,dz \qquad f(x) = F'(x)$$

The function $f(x)$ is called ***probability density function (pdf)*** *of $F(x)$.*

Proof Omitted: see [2] p. 288. ■

It is easy to show that, taken a non negative, Lebesgue integrable and 1-normalized function $f(x)$, the function

$$F(x) = \int_{-\infty}^{x} f(z)\,dz$$

always is an *ac cdf* The Radon-Nikodym theorem states the remarkable fact that also the reverse holds: every *ac cdf* $F(x)$ is the primitive function of a suitable *pdf* $f(x)$, so that every *ac cdf* can be given through a *pdf*, which is unique but for its values on a Lebesgue negligible set of points. It is possible to show that an *ac cdf* is also continuous and derivable (but for a Lebesgue negligible set of points), and that in this case the *pdf* is nothing else than its derivative

$$f(x) = F'(x)$$

There are on the other hand (a simple example will be discussed in the Sect. 2.2.4) *cdf*'s $F(x)$ which are continuous but not *ac*, so that the existence of a *pdf* is not a consequence of the simple continuity of a *cdf*. Taking then into account (2.12) and the continuity of $F(x)$, we can now calculate the probability of an interval $[a, b]$ from a *pdf* $f(x)$ as the integral

$$\boldsymbol{P}\{[a, b]\} = F(b) - F(a) = \int_{a}^{b} f(t)\,dt$$

It is apparent on the other hand that a discrete *cdf* can never be *ac* because it is not even continuous: in this case we can never speak of a *pdf*, and we must restrict ourselves to the use of the *cdf*

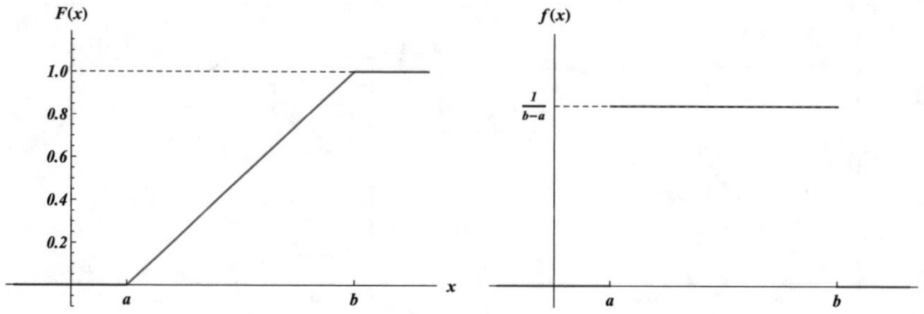

Fig. 2.5 *cdf* and *pdf* of the uniform distribution $\mathfrak{U}(a, b)$

Example 2.16 Uniform distributions: Take first the family of the **uniform laws** on an interval $[a, b]$ denoted as $\mathfrak{U}(a, b)$. The *cdf* is

$$F(x) = \begin{cases} 0 & \text{if } x < a \\ \frac{x-a}{b-a} & \text{if } a \le x \le b \\ 1 & \text{if } b < x \end{cases}$$

as displayed in the Fig. 2.5. This *cdf* defines on $(\boldsymbol{R}, \mathcal{B}(\boldsymbol{R}))$ a probability $\boldsymbol{P}$ concentrated on $[a, b]$ that to every interval $[x, y] \subseteq [a, b]$ gives the probability

$$\boldsymbol{P}\{[x, y]\} = \frac{y - x}{b - a}$$

On the other hand intervals lying outside $[a, b]$ have zero probability, while, since $F(x)$ is continuous, $\boldsymbol{P}\{x\} = 0$ for every event reduced to the point x. The *pdf* is deduced by derivation

$$f(x) = \begin{cases} \frac{1}{b-a} & \text{if } a \le x \le b \\ 0 & \text{else} \end{cases} \tag{2.14}$$

and is displayed in the Fig. 2.5 along with its *cdf*. These behaviors also justify the name of these laws because the probability of every interval $[x, y] \subseteq [a, b]$ depends only on its amplitude $y - x$ and not on its position inside $[a, b]$. For short: all the locations inside $[a, b]$ are uniformly weighted.

Example 2.17 Gaussian (normal) distributions: The family of the **Gaussian (normal) laws** $\mathfrak{N}(b, a^2)$ is characterized by the *pdf*

$$f(x) = \frac{1}{a\sqrt{2\pi}} e^{-(x-b)^2/2a^2} \qquad a > 0, \ b \in \boldsymbol{R} \tag{2.15}$$

displayed with its *cdf* in the Fig. 2.6. The so called degenerate case $a = 0$, that is here excluded, needs a particular discussion developed in the Sect. 4.2.2. The Gaussian *pdf* shows a typical bell-like shape with the maximum in $x = b$. The two flexes in

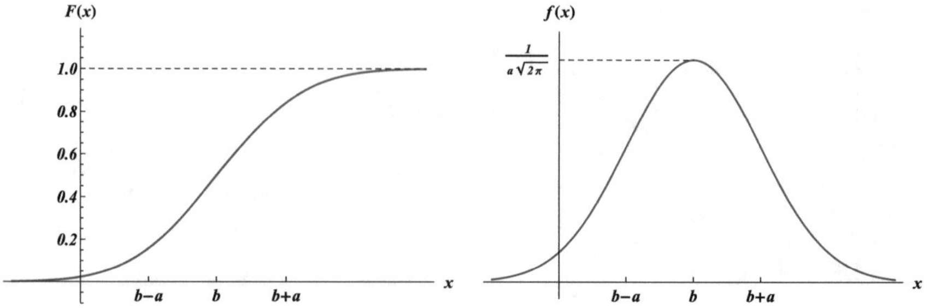

Fig. 2.6 *cdf* and *pdf* of the normal distribution $\mathfrak{N}(b, a^2)$

$x = b \pm a$ give a measure of the width that hance depends on the parameter a. We will speak of **standard normal law** when $b = 0$ and $a = 1$, namely when the *pdf* is

$$\phi(x) = \frac{1}{\sqrt{2\pi}}\, e^{-x^2/2}$$

Both the standard and non standard Gaussian *cdf*, also called **error functions**, respectively are

$$\Phi(x) = \frac{1}{\sqrt{2\pi}} \int_{-\infty}^{x} e^{-z^2/2}\, dz \qquad F(x) = \frac{1}{a\sqrt{2\pi}} \int_{-\infty}^{x} e^{-(z-b)^2/2a^2}\, dz \tag{2.16}$$

and are shown in the Fig. 2.6: they can not be given as finite combinations of elementary functions, but have many analytical expressions and can always be calculated numerically.

Example 2.18 Exponential distributions: The family of the **exponential laws** $\mathfrak{E}(a)$ has the *pdf*

$$f(x) = a\, e^{-ax}\vartheta(x) = \begin{cases} a\, e^{-ax} & \text{if } x \geq 0 \\ 0 & \text{if } x < 0 \end{cases} \qquad a > 0 \tag{2.17}$$

while the corresponding *cdf* is

$$F(x) = (1 - e^{-ax})\vartheta(x) = \begin{cases} 1 - e^{-ax} & \text{if } x \geq 0 \\ 0 & \text{if } x < 0 \end{cases}$$

both represented in the Fig. 2.7.

Example 2.19 Laplace distributions: We call **Laplace laws**, or even bilateral exponentials, denoted as $\mathfrak{L}(a)$, the laws with *pdf*

$$f(x) = \frac{a}{2}\, e^{-a|x|} \qquad a > 0 \tag{2.18}$$

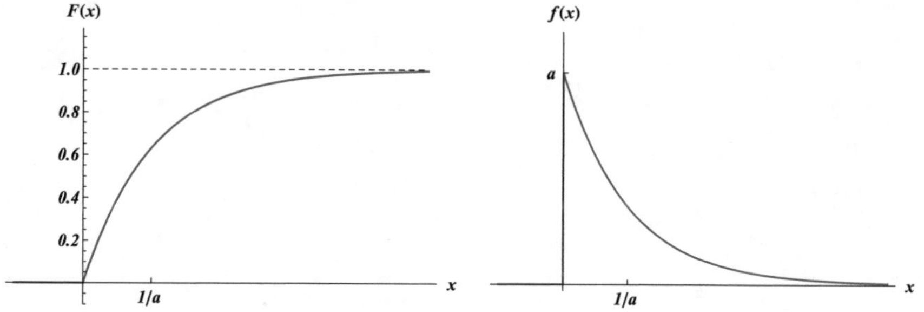

Fig. 2.7 *cdf* and *pdf* of the exponential distribution $\mathfrak{E}(a)$

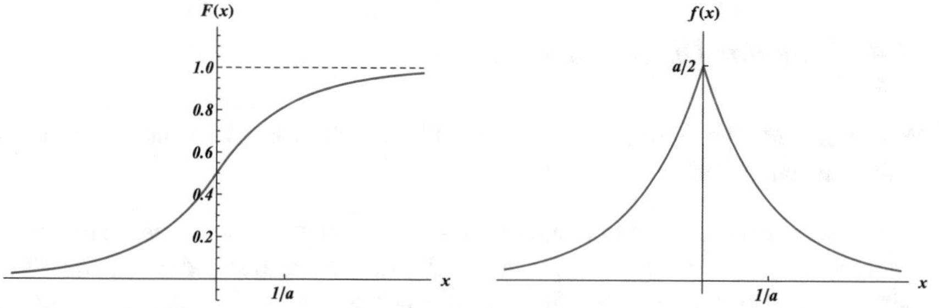

Fig. 2.8 *cdf* and *pdf* of the Laplace distribution $\mathfrak{L}(a)$

and *cdf*

$$F(x) = \frac{1}{2} + \frac{|x|}{x}\frac{1 - e^{-a|x|}}{2}$$

represented in the Fig. 2.8.

Example 2.20 Cauchy distributions: Finally the family of the **Cauchy laws** $\mathfrak{C}(b, a)$ has the *pdf*

$$f(x) = \frac{1}{\pi}\frac{a}{a^2 + (x - b)^2} \qquad a > 0 \tag{2.19}$$

and the *cdf*

$$F(x) = \frac{1}{2} + \frac{1}{\pi}\arctan\frac{x - b}{a}$$

both represented in the Fig. 2.9. It is easy to see from the Figs. 2.6 and 2.9, that the qualitative behavior of the $\mathfrak{N}(b, a^2)$ and $\mathfrak{C}(b, a)$ *pdf*'s are roughly similar: both are bell shaped curves, symmetrically centered around $x = b$ with a width ruled by $a > 0$. They however essentially differ for the velocities of their queues vanishing: while the normal *pdf* asymptotically vanishes rather quickly, the Cauchy *pdf* goes slowly to zero as x^{-2}. As a consequence the central body of the Cauchy *pdf* is thinner w.r.t. the normal function, while its queues are correspondingly fatter.

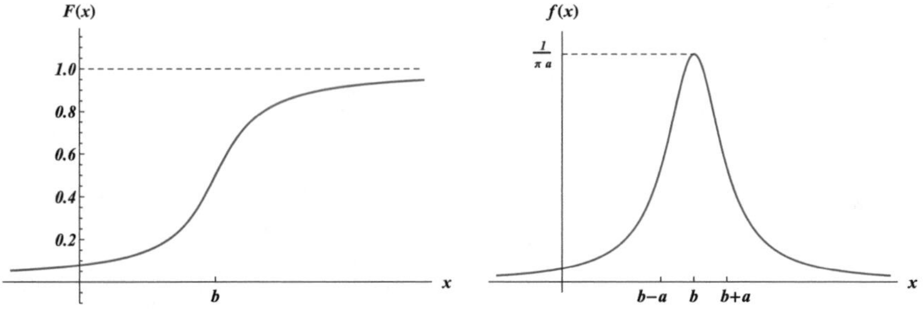

Fig. 2.9 *cdf* and *pdf* of the Cauchy distribution $\mathfrak{C}(b, a)$

2.2.4 Singular Distributions

Definition 2.21 We say that $\boldsymbol{P}$ is a **singular distribution** when its *cdf* $F(x)$ is continuous, but not *ac*

We have seen that the probability measures which are *ac* w.r.t. the Lebesgue measure, namely that with an *ac cdf* $F(x)$, have a *pdf* $f(x)$. We also stated that an *ac* $F(x)$ is also continuous, while instead the reverse is not in general true: there are—but we will neglect here to produce the classical examples—*cdf* $F(x)$ which are continuous (and hence they are not discrete) but not *ac* (and hence have no *pdf*). It is important then to introduce the previous definition to point out that a singular law can be given neither as a discrete distribution (by means of the numbers p_k), nor through a *pdf* $f(x)$: the unique way to define it is to produce a suitable continuous *cdf* $F(x)$ that certainly exists. In the following however we will restrict ourselves to the discrete and *ac* distributions, or—as we will see in the next section—to their *mixtures*, so that the singular distributions will play here only a marginal role.

2.2.5 Mixtures

Definition 2.22 We say that a distribution $\boldsymbol{P}$ is a **mixture** when its *cdf* $F(x)$ is a convex combination of other *cdf*'s, namely when it can be represented as

$$F(x) = \sum_{k=1}^{n} p_k F_k(x), \qquad 0 \le p_k \le 1, \qquad \sum_{k=1}^{n} p_k = 1$$

where $F_k(x)$ for $k = 1, \dots, n$ are arbitrary *cdf*'s

When the $F_k(x)$ are all *ac* with *pdf*'s $f_k(x)$ it is easy to understand that also the mixture $F(x)$ comes out to be *ac* with *pdf*

$$f(x) = \sum_{k=1}^{n} p_k f_k(x)$$

It is not forbidden, however, to have mixtures composed of every possible kind of *cdf*, and the following important result puts in evidence that the three types of distributions so far introduced (discrete, absolutely continuous and singular), along with their mixtures, in fact exhaust all the available possibilities.

Theorem 2.23 Lebesgue-Nikodym theorem: *Every* $\boldsymbol{P}$ *on* $(\boldsymbol{R}, \mathcal{B}(\boldsymbol{R}))$ *can be represented as a mixture of discrete, ac and singular probabilities, namely its cdf* $F(x)$ *always is a convex combination*

$$F(x) = p_1 F_1(x) + p_2 F_2(x) + p_3 F_3(x)$$

where F_1 *is discrete,* F_2 *is ac,* F_3 *is singular, while* p_1, p_2, p_3 *are non negative numbers such that* $p_1 + p_2 + p_3 = 1$.

Proof Omitted: see [2] p. 290. ■

Example 2.24 Mixtures: To elucidate these ideas take for instance the *cdf* displayed in the Fig. 2.10: it is the mixture of a normal $\mathfrak{N}(b, a^2)$ and a Bernoulli $\mathfrak{B}(1; p)$, with arbitrary coefficients p_1, p_2. This $F(x)$ has discontinuities in $x = 0$ and $x = 1$ because of its Bernoulli component, but wherever it is continuous it is not constant (as for a purely discrete distribution) because of its Gaussian component. Remark that in this example the distribution—without being singular—can be given neither as a discrete distribution on 0 and 1, nor by means of a *pdf* $f(x)$: its unique correct representation can be given through its *cdf* $F(x)$.

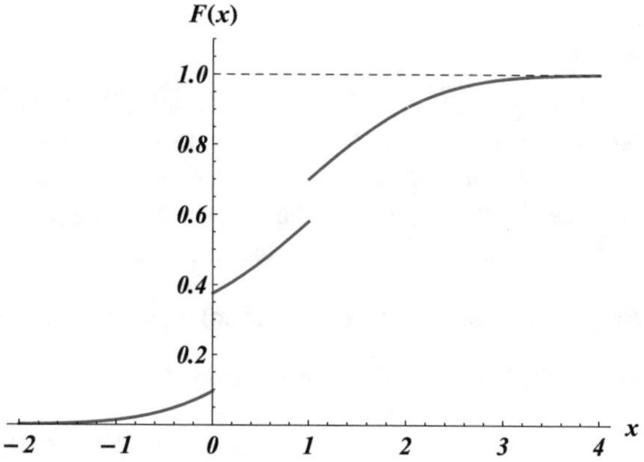

Fig. 2.10 *cdf* of the mixture of a Bernoulli and a Gaussian

2.3 Distributions on $\boldsymbol{R}^n$

In the case of $(\boldsymbol{R}^n, \mathcal{B}(\boldsymbol{R}^n))$ we can extend with a few changes the definitions adopted for $(\boldsymbol{R}, \mathcal{B}(\boldsymbol{R}))$ in the Sect. 2.2, the relevant innovation being the interrelationship between the *marginal distributions* and their (possible) common *joint, multivariate distribution*.

2.3.1 Multivariate Distribution Functions

In analogy with Sezione 2.2.1 take first $\boldsymbol{P}$ as a given probability on $(\boldsymbol{R}^n, \mathcal{B}(\boldsymbol{R}^n))$, and define the n-variate function

$$F(\boldsymbol{x}) = F(x_1, \dots, x_n) = \boldsymbol{P}\{(-\infty, x_1] \times \cdots \times (-\infty, x_n]\} \tag{2.20}$$

where $\boldsymbol{x} = (x_1, \dots, x_n)$. Within the synthetic notation

$$\begin{aligned} \Delta_k F(\boldsymbol{x}) &= F(x_1, \dots, x_k + \Delta x_k, \dots, x_n) - F(x_1, \dots, x_k, \dots, x_n) \\ (\boldsymbol{x}, \boldsymbol{x} + \Delta \boldsymbol{x}] &= (x_1, x_1 + \Delta x_1] \times \cdots \times (x_n, x_n + \Delta x_n] \end{aligned}$$

with $\Delta x_k \geq 0$, it is then possible to show that

$$\boldsymbol{P}\{(\boldsymbol{x}, \boldsymbol{x} + \Delta \boldsymbol{x}]\} = \Delta_1 \dots \Delta_n F(\boldsymbol{x})$$

For instance, in the case $n = 2$ we have

$$\begin{aligned} \boldsymbol{P}\{(\boldsymbol{x}, \boldsymbol{x} + \Delta \boldsymbol{x}]\} &= \Delta_1 \Delta_2 F(\boldsymbol{x}) \\ &= \big[F(x_1 + \Delta x_1, x_2 + \Delta x_2) - F(x_1 + \Delta x_1, x_2)\big] \\ &\qquad - \big[F(x_1, x_2 + \Delta x_2) - F(x_1, x_2)\big] \end{aligned}$$

as it is easy to see from Fig. 2.11. Remark that, at variance with the case $n = 1$, the probability $\boldsymbol{P}\{(\boldsymbol{x}, \boldsymbol{x} + \Delta \boldsymbol{x}]\}$ of a Cartesian product of intervals does not coincide with the simple difference $F(\boldsymbol{x} + \Delta \boldsymbol{x}) - F(\boldsymbol{x})$, but it is a combination of 2^n terms produced by the iteration of the Δ_k operator. The properties of these $F(\boldsymbol{x})$ generalize that of the case $n = 1$ given in the Sect. 2.2.1.

Proposition 2.25 *The function $F(\boldsymbol{x})$ defined in* (2.20) *has the following properties:*

1. *For every $\Delta x_k \geq 0$ with $k = 1, \dots, n$ it is always*

$$\Delta_1 \dots \Delta_n F(\boldsymbol{x}) \geq 0$$

so that $F(\boldsymbol{x})$ comes out to be non decreasing in every variable x_k

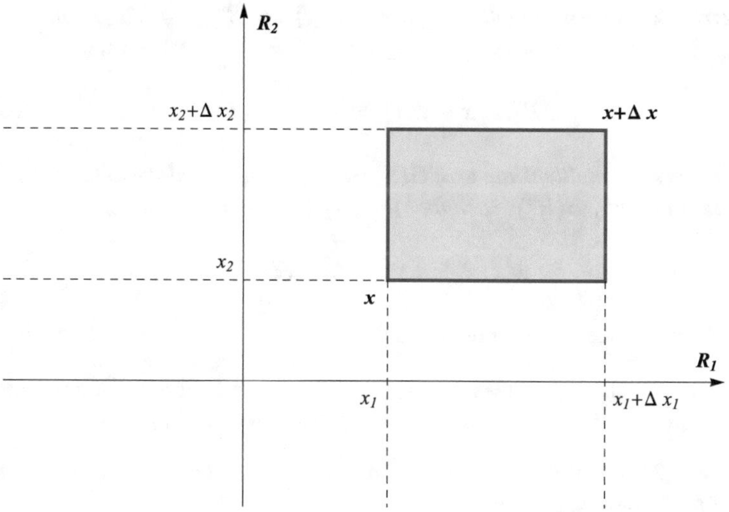

Fig. 2.11 Probability for Cartesian products of intervals

2. *We always have*

$$\lim_{x\to+\infty} F(x) = 1 \qquad\qquad \lim_{x\to-\infty} F(x) = 0$$

where it is understood that the limit $x \to +\infty$ *means that* ***every*** x_k *goes to* $+\infty$, *while the* $x \to -\infty$ *means that* ***at least one*** *among the* x_k *goes to* $-\infty$

3. $F(x)$ *is always continuous* ***from above***

$$\lim_{x_k \downarrow x} F(x_k) = F(x)$$

where it is understood that $x_k \downarrow x$ *means that every component of the sequence* x_k *goes* ***decreasing*** *to the corresponding component of* x.

Proof Property 1. results from the positivity of every probability; property 2. on the other hand comes from the remark that the set

$$(-\infty, +\infty] \times \ldots \times (-\infty, +\infty] = R^n$$

coincides with the whole sample space, while every Cartesian product containing even one empty factor, is itself empty. The argument for the last property is similar to that of the Proposition 2.7 and we will neglect it for short . ■

Definition 2.26 We call **multivariate distribution function** on R^n every function $F(x)$ satisfying 1, 2 and 3; we call on the other hand **generalized, multivariate distribution function** on R^n every function $G(x)$ satisfying 1 and 3, but not necessarily 2.

Theorem 2.27 *Given a multivariate cdf $F(\boldsymbol{x})$ on $\boldsymbol{R}^n$, it exists a unique $\boldsymbol{P}$ on $(\boldsymbol{R}^n, \mathcal{B}(\boldsymbol{R}^n))$ such that for every $\Delta x_k \geq 0$ with $k = 1, \dots, n$ we have*

$$\boldsymbol{P}\{(\boldsymbol{x}, \boldsymbol{x} + \Delta\boldsymbol{x}]\} = \Delta_1 \dots \Delta_n F(\boldsymbol{x})$$

Similarly, given a multivariate gcdf $G(\boldsymbol{x})$ on $\boldsymbol{R}^n$, it exists a unique Lebesgue-Stieltjes measure μ on $(\boldsymbol{R}^n, \mathcal{B}(\boldsymbol{R}^n))$ such that for $\Delta x_k \geq 0,\ k = 1, \dots, n$ it is

$$\mu(\boldsymbol{x}, \boldsymbol{x} + \Delta\boldsymbol{x}] = \Delta_1 \dots \Delta_n G(\boldsymbol{x})$$

Proof Omitted: see also Proposition 2.9. ■

In short, also in the n-variate case, a probability $\boldsymbol{P}$ or a Lebesgue–Stieltjes measure μ are uniquely defined respectively by a *cdf* $F(\boldsymbol{x})$ or by a *gcdf* $G(\boldsymbol{x})$.

Example 2.28 Given the univariate *cdf* of a uniform law on [0, 1], and the univariate *gcdf* of the Lebesgue measure

$$F_1(x) = \begin{cases} 0, & \text{if } x < 0; \\ x, & \text{if } 0 \leq x \leq 1; \\ 1, & \text{if } 1 < x, \end{cases} \qquad\qquad G_1(x) = x$$

it is easy to show that

$$\begin{aligned} F(\boldsymbol{x}) &= F_1(x_1) \cdot \ldots \cdot F_1(x_n) \\ G(\boldsymbol{x}) &= G_1(x_1) \cdot \ldots \cdot G_1(x_n) = x_1 \cdot \ldots \cdot x_n \end{aligned}$$

respectively are the *cdf* of a uniform law on the hypercube $[0, 1]^n$, and the *gcdf* of the Lebesgue measure on $\boldsymbol{R}^n$.

The previous example can be generalized: if $F_1(x), \dots, F_n(x)$ are n *cdf* on $\boldsymbol{R}$, it is easy to show that

$$F(\boldsymbol{x}) = F_1(x_1) \cdot \ldots \cdot F_n(x_n)$$

always is a *cdf* on $\boldsymbol{R}^n$. The reverse, instead, is not true in general: a *cdf* on $\boldsymbol{R}^n$ can not always be factorized in the product of n *cdf* on $\boldsymbol{R}$; this happens only under particular circumstances to be discussed later.

2.3.2 *Multivariate Densities*

When $\boldsymbol{P}$ is *ac* w.r.t. the Lebesgue measure on $(\boldsymbol{R}^n, \mathcal{B}(\boldsymbol{R}^n))$, namely when $F(\boldsymbol{x})$ is *ac*, a generalization of the Radon–Nikodym theorem 2.15 entails the existence of a non negative, normalized ***multivariate density function*** $f(\boldsymbol{x})$

$$\int_{R^n} f(x_1, \ldots, x_n)\, dx_1 \ldots dx_n = \int_{R^n} f(\boldsymbol{x})\, d^n\boldsymbol{x} = 1$$

and in this case we always find

$$F(\boldsymbol{x}) = \int_{-\infty}^{x_1} \ldots \int_{-\infty}^{x_n} f(\boldsymbol{z})\, d^n\boldsymbol{z} \qquad f(\boldsymbol{x}) = \frac{\partial^n F(\boldsymbol{x})}{\partial x_1 \ldots \partial x_n} \tag{2.21}$$

while the probability of the Cartesian products of intervals are given as

$$\boldsymbol{P}\left\{(a_1, b_1] \times \cdots \times (a_n, b_n]\right\} = \int_{a_1}^{b_1} \ldots \int_{a_n}^{b_n} f(\boldsymbol{x})\, d^n\boldsymbol{x}$$

Example 2.29 Multivariate Gaussian (normal) laws: The family of the multivariate normal laws $\mathfrak{N}(\boldsymbol{b}, \mathbb{A})$ is characterized by the vectors of real numbers $\boldsymbol{b} = (b_1, \ldots, b_n)$, and by the symmetric ($a_{ij} = a_{ji}$), and positive definite[1] matrices $\mathbb{A} = \|a_{ij}\|$: the statistical meaning of $\boldsymbol{b}$ and $\mathbb{A}$ will be discussed in the Sect. 3.34. Since $\mathbb{A}$ is positive, its inverse $\mathbb{A}^{-1}$ always exists and $\mathfrak{N}(\boldsymbol{b}, \mathbb{A})$ has a multivariate *pdf*:

$$f(\boldsymbol{x}) = f(x_1, \ldots, x_n) = \sqrt{\frac{|\mathbb{A}^{-1}|}{(2\pi)^n}}\, e^{-\frac{1}{2}(\boldsymbol{x}-\boldsymbol{b})\cdot\mathbb{A}^{-1}(\boldsymbol{x}-\boldsymbol{b})} \tag{2.22}$$

where $|\mathbb{A}^{-1}|$ is the determinant of $\mathbb{A}^{-1}$, $\boldsymbol{x} \cdot \boldsymbol{y} = \sum_k x_k y_k$ is the Euclidean scalar product between the vectors $\boldsymbol{x}$ e $\boldsymbol{y}$, and between vectors and matrices the usual rows by columns product is adopted, so that for instance

$$\boldsymbol{x} \cdot \mathbb{A}\boldsymbol{y} = \sum_{i,j=1}^{n} a_{ij} x_i y_j$$

On the other hand (in analogy with the case $a = 0$ when $n = 1$) the laws $\mathfrak{N}(\boldsymbol{b}, \mathbb{A})$ with a singular, non invertible $\mathbb{A}$ can be defined, but have no *pdf*: they will be discussed in detail in the Sect. 4.2.2. The multivariate, normal *pdf* (2.22) is then a generalization of the univariate case presented in the Sect. 2.2.3: when $n = 1$ the *pdf* of $\mathfrak{N}(b, a^2)$ has just two numerical parameters, b and $a \geq 0$; in the multivariate case, instead, we need a vector $\boldsymbol{b}$ and a symmetric, non-negative matrix $\mathbb{A}$. Remark also that for

---

[1] A matrix $\mathbb{A}$ is ***non-negative definite*** if, however taken a vector of real numbers $\boldsymbol{x} = (x_1, \ldots, x_n)$, it is always

$$\boldsymbol{x} \cdot \mathbb{A}\boldsymbol{x} = \sum_{i,j=1}^{n} a_{ij} x_i x_j \geq 0$$

and it is ***positive definite*** if this sum is always strictly positive (namely non zero). If $\mathbb{A}$ is positive, it is also ***non singular***, namely its determinant $|\mathbb{A}| > 0$ does not vanish, and hence it has an inverse $\mathbb{A}^{-1}$.

$n = 2$, defining $a_k = \sqrt{a_{kk}} > 0, k = 1, 2$, and $r = a_{12}/\sqrt{a_{11}a_{22}}$ with $|r| < 1$, $\mathbb{A}$ and its inverse are

$$\mathbb{A} = \begin{pmatrix} a_1^2 & a_1a_2r \\ a_1a_2r & a_2^2 \end{pmatrix} \qquad \mathbb{A}^{-1} = \frac{1}{(1-r^2)a_1^2a_2^2}\begin{pmatrix} a_2^2 & -a_1a_2r \\ -a_1a_2r & a_1^2 \end{pmatrix} \tag{2.23}$$

and the ***pdf*** **bivariate normal** takes the form

$$f(x_1, x_2) = \frac{e^{-\frac{1}{2(1-r^2)}\left[\frac{(x_1-b_1)^2}{a_1^2} - 2r\frac{(x_1-b_1)(x_2-b_2)}{a_1a_2} + \frac{(x_2-b_2)^2}{a_2^2}\right]}}{2\pi a_1a_2\sqrt{1-r^2}} \tag{2.24}$$

2.3.3 Marginal Distributions

For a given *cdf* $F(\boldsymbol{x})$ on $\boldsymbol{R}^n = \boldsymbol{R}_1 \times \cdots \times \boldsymbol{R}_n$ it is easy to show that the $n-1$ variables function

$$F^{(1)}(x_2, \ldots, x_n) = F(+\infty, x_2, \ldots, x_n) = \lim_{x_1 \to +\infty} F(x_1, x_2 \ldots, x_n) \tag{2.25}$$

again is a *cdf* on $\boldsymbol{R}^{n-1} = \boldsymbol{R}_2 \times \cdots \times \boldsymbol{R}_n$ because it still complies with the properties 1, 2 and 3 listed in the Sect. 2.3.1. This is true in fact whatever x_i we choose to perform the limit; by choosing however different coordinates we get *cdf*'s $F^{(i)}$ which are in general different from each other. To avoid ambiguities we then adopt a notation with upper indices telling the *removed coordinates*. This operation can, moreover, be performed on arbitrary $m < n$ variables: we always get *cdf*'s on a suitable $\boldsymbol{R}^{n-m}$. For instance

$$F^{(1,2)}(x_3, \ldots, x_n) = F(+\infty, +\infty, x_3, \ldots, x_n) \tag{2.26}$$

is a *cdf* on $\boldsymbol{R}^{n-2} = \boldsymbol{R}_3 \times \cdots \times \boldsymbol{R}_n$. At the end of this procedure we find n *cdf*'s with a single variable, as for instance the *cdf* on $\boldsymbol{R}_1$

$$F^{(2,\ldots,n)}(x_1) = F(x_1, +\infty, \ldots, +\infty)$$

All the *cdf*'s deduced from a given multivariate *cdf* F are called ***marginal distribution function***, and the operation to get them is called ***marginalization***. From the Theorem 2.27 we can extend these remarks also to the probability measures: if $\boldsymbol{P}$ is the probability on $(\boldsymbol{R}^n, \mathcal{B}(\boldsymbol{R}^n))$ associated to $F(\boldsymbol{x})$, we can define the marginal probabilities on $\left(\boldsymbol{R}^k, \mathcal{B}(\boldsymbol{R}^k)\right)$ with $k < n$ associated to the corresponding marginal *cdf*'s. The relation among $\boldsymbol{P}$ and its marginals is then for example

$$\begin{aligned} \boldsymbol{P}^{(2,\ldots,n)}\{(-\infty, x_1]\} &= F^{(2,\ldots,n)}(x_1) \\ &= F(x_1, +\infty, \ldots, +\infty) = \boldsymbol{P}\{(-\infty, x_1] \times \boldsymbol{R}_2 \times \cdots \times \boldsymbol{R}_n\} \end{aligned}$$

If the initial multivariate *cdf* F is *ac* also its marginals will be *ac*, and from (2.21) we deduce for instance that the *pdf*'s of $F^{(1)}$, $F^{(1,2)}$ and $F^{(2,\dots,n)}$ respectively are

$$f^{(1)}(x_2,\dots,x_n) = \int_{-\infty}^{+\infty} f(x_1,x_2,\dots,x_n)\,dx_1 \tag{2.27}$$

$$f^{(1,2)}(x_3,\dots,x_n) = \int_{-\infty}^{+\infty}\int_{-\infty}^{+\infty} f(x_1,x_2,x_3,\dots,x_n)\,dx_1dx_2 \tag{2.28}$$

$$f^{(2,\dots,n)}(x_1) = \int_{-\infty}^{+\infty}\dots\int_{-\infty}^{+\infty} f(x_1,x_2,\dots,x_n)\,dx_2\dots dx_n \tag{2.29}$$

For a *pdf*, in other words, the marginalization is performed by integrating on the variables that are to be removed.

Starting hence from a multivariate *cdf* (or *pdf*) on $\boldsymbol{R}^n$ we can always deduce, in an unambiguous way, an entire hierarchy of marginal *cdf*'s with an ever smaller number of variables, until we get n univariate *cdf*'s. It is natural to ask then if this path can also be taken backwards: given a few (either univariate or multivariate) *cdf*'s, is it possible to *unambiguously* find a multivariate *cdf* such that the initial *cdf*'s are its marginals? The answer is, in general, surprisingly *negative, at least for what concerns unicity*, and deserves a short discussion. Take first n arbitrary univariate *cdf*'s $F_1(x),\dots,F_n(x)$: it is easy to see that

$$F(\boldsymbol{x}) = F_1(x_1)\cdot\ldots\cdot F_n(x_n)$$

is again a multivariate *cdf* (see the end of the Sect. 2.3.1) whose univariate marginals are the $F_k(x)$. From the previous marginalization rules we indeed have for instance

$$F^{(2,\dots,n)}(x_1) = F_1(x_1)F_2(+\infty)\dots F_n(+\infty) = F_1(x_1)$$

If moreover the given *cdf*'s are also *ac* with *pdf*'s $f_k(x)$, the product multivariate F will also be *ac* with *pdf*

$$f(\boldsymbol{x}) = f_1(x_1)\cdot\ldots\cdot f_n(x_n)$$

while its marginal *pdf*'s are exactly the $f_k(x)$. It is easy to check, however, by means of a few elementary counterexamples that the previous *product cdf* is *far from the unique* multivariate *cdf* allowing the F_k as its marginals.

Example 2.30 Multivariate distributions and their marginals: Take the following pair of bivariate *pdf*'s which are uniform on the domains[2] represented in the Fig. 2.12

[2]Since the laws with the *pdf*'s f and g are *ac*, the boundaries of the chosen domains have zero measure, and hence we can always take such domains as *closed* without risk of errors.

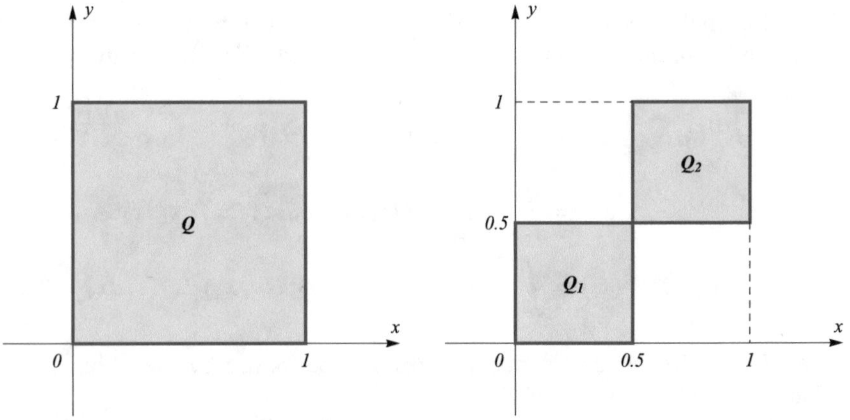

Fig. 2.12 Uniform *pdf* on different domains

$$f(x, y) = \begin{cases} 1, \text{ if } (x, y) \in Q, \\ 0, \text{ if } (x, y) \notin Q, \end{cases} \qquad Q = [0, 1] \times [0, 1]$$

$$g(x, y) = \begin{cases} 2, \text{ if } (x, y) \in Q_1 \cup Q_2, \\ 0, \text{ if } (x, y) \notin Q_1 \cup Q_2, \end{cases} \qquad \begin{aligned} Q_1 &= \left[0, {}^1/_2\right] \times \left[0, {}^1/_2\right] \\ Q_2 &= \left[{}^1/_2, 1\right] \times \left[{}^1/_2, 1\right] \end{aligned}$$

It is easy to show now by elementary integrations that first of all the two univariate marginals of f are uniform $\mathfrak{U}(0, 1)$ respectively with the *pdf*'s

$$f_1(x) = f^{(2)}(x) = \int_{-\infty}^{+\infty} f(x, y)\, dy = \begin{cases} 1, \text{ if } x \in [0, 1], \\ 0, \text{ if } x \notin [0, 1], \end{cases}$$

$$f_2(y) = f^{(1)}(y) = \int_{-\infty}^{+\infty} f(x, y)\, dx = \begin{cases} 1, \text{ if } y \in [0, 1], \\ 0, \text{ if } y \notin [0, 1], \end{cases}$$

and then that they also exactly coincide with the corresponding marginals g_1 and g_2 of g, so that

$$f_1(x) = g_1(x) \qquad f_2(y) = g_2(y)$$

This apparently shows that different, multivariate *pdf*'s can have the same marginals *pdf*'s, and hence that if we just have the marginals we can not **in a unique way** retrace back the multivariate *pdf* engendering them. It is also easy to see, moreover, that in our example

$$f(x, y) = f_1(x) f_2(y) \qquad g(x, y) \neq g_1(x) g_2(y) \tag{2.30}$$

namely that, as already remarked, the product turns out to be a possible bivariate *pdf* with the given marginals, but also that this is not the only possibility.

Example 2.31 Marginals of multivariate Gaussian laws: An elementary, but tiresome integration of the type (2.29)—that we will neglect—explicitly gives the univariate marginals *pdf*'s of a multivariate Gaussian $\mathfrak{N}(\boldsymbol{b}, \mathbb{A})$ (2.22): it turns out that such marginals are again all Gaussian $\mathfrak{N}(b_k, a_k^2)$ as in (2.15), with $a_k^2 = a_{kk}$ and that their *pdf*'s are

$$f_k(x_k) = \frac{1}{a_k\sqrt{2\pi}} e^{-(x_k-b_k)^2/2a_k^2} \tag{2.31}$$

It is easy to understand, however, that in general the product of these *pdf*'s—which still is a multivariate normal *pdf*—does not coincide with the initial multivariate *pdf* (2.22), unless $\mathbb{A}$ is a diagonal matrix: in the simple product, indeed, we would not find the off-diagonal terms of the quadratic form at the exponent of (2.22). Remark in particular that, from the discussion in the Example 2.29, it turns out that the matrix $\mathbb{A}$ of a **bivariate** normal is diagonal *iff* $r = 0$: this point will be resumed in the discussion of the forthcoming Example 3.34.

2.3.4 Copulas

The previous remarks prompt a discussion of the following two interrelated problems:

1. what is the general relation between an n-variate *cdf* $F(x_1, \ldots, x_n)$ and its n univariate marginals $F_k(x)$?
2. given n univariate *cdf*'s $F_k(x)$, do they exist (one or more) n-variate *cdf*'s $F(x_1, \ldots, x_n)$ having the F_k as their marginals? And, if *yes*, how and in how many ways could we retrieve them?

Definition 2.32 We say that a function $C(u, v)$ defined on $[0, 1] \times [0, 1]$ and taking values in $[0, 1]$ is a **copula** when it has the following properties:

1. $C(u, 0) = C(0, v) = 0, \quad \forall\, u, v \in [0, 1]$
2. $C(u, 1) = u, \;\; C(1, v) = v, \quad \forall\, u, v \in [0, 1]$
3. $C(u_2, v_2) - C(u_2, v_1) - C(u_1, v_2) + C(u_1, v_1) \geq 0, \quad \forall\, u_1 \leq u_2,\; v_1 \leq v_2$

In short a copula is the restriction to $[0, 1] \times [0, 1]$ of a *cdf* with uniform marginals $\mathfrak{U}(0, 1)$. Typical examples are

$$\begin{aligned}
C_M(u, v) &= u \wedge v = \min\{u, v\} \\
C_m(u, v) &= (u + v - 1)^+ = \max\{u + v - 1, 0\} \\
C_0(u, v) &= uv \\
C_\theta(u, v) &= (u^{-\theta} + v^{-\theta} - 1)^{-\frac{1}{\theta}} \qquad \theta > 0 \qquad \text{(Clayton)}
\end{aligned}$$

while many others exist in the literature along with their combinations (see [3]). It is also known that every copula $C(u, v)$ falls between the *Fréchet-Höffding bounds*

$$C_m(u, v) \leq C(u, v) \leq C_M(u, v)$$

Theorem 2.33 Sklar theorem (bivariate):

- *If $H(x, y)$ is a bivariate cdf and $F(x) = H(x, +\infty)$, $G(y) = H(+\infty, y)$ are its marginals, there is always a copula $C(u, v)$ such that*

$$H(x, y) = C\,[F(x), G(y)] \tag{2.32}$$

this copula is unique if F and G are continuous; otherwise C is unique only on the points (u, v) which are possible values of $(F(x), G(y))$;
- *if $F(x)$ and $G(y)$ are two cdf, and $C(u, v)$ is a copula, then $H(x, y)$ defined as in* (2.32) *always is a bivariate cdf having F and G as its marginals.*

Proof Omitted: see [3] p. 18. ■

In short the Sklar theorem states that every bivariate *cdf* comes from the application of a suitable copula to its marginals, and that viceversa the application of an arbitrary copula to any pair of univariate *cdf*'s always results in a bivariate *cdf* with the given distributions as marginals. In particular the *product* bivariate of two univariate *cdf*'s comes from the application of the copula C_0, and hence is just one among many other available possibilities. Remark finally that in general the bivariate *cdf* resulting from the application of a copula may be not *ac* even when the two univariate *cdf* are *ac*.

Example 2.34 Cauchy bivariate distributions: Take two Cauchy distributions $\mathfrak{C}(0, 1)$ respectively with *cdf* and *pdf*

$$F(x) = \frac{1}{2} + \frac{1}{\pi}\arctan x \qquad f(x) = \frac{1}{\pi}\frac{1}{1+x^2}$$
$$G(y) = \frac{1}{2} + \frac{1}{\pi}\arctan y \qquad g(y) = \frac{1}{\pi}\frac{1}{1+y^2}$$

The product copula C_0 gives the bivariate *cdf*

$$H_0(x, y) = \left(\frac{1}{2} + \frac{1}{\pi}\arctan x\right)\left(\frac{1}{2} + \frac{1}{\pi}\arctan y\right)$$

which is again *ac* with *pdf*

$$h_0(x, y) = \frac{1}{\pi}\frac{1}{1+x^2}\cdot\frac{1}{\pi}\frac{1}{1+y^2} = f(x)g(y)$$

while the simplest Clayton copula, that with $\theta = 1$

$$C_1(u, v) = (u^{-1} + v^{-1} - 1)^{-1} = \frac{uv}{u + v - uv}$$

would give a different *cdf* $H_1(x, y)$ (we neglect it for brevity) which is still *ac* with the *pdf*

$$h_1(x, y) = \frac{32\pi^2(\pi + 2\arctan x)(\pi + 2\arctan y)}{(1 + x^2)(1 + y^2)[2\arctan x(\pi - 2\arctan y) + \pi(3\pi + 2\arctan y)]^3}$$

On the other hand an application of the extremal Frchet-Höffding copulas C_M and C_m would give rise to different *cdf* which no longer are *ac*, but we will not make explicit reference to them.

The Sklar theorem 2.33 can be generalized to all the multivariate *cdf*'s $H(\boldsymbol{x}) = H(x_1, \ldots, x_n)$ that turn out to be deducible from the application of suitable multivariate copulas $C(\boldsymbol{u}) = C(u_1, \ldots, u_n)$ to univariate *cdf*'s $F_1(x_1), \ldots, F_n(x_n)$. It is also possible to show that even in this case n arbitrary univariate *cdf*'s can always—and in several, different ways, according to the chosen copula—be combined in multivariate *cdf*'s.

A radically different problem arises instead when we try to combine *multivariate* marginals into higher order multivariate *cdf*'s. We will not indulge into details, and we will just restrict us to remark that, given for instance a trivariate *cdf* $F(x, y, z)$, we can always find its three bivariate marginals

$$F^{(1)}(y, z) = F(+\infty, y, z), \quad F^{(2)}(x, z) = F(x, +\infty, z), \quad F^{(3)}(x, y) = F(x, y, +\infty)$$

and that it would be possible—albeit not trivial—to find a way of reassembling F from these marginals by means of suitable copulas. The reverse problem, however, at variance with the case of the Sklar theorem, not always has a solution, because we can not always hope to find a *cdf* $F(x, y, z)$ endowed with three *arbitrarily given* bivariate marginals *cdf* $F_1(y, z)$, $F_2(x, z)$ and $F_3(x, y)$. At variance with the case of the univariate marginals, indeed, first of all a problem of *compatibility* among the given *cdf*'s arises. For instance it is apparent that—in order to be deducible as marginals of the same trivariate $F(x, y, z)$—they must at least agree on the univariate marginals deduced from them, namely we should have

$$F_1(+\infty, z) = F_2(+\infty, z), \quad F_1(y, +\infty) = F_3(+\infty, y), \quad F_2(x, +\infty) = F_3(x, +\infty)$$

while even this (only necessary) condition can not be ensured if F_1, F_2 e F_3 are totally arbitrary. In short, the choice of the multivariate marginal *cdf*'s must be made according to some suitable *consistency* criterion, arguably not even restricted just to the simplest one previously suggested: a short discussion on this point can be found in the Appendix A.

2.4 Distributions on R^∞ and R^T

The extension of the previous results to the case of the space $(\boldsymbol{R}^\infty, \mathcal{B}(\boldsymbol{R}^\infty))$ of the real sequences is understandably less straightforward because, while on a $(\boldsymbol{R}^n, \mathcal{B}(\boldsymbol{R}^n))$ we can always give a probability by means of an n-variables *cdf*, this is not possible for $(\boldsymbol{R}^\infty, \mathcal{B}(\boldsymbol{R}^\infty))$ because it would be meaningless to have a *cdf* with an *infinite* number of variables. To give a probability on $(\boldsymbol{R}^\infty, \mathcal{B}(\boldsymbol{R}^\infty))$ we must hence use different tools.

To this end remark first that if a probability $\boldsymbol{P}$ is given on $(\boldsymbol{R}^\infty, \mathcal{B}(\boldsymbol{R}^\infty))$, we could inductively deduce a whole family of probabilities on the finite dimensional spaces that we get by selecting an arbitrary, but finite, number of sequence components. As a matter of fact n arbitrary components of the sequences in $(\boldsymbol{R}^\infty, \mathcal{B}(\boldsymbol{R}^\infty))$ always are a point in a space $(\boldsymbol{R}^n, \mathcal{B}(\boldsymbol{R}^n))$: to give a probability on this $(\boldsymbol{R}^n, \mathcal{B}(\boldsymbol{R}^n))$ from the given $\boldsymbol{P}$ it would be enough to take $B \in \mathcal{B}(\boldsymbol{R}^n)$ as the basis of a *cylinder* in $\mathcal{B}(\boldsymbol{R}^\infty)$ (see Example 1.7), and then give to B the probability that $\boldsymbol{P}$ gives to the cylinder. We get in this way an entire family of finite probability spaces which are **consistent** (see also Appendix A), in the sense that the *cdf*'s in a $\left(\boldsymbol{R}^k, \mathcal{B}(\boldsymbol{R}^k)\right)$ which is subspace of an $(\boldsymbol{R}^n, \mathcal{B}(\boldsymbol{R}^n))$ with $k \leq n$ are derived by marginalization of the extra components through the usual relations (2.25) and (2.26).

This prompt the idea of defining a probability on $(\boldsymbol{R}^\infty, \mathcal{B}(\boldsymbol{R}^\infty))$ through the *reverse procedure*: give first a family of probabilities on all the finite subspaces $(\boldsymbol{R}^n, \mathcal{B}(\boldsymbol{R}^n))$, and then extend them to all $(\boldsymbol{R}^\infty, \mathcal{B}(\boldsymbol{R}^\infty))$. In order to get a successful procedure, however, these finite probabilities can not be given in a totally arbitrary way: they must indeed be a ***consistent family of probabilities***, in the sense of the previously discussed *consistence*. The subsequent theorem encodes this important result.

Theorem 2.35 Kolmogorov theorem on $\boldsymbol{R}^\infty$: *Given a consistent family of finite probability spaces $(\boldsymbol{R}^n, \mathcal{B}(\boldsymbol{R}^n), \boldsymbol{P}_n)$ there is always a unique probability $\boldsymbol{P}$ on $(\boldsymbol{R}^\infty, \mathcal{B}(\boldsymbol{R}^\infty))$ which is an extension of the given family.*

Proof Omitted: see [1] p. 163. ■

Example 2.36 Bernoulli sequences: The simplest way to meet the conditions of the Theorem 2.35 is to take a sequence of univariate *cdf*'s $G_k(x)$, $k \in N$ and to define then another sequence of multivariate *cdf*'s as

$$F_n(x_1, \ldots, x_n) = G_1(x_1) \cdot \ldots \cdot G_n(x_n) \qquad n \in N$$

According to the Theorem 2.27 we can then define on $(\boldsymbol{R}^n, \mathcal{B}(\boldsymbol{R}^n))$ the probabilities $\boldsymbol{P}_n$ associated to F_n, and we can check that these $\boldsymbol{P}_n$ are a consistent family of probabilities: according to the Kolmogorov theorem 2.35 it exists then a unique probability $\boldsymbol{P}$ on $(\boldsymbol{R}^\infty, \mathcal{B}(\boldsymbol{R}^\infty))$ such that

$$\boldsymbol{P}\left\{x \in \boldsymbol{R}^\infty : (x_1, \ldots, x_n) \in B\right\} = \boldsymbol{P}_n\{B\} \qquad \forall B \in \mathcal{B}(\boldsymbol{R}^n)$$

and in particular

$$\boldsymbol{P}\left\{x \in \boldsymbol{R}^\infty : x_1 \leq a_1, \ldots, x_n \leq a_n\right\} = F_n(a_1, \ldots, a_n) = G_1(a_1) \cdot \ldots \cdot G_n(a_n)$$

If for example all the $G_n(x)$ are identical Bernoulli distributions $\mathfrak{B}(1; p)$ so that

$$G_n(x) = \begin{cases} 0, & \text{if } x < 0; \\ 1-p, & \text{if } 0 \le x < 1; \\ 1, & \text{if } 1 \le x, \end{cases}$$

we can define $\boldsymbol{P}$ of x_j sequences taking values $a_j = 0, 1$, so that for every $k = 0, 1, \dots, n$

$$\boldsymbol{P}\left\{ x \in \boldsymbol{R}^\infty \,:\, x_1 = a_1, \dots, x_n = a_n, \text{ with } \sum_{j=1}^n a_j = k \right\} = p^k q^{n-k}$$

Such a $\boldsymbol{P}$ extends to the (uncountable) space of infinite sequences of draws (**Bernoulli sequences**) the binomial distributions defined by (2.6) and (2.7) on the finite spaces of n-tuples of draws, as shown in the Sect. 2.1.2: this extension is crucial in order to be able to define limits for an infinite number of drawings as will be seen in the Appendix F.

Take finally the space $\big(\boldsymbol{R}^T, \mathcal{B}(\boldsymbol{R}^T)\big)$ and suppose first, as for $(\boldsymbol{R}^\infty, \mathcal{B}(\boldsymbol{R}^\infty))$, to have a probability $\boldsymbol{P}$ on it. This allows again (by adopting the usual cylinder procedure of the Example 1.7) to get an entire family of probabilities on the finite dimensional subspaces which are consistent as in the case of $\boldsymbol{R}^\infty$. We ask then if, by starting backward from a consistent family of probability spaces, we can extend it again to a $\boldsymbol{P}$ on $\big(\boldsymbol{R}^T, \mathcal{B}(\boldsymbol{R}^T)\big)$: this would ensure the definition of a probability on the *(uncountably) infinite dimensional* space $\big(\boldsymbol{R}^T, \mathcal{B}(\boldsymbol{R}^T)\big)$ by giving an infinite consistent family of probabilities on *finite dimensional* spaces. The positive answer to this question is in the following theorem.

Theorem 2.37 Kolmogorov theorem on $\boldsymbol{R}^T$: *Take $S = \{t_1, \dots, t_n\}$ arbitrary finite subset of T, and a consistent family of probability spaces $(\boldsymbol{R}^n, \mathcal{B}(\boldsymbol{R}^n), \boldsymbol{P}_S)$: then there always exists a unique probability $\boldsymbol{P}$ on $\big(\boldsymbol{R}^T, \mathcal{B}(\boldsymbol{R}^T)\big)$ which turns out to be an extension of the given family.*

Proof Omitted: see [1] p. 167. ■

Example 2.38 Wiener measure: Consider $T = [0, +\infty)$, so that $\boldsymbol{R}^T$ will turn out to be the set of the functions $(x_t)_{t \ge 0}$, and take the following family of Gaussian $\mathfrak{N}(0, t)$ *pdf*'s

$$\varphi_t(x) = \frac{e^{-x^2/2t}}{\sqrt{2\pi t}} \qquad t > 0$$

Given then $S = \{t_1, \dots, t_n\}$ with $0 < t_1 < \dots < t_n$, and a Borel set in $\boldsymbol{R}^n$, for instance $B = A_1 \times \dots \times A_n \in \mathcal{B}(\boldsymbol{R}^n)$, define $\boldsymbol{P}_S$ by giving to B the probability

$$\boldsymbol{P}_S\{B\} = \int_{A_n} \dots \int_{A_1} \varphi_{t_n - t_{n-1}}(x_n - x_{n-1}) \dots \varphi_{t_2 - t_1}(x_2 - x_1)\varphi_{t_1}(x_1)\, dx_1 \dots dx_n$$

It is possible to check then that, with every possible S, the family $(\boldsymbol{R}^n, \mathcal{B}(\boldsymbol{R}^n), \boldsymbol{P}_S)$ is consistent, and hence according to the Theorem 2.37 there exists a unique probability $\boldsymbol{P}$ defined on $\big(\boldsymbol{R}^T, \mathcal{B}(\boldsymbol{R}^T)\big)$ as an extension of the given family. This probability is also called **Wiener measure** and plays a crucial role in the theory of the stochastic processes. Its meaning could be intuitively clarified as follows: if $(x_t)_{t\geq 0}$ is the generic trajectory of a point particle, the cylinder of basis $B = A_1 \times \cdots \times A_n$ will be the bundle of the trajectories starting from $x = 0$, and passing through the windows $A_1, \ldots, A_n$ at the times $t_1 < \cdots < t_n$. The $\varphi_{t_k - t_{k-1}}(x_k - x_{k-1})\, dx_k$ are moreover the (Gaussian) probabilities that the particle, starting from x_{k-1} at the time t_{k-1}, will be in $[x_k, x_k + dx_k]$ after a delay $t_k - t_{k-1}$, while the product of these *pdf*'s appearing in the definition indicates the displacements independence in the time intervals $[0, t_1], [t_1, t_2], \ldots, [t_{n-1}, t_n]$. The multiple integral of the definition, finally, allows to calculate the probability attributed to the bundle of trajectories that, at the times $t_1 < \cdots < t_n$, go through the windows $A_1, \ldots, A_n$.

References

1. Shiryaev, A.N.: Probability. Springer, New York (1996)
2. Métivier, M.: Notions Fondamentales de la Théorie des Probabilités. Dunod, Paris (1972)
3. Nelsen, R.B.: An Introduction to Copulas. Springer, New York (1999)

Chapter 3
Random Variables

3.1 Random Variables

3.1.1 Measurability

Definition 3.1 Given the probabilizable spaces $(\Omega, \mathcal{F})$ e $(\boldsymbol{R}, \mathcal{B}(\boldsymbol{R}))$, a function $X : \Omega \to \boldsymbol{R}$ is said to be $\mathcal{F}$-**measurable**—or simply **measurable** when there is no ambiguity—if (see also Fig. 3.1)

$$X^{-1}(B) = \{X \in B\} = \big\{\omega \in \Omega : X(\omega) \in B\big\} \in \mathcal{F}, \qquad \forall B \in \mathcal{B}(\boldsymbol{R}).$$

while to mention the involved σ-algebras we often write

$$X : (\Omega, \mathcal{F}) \to (\boldsymbol{R}, \mathcal{B}(\boldsymbol{R}))$$

In probability a measurable X is also called **random variable** (*rv*), and when $(\Omega, \mathcal{F})$ coincides with $(\boldsymbol{R}^n, \mathcal{B}(\boldsymbol{R}^n))$ it is called **Borel function**.

Remark that in the previous definition no role whatsoever is played by the probability measures: X is a *rv* as a result of its measurability only. On the other hand it is indispensable to single out the two involved σ-algebras without which our definition would be meaningless

Example 3.2 Indicators, and degenerate and simple *rv*'s: The simplest *rv*'s are the **indicators** $I_A(\omega)$ **of an event** $A \in \mathcal{F}$ defined as

$$I_A(\omega) = \begin{cases} 1, & \text{if } \omega \in A, \\ 0, & \text{if } \omega \notin A, \end{cases}$$

N. Cufaro Petroni, *Probability and Stochastic Processes for Physicists*, UNITEXT for Physics, https://doi.org/10.1007/978-3-030-48408-8_3

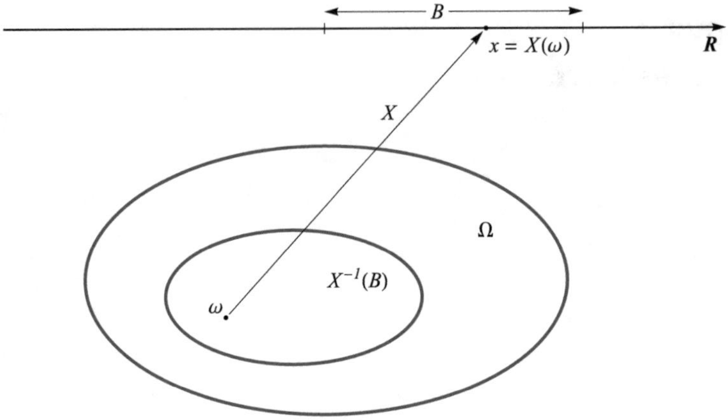

Fig. 3.1 Graphic depiction of the random variable definition

which apparently are measurable w.r.t. $\mathcal{F}$ since $A \in \mathcal{F}$. The indicators have several properties: for instance it is easy to check that $\forall\omega \in \Omega$

$$I_{\emptyset}(\omega) = 0 \qquad I_{\Omega}(\omega) = 1 \qquad I_A(\omega) + I_{\bar{A}}(\omega) = 1$$
$$I_{AB}(\omega) = I_A(\omega)\, I_B(\omega) \qquad I_{A\cup B}(\omega) = I_A(\omega) + I_B(\omega) - I_A(\omega)\, I_B(\omega)$$

Aside from the indicators, a relevant role is played by the **simple *rv*'s**, namely those taking just a finite number of values $x_k,\ k = 1, \dots, n$ according to

$$X(\omega) = \sum_{k=1}^{n} x_k I_{D_k}(\omega)$$

where the $D_k \in \mathcal{F}$ are a finite decomposition of Ω: in short the simple *rv* X takes the value x_k on the $\omega \in D_k$ with $k = 1, \dots, n$. It is not excluded finally the case of a **degenerate (constant) *rv*** which takes just one value b on Ω.

In short every measurable function $X : \Omega \to \boldsymbol{R}$ is a *rv* and can be considered first as a simple way to ascribe numerical values in $\boldsymbol{R}$ to every $\omega \in \Omega$. Every procedure that for example awards money winnings to the sides of a dice (the measurability is trivially met in these simple cases) can be considered as a *rv*. The rationale to require the measurability in the general case will be made clear in the next section.

3.1.2 Laws and Distributions

The measurability comes into play when we take a probability $\boldsymbol{P}$ on $(\Omega, \mathcal{F})$: the fact that $\{X \in B\} \in \mathcal{F}$, $\forall B \in \mathcal{B}(\boldsymbol{R})$ enables indeed the *rv* X to induce on $(\boldsymbol{R}, \mathcal{B}(\boldsymbol{R}))$ a new probability as explained in the following definition

Definition 3.3 Given a probability space $(\Omega, \mathcal{F}, \boldsymbol{P})$ and a *rv* $X : (\Omega, \mathcal{F}) \to (\boldsymbol{R}, \mathcal{B}(\boldsymbol{R}))$, the **law or distribution of** X is the new probability $\boldsymbol{P}_X$ induced by X on $(\boldsymbol{R}, \mathcal{B}(\boldsymbol{R}))$ through the relation

$$\boldsymbol{P}_X\{B\} = \boldsymbol{P}\{X \in B\}, \qquad B \in \mathcal{B}(\boldsymbol{R})$$

while the *cdf* $F_X(x)$ of $\boldsymbol{P}_X$, namely

$$F_X(x) = \boldsymbol{P}_X\{(-\infty, x]\} = \boldsymbol{P}\{X \le x\} = \boldsymbol{P}\{\omega \in \Omega : X(\omega) \le x\}, \qquad x \in \boldsymbol{R}$$

is called **distribution function *(cdf)* of** X, and coincides with the probability that *rv* X be smaller than, or equal to x. We will finally adopt the shorthand notation

$$X \sim \boldsymbol{P}_X$$

to indicate that $\boldsymbol{P}_X$ is the distribution of the *rv* X

It is apparent that two *rv*'s X and Y, different in the sense of the Definition 3.1, in general have differen laws $\boldsymbol{P}_X$ e $\boldsymbol{P}_Y$, but we must remark at once that it is not impossible for them to have the same law and hence to be *identically distributed.* For instance, on the probability space of a fair dice, we can define the following two *rv*'s: X taking value 1 on the first four sides of the dice (and 0 on the other two), and Y taking value 0 on the first two sides (and 1 on the remaining four). Albeit different as functions of ω, X and Y take the same values (0 e 1) with the same probabilities (respectively $^1/_3$ and $^2/_3$), and hence they have the same distribution. On the other hand, a given *rv* X can have several different laws according to the different probabilities $\boldsymbol{P}$ that we may define on $(\Omega, \mathcal{F})$: remember for instance that every conditioning modifies the probability on $(\Omega, \mathcal{F})$. These remarks show that the law—even though that only is practically accessible to our observations—does not define the *rv*, but it only gives its statistical behavior. It is relevant then to add a few words about what it possibly means *to be equal* for two or more *rv*'s

Definition 3.4 Two *rv*'s X and Y defined on the same $(\Omega, \mathcal{F}, \boldsymbol{P})$ are

- **indistinguishable**, and we simply write $X = Y$, when

$$X(\omega) = Y(\omega) \qquad \forall\omega \in \Omega$$

- **identical *P*-a.s.**, and we also write $X \overset{as}{=} Y$, when

$$\boldsymbol{P}\{X \neq Y\} = \boldsymbol{P}\{\omega \in \Omega : X(\omega) \neq Y(\omega)\} = 0$$

- **identically distributed** ***(id)***, and we also write $X \stackrel{d}{=} Y$, when their laws coincide namely if $\boldsymbol{P}_X = \boldsymbol{P}_Y$ so that

$$F_X(x) = F_Y(x) \qquad \forall x \in \boldsymbol{R}$$

It is apparent that indistinguishable *rv*'s also are identical $\boldsymbol{P}$-a.s., and that *rv*'s identical $\boldsymbol{P}$-a.s. also are *id*: the reverse instead does not hold in general as could be shown with a few counterexamples that we will neglect. We will give now a classification of the *rv*'s according to their laws

3.1.2.1 Discrete *rv*'s

The *discrete rv*'s are of the type

$$X(\omega) = \sum_k x_k I_{D_k}(\omega)$$

where k ranges on a finite or countable set of integers, while the events $D_k = \{X = x_k\}$ are a decomposition of Ω. It is easy to see that the distribution $\boldsymbol{P}_X$ is in this case a discrete probability on the—at most countable—set of numbers $x_k \in \boldsymbol{R}$, so that its *cdf* is *discrete* in the sense discussed in the Sect. 2.2.1. If then we take

$$p_k = \boldsymbol{P}_X\{x_k\} = \boldsymbol{P}\{X = x_k\} = F_X(x_k) - F_X(x_k^-)$$

we also have

$$\boldsymbol{P}_X\{B\} = \sum_{k\,:\,x_k \in B} p_k, \qquad B \in \mathcal{B}(\boldsymbol{R})$$

Apparently the *simple rv*'s previously introduced are discrete *rv*'s taking just a finite number of values, and so are also the ***P****-a.s. degenerate rv*'s taking just the value b with probability 1, namely with $\boldsymbol{P}_X\{b\} = 1$. The discrete *rv*'s are identified according to their laws, namely we will have for example degenerate *rv*'s $\boldsymbol{\delta}_b$, Bernoulli $\mathfrak{B}(1; p)$, binomial $\mathfrak{B}(n; p)$ and Poisson $\mathfrak{P}(\alpha)$ whose distributions have already been discussed in the Sect. 2.2.2.

3.1.2.2 Continuous and *ac rv*'s

A *rv* is said *continuous* if its *cdf* $F_X(x)$ is a continuous function of x, and in particular it is said *absolutely continuous* (*ac*) if its F_X is *ac*, namely if there exists a non negative *probability density* (*pdf*) $f_X(x)$, normalized and such that

$$F_X(x) = \int_{-\infty}^{x} f_X(y)\,dy$$

Remember that not every continuous *rv* also is *ac*: there are indeed continuous, but singular *rv*'s (see Sect. 2.2.4). These *rv*'s are however devoid of any practical value, so that often in the applied literature the *ac rv*'s are just called *continuous*. All the remark and examples of the Sect. 2.2.3 can of course be extended to the *ac rv*'s that will be classified again according to their laws: we will have then uniform *rv*'s $\mathfrak{U}(a, b)$, Gaussian (normal) $\mathfrak{N}(b, a^2)$, exponential $\mathfrak{E}(a)$, Laplace $\mathfrak{L}(a)$, Cauchy $\mathfrak{C}(b, a)$ and so on. We finally remember here—but this will be reconsidered in further detail later—how to calculate $\boldsymbol{P}_X(B)$ from a *pdf* $f_X(x)$: for $B = (-\infty, x]$ from the given definitions we get first

$$\boldsymbol{P}_X\{(-\infty, x]\} = \boldsymbol{P}\{X \le x\} = F_X(x) = \int_{-\infty}^{x} f_X(t)\, dt = \int_{(-\infty,x]} f_X(t)\, dt$$

so that for $B = (a, b]$ we have

$$\boldsymbol{P}_X\{(a, b]\} = \boldsymbol{P}\{a < X \le b\} = F_X(b) - F_X(a) = \int_{a}^{b} f_X(t)\, dt = \int_{(a,b]} f_X(t)\, dt$$

With an intuitive generalization we can extend by analogy this result to an arbitrary $B \in \mathcal{B}(\boldsymbol{R})$ so that

$$\boldsymbol{P}_X\{B\} = \int_B dF_X(x) = \int_B f_X(x)\, dx$$

Its precise meaning will be presented however in the discussion of the Corollary 3.23.

3.1.3 Generating New *rv's*

We have shown that a *rv* X on $(\Omega, \mathcal{F}, \boldsymbol{P})$ is always equipped with a distribution given by means of a *cdf* $F_X(x)$, but in fact also the reverse holds. If we have indeed just a *cdf* $F(x)$, but neither a *rv* or a probability space, we can always consider $\Omega = \boldsymbol{R}$ as sample space, and the *rv* X defined as the *identical map* $X : \boldsymbol{R} \to \boldsymbol{R}$ that to every $x = \omega \in \Omega$ associates the same number $x \in \boldsymbol{R}$. If then on $(\Omega, \mathcal{F}) = (\boldsymbol{R}, \mathcal{B}(\boldsymbol{R}))$ we define a $\boldsymbol{P}$ by means of the given *cdf* $F(x)$, it is easy to acknowledge that the *cdf* of X exactly coincides with $F(x)$

Definition 3.5 If on $(\Omega, \mathcal{F}) = (\boldsymbol{R}, \mathcal{B}(\boldsymbol{R}))$ we take the probability $\boldsymbol{P}$ defined through a given *cdf* $F(x)$, the identical map X from $(\boldsymbol{R}, \mathcal{B}(\boldsymbol{R}), \boldsymbol{P})$ to $(\boldsymbol{R}, \mathcal{B}(\boldsymbol{R}))$ will be called **canonical *rv***, and its *cdf* will coincide with the given $F(x)$

3.1.3.1 Functions of *rv*'s

Proposition 3.6 *Given a rv $X(\omega)$ $(\omega \in \Omega)$ and a Borel function $\phi(x)$ $(x \in \boldsymbol{R})$ defined (at least) on the range of X, the compound function*

$$\phi\big[X(\omega)\big] = Y(\omega)$$

is measurable again and hence is a rv

Proof Omitted: see [1] p. 172 ■

According to the previous proposition, every $Y = \phi(X)$ is a *rv* if X is a *rv* and ϕ a Borel function: for instance X^n, $|X|$, $\cos X$, e^X and so on are all *rv*'s. On the other hand it is important to remark that, given two *rv*'s X and Y, does not necessarily exist a Borel function ϕ such that Y coincides with $\phi(X)$: this could be shown by means of simple examples that we will neglect. To further discuss this point let us introduce a new notion: if $X : (\Omega, \mathcal{F}) \to (\boldsymbol{R}, \mathcal{B}(\boldsymbol{R}))$ is a *rv* it is possible to show that the following family of subsets of Ω

$$\mathcal{F}_X = \big(X^{-1}(B)\big)_{B\in\mathcal{B}(\boldsymbol{R})} \subseteq \mathcal{F}$$

again is a σ-algebra taking the name of ***σ-algebra generated by*** X. Apparently $\mathcal{F}_X$ is also the *smallest* σ-algebra of Ω subsets w.r.t. which X turns out to be measurable. Remark that given two *rv*'s X and Y their respective σ-algebras $\mathcal{F}_X$ and $\mathcal{F}_Y$ in general neither coincide, nor are are included in one another. The following result is then particularly relevant

Theorem 3.7 *Given two rv's X and Y, if Y turns out to be $\mathcal{F}_X$-measurable (namely if $\mathcal{F}_Y \subseteq \mathcal{F}_X$) then there exists a Borel function ϕ such that $Y(\omega) = \phi\big[X(\omega)\big]\ \forall\,\omega \in \Omega$.*

Proof Omitted: see [1] p. 174 ■

In short, the *rv* Y happens to be a function of another *rv* X if (and only if) all the events in its σ-algebra $\mathcal{F}_Y$ also are events in $\mathcal{F}_X$, namely when every statement about Y (the events in $\mathcal{F}_Y$) can be reformulated as an equivalent statement about X (namely are also events in $\mathcal{F}_X$)

3.1.3.2 Limits of Sequences of *rv*'s

An alternative procedure to generate *rv*'s consists in taking *limits of sequences* of *rv*'s $(X_n)_{n\in\boldsymbol{N}}$. To this end we will preliminarily introduce a suitable definition of ***convergence***, even if this point will be subsequently considered in more detail in the Sect. 4.1. Remark first that in the following the qualification *convergent* will be attributed to both the sequences convergent to finite limits, and that divergent either to $+\infty$ or to $-\infty$; the wording *non convergent* will be instead reserved for the sequences that do not admit a limit whatsoever. Remember moreover that the notion of convergence that we define here is just the first among the several (non equivalent) that will be introduced later

A sequence of *rv*'s $(X_n)_{n\in\boldsymbol{N}}$ is a sequence not of numbers, but of functions of $\omega \in \Omega$. Only when we fix an ω the elements X_n of the sequence will take the particular values x_n and we get the numerical sequence $(x_n)_{n\in\boldsymbol{N}}$ as the *sample trajectory* of our sequence of *rv*'s. By choosing on the other hand a different ω' we will get a different numerical sequence $(x'_n)_{n\in N}$, and so on with ω''... We can then think of the sequence of *rv*'s $(X_n)_{n\in N}$ as the set of all its possible samples $(x_n)_{n\in N}$ obtained according to the chosen $\omega \in \Omega$

Definition 3.8 We say that a sequence of *rv*'s $(X_n)_{n\in N}$ **pointwise convergent** when all the numerical sequences $x_n = X_n(\omega)$ converge for every $\omega \in \Omega$

Of course, when our sequence $(X_n)_{n\in N}$ is pointwise convergent every different sample $(x_n)_{n\in N}$ either converges toward a different number x, or diverges toward $\pm\infty$. As a consequence the limit is a new function $X(\omega)$ of $\omega \in \Omega$, and the following results state indeed that such a function again is a *rv*, while in the reverse every *rv* X can be recovered as a limit of suitable simple *rv*'s

Proposition 3.9 *If the sequence of rv's $(X_n)_{n\in N}$ is pointwise convergent toward $X(\omega)$, then X too is measurable, namely it is a rv*

Proof Omitted: see [1] p. 173 ■

Theorem 3.10 Lebesgue theorem: *If X is a non negative rv ($X \geq 0$), we can always find a sequence $(X_n)_{n\in N}$ of simple, non decreasing rv's*

$$0 \leq X_n \leq X_{n+1} \leq X, \qquad \forall\omega \in \Omega, \quad \forall n \in N$$

which is pointwise convergent (from below) toward X, and we will also write $X_n \uparrow X$. If instead X is an arbitrary rv we can always find a sequence $(X_n)_{n\in N}$ of simple rv's with

$$|X_n| \leq |X|, \qquad \forall\omega \in \Omega, \quad \forall n \in N$$

which is pointwise convergent toward X

Proof Omitted: see [1] p. 173 ■

From the previous results we can prove that, if X and Y are *rv*'s, then also $X \pm Y$, XY, X/Y ... are *rv*'s, provided that they do not take one of the indeterminate forms $\infty - \infty$, ∞/∞, $0/0$. Remark finally that sometimes, in order to take into account the possible $\pm\infty$ limits for some ω, we will suppose that our *rv*'s can also take the values $+\infty$ and $-\infty$. In this case we speak of ***extended rv*'s** which however, with some caution, have the same properties of the usual *rv*'s.

3.2 Random Vectors and Stochastic Processes

3.2.1 Random Elements

The notion of a *rv* as a measurable function $X : (\Omega, \mathcal{F}) \to (\boldsymbol{R}, \mathcal{B}(\boldsymbol{R}))$ can be generalized by allowing values in spaces different from $(\boldsymbol{R}, \mathcal{B}(\boldsymbol{R}))$. The sole property

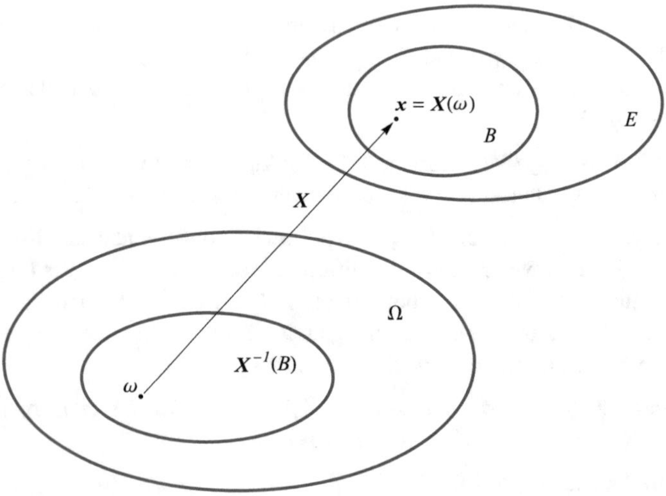

Fig. 3.2 Graphic depiction of the random element definition

of $(\boldsymbol{R}, \mathcal{B}(\boldsymbol{R}))$ which is relevant for the definition is indeed to be a probabilizable space: we can then suppose that our functions take values in more general spaces, and we will speak of *random elements* that, when not reduced to a simple *rv*, will be denoted as $\boldsymbol{X}$

Definition 3.11 Take two probabilizable spaces $(\Omega, \mathcal{F})$ and $(\mathrm{E}, \mathcal{E})$: we say that a function $\boldsymbol{X} : (\Omega, \mathcal{F}) \to (\mathrm{E}, \mathcal{E})$ is a **random element** when it is $\mathcal{F}/\mathcal{E}$-measurable, namely when (see also Fig. 3.2)

$$\boldsymbol{X}^{-1}(B) \in \mathcal{F}, \qquad \forall\, B \in \mathcal{E}$$

3.2.1.1 Random Vectors

Suppose first that $(\mathrm{E}, \mathcal{E}) = (\boldsymbol{R}^n, \mathcal{B}(\boldsymbol{R}^n))$ with $\boldsymbol{R}^n = \boldsymbol{R}_1 \times \cdots \times \boldsymbol{R}_n$ Cartesian product of n replicas of the real line: the values taken by the random element $\boldsymbol{X}$ are then the n-tuples of real numbers

$$\boldsymbol{x} = (x_1, \ldots, x_n) = (x_k)_{k=1,\ldots,n} \in \boldsymbol{R}^n$$

and in this case we say that $\boldsymbol{X}$ is a ***random vector (r-vec)*** taking values in $\boldsymbol{R}^n$. It would be easy to see that it can be equivalently represented as an n-tuple $(X_k)_{k=1,\ldots,n}$ of *rv*'s X_k each taking values in $\big(\boldsymbol{R}_k, \mathcal{B}(\boldsymbol{R}_k)\big)$ and called ***components*** of the *r-vec*: in short to take a *r-vec* $\boldsymbol{X} = (X_k)_{k=1,\ldots,n}$ is equivalent to take n *rv*'s as its components. In particular consider the case $(\mathrm{E}, \mathcal{E}) = \big(\boldsymbol{C}, \mathcal{B}(\boldsymbol{C})\big)$ where $\boldsymbol{C}$ is the set of the complex numbers $z = x + iy$: since $\boldsymbol{C}$ and $\boldsymbol{R}^2$ are isomorphic, a ***complex rv*** $\boldsymbol{Z}$ will be nothing else than a *r-vec* with $n = 2$ whose components are its real and imaginary parts according to $\boldsymbol{Z}(\omega) = X(\omega) + iY(\omega)$, where X and Y are real *rv*'s

3.2.1.2 Random Sequences

When instead $(\mathrm{E}, \mathcal{E}) = (\boldsymbol{R}^\infty, \mathcal{B}(\boldsymbol{R}^\infty))$ the values of the random element are sequences of real number

$$\boldsymbol{x} = (x_1, \ldots, x_n, \ldots) = (x_n)_{n \in \boldsymbol{N}} \in \boldsymbol{R}^\infty$$

and $\boldsymbol{X}$ is called ***random sequence (r-seq)***. In this case we can equivalently say that $\boldsymbol{X}$ coincides with a ***sequence of rv*'s** $(X_n)_{n \in \boldsymbol{N}}$, a notion already introduced in the previous section while discussing of convergence

3.2.1.3 Stochastic Processes

If finally $(\mathrm{E}, \mathcal{E}) = \big(\boldsymbol{R}^T, \mathcal{B}(\boldsymbol{R}^T)\big)$, with T a (neither necessarily finite nor countable) subset of $\boldsymbol{R}$, the random element $\boldsymbol{X}$—now called ***stochastic process (sp)***—associates to every $\omega \in \Omega$ a function $(x_t)_{t \in T}$, also denoted as $x(t)$ and called ***sample*** or ***trajectory***. Even here it could be shown that a *sp* can be considered as a family $(X_t)_{t \in T}$ of *rv*'s X_t, also denoted as $X(t)$. A *sp* $\boldsymbol{X}$ can then be considered either as a map associating to $\omega \in \Omega$ a whole function (trajectory) $(x_t)_{t \in T} \in \boldsymbol{R}^T$, or as a map associating to every $t \in T$ a *rv* $X_t : (\Omega, \mathcal{F}) \to (\boldsymbol{R}, \mathcal{B}(\boldsymbol{R}))$. In short the components $X_t(\omega) = X(\omega; t)$ of a *sp* are functions of two variables t and ω, and the different notations are adopted in order to emphasize one or both these variables. Further details about the *sp*'s will be provided in the Second Part of these lectures. Remark finally that if in particular $T = \{1, 2, \ldots\} = \boldsymbol{N}$ is the set of the natural numbers then the *sp* boils down to a sequence of *rv*'s $(X_n)_{n \in \boldsymbol{N}}$ and is also called *discrete time stochastic process.*

3.2.2 *Joint and Marginal Distributions and Densities*

When on $(\Omega, \mathcal{F})$ we take a probability $\boldsymbol{P}$, the usual procedures will allow to induce probabilities also on the spaces $(\mathrm{E}, \mathcal{E})$ in order to define laws and distributions of the random elements

3.2.2.1 Laws of Random Vectors

Given a *r-vec* $\boldsymbol{X} = (X_k)_{k=1,\ldots,n}$, we call ***joint law (or joint distribution)*** of its components X_k, the $\boldsymbol{P}_{\boldsymbol{X}}$ defined on $(\boldsymbol{R}^n, \mathcal{B}(\boldsymbol{R}^n))$ through

$$\boldsymbol{P}_X\{B\} = \boldsymbol{P}\{X^{-1}(B)\} = \boldsymbol{P}\{\boldsymbol{X} \in B\} = \boldsymbol{P}\{(X_1, \ldots, X_n) \in B\}$$

where B is an arbitrary element of $\mathcal{B}(\boldsymbol{R}^n)$; we call instead ***marginal laws (or marginal distributions)*** of the components X_k, the $\boldsymbol{P}_{X_k}$ defined on $(\boldsymbol{R}_k, \mathcal{B}(\boldsymbol{R}_k))$ through

$$\boldsymbol{P}_{X_k}\{A\} = \boldsymbol{P}\{X_k^{-1}(A)\} = \boldsymbol{P}\{X_k \in A\}$$

where A is an arbitrary element of $\mathcal{B}(\boldsymbol{R}_k)$. We consequently call ***joint distribution function*** (joint *cdf*) of the *r-vec* $\boldsymbol{X}$ the *cdf* of $\boldsymbol{P}_X$, namely

$$\begin{aligned} F_X(\boldsymbol{x}) &= F_X(x_1, \ldots, x_n) = \boldsymbol{P}_X\{(-\infty, x_1] \times \cdots \times (-\infty, x_n]\} \\ &= \boldsymbol{P}\{X_1 \leq x_1, \ldots, X_n \leq x_n\} \end{aligned}$$

with $\boldsymbol{x} = (x_1, \ldots, x_n) \in \boldsymbol{R}^n$, and ***marginal distribution function*** (marginal *cdf*) of its components X_k the *cdf*'s of the $\boldsymbol{P}_{X_k}$, namely

$$F_{X_k}(x_k) = \boldsymbol{P}\{X_k \leq x_k\} = \boldsymbol{P}_{X_k}\{(-\infty, x_k]\} = \boldsymbol{P}_X\{\boldsymbol{R}_1 \times \cdots \times (-\infty, x_k] \times \cdots \times \boldsymbol{R}_n\}$$

with $x_k \in \boldsymbol{R}_k$ and $k = 1, \ldots, n$.

3.2.2.2 Laws of Random Sequences

The notions about the *r-vec*'s can be easily extended to the *r-seq*'s $\boldsymbol{X} = (X_n)_{n \in \boldsymbol{N}}$ by selecting a finite subset of components constituting a *r-vec*: taken indeed a finite subset of indices $\{k_1, \ldots, k_m\}$, we consider the finite-dimensional, joint law or distribution of the *r-vec* $(X_{k_1}, \ldots, X_{k_m})$. By choosing the indices $\{k_1, \ldots, k_m\}$ in all the possible ways we will have then a (consistent, see Sect. 2.4) family of finite-dimensional distributions with their finite-dimensional *cdf*'s

$$F(x_1, k_1; \ldots; x_m, k_m) = \boldsymbol{P}\{X_{k_1} \leq x_1; \ldots; X_{k_m} \leq x_m\}$$

These finite-dimensional distributions (and their *cdf*'s) of the components X_k are also called ***joint, marginal laws or distributions***. We must just remember at this point that, according to the Theorem 2.35, to have this consistent family of finite-dimensional distributions is tantamount to define the probability $\boldsymbol{P}_X$ on the whole space $(\boldsymbol{R}^\infty, \mathcal{B}(\boldsymbol{R}^\infty))$ of our *r-seq* samples

3.2.2.3 Laws of Stochastic Processes

In a similar way we can manage the case of a *sp* $\boldsymbol{X} = (X_t)_{t \in T}$: given indeed a finite set $\{t_1, \ldots, t_m\}$ of instants in T we take the finite-dimensional, joint law or distribution of the *r-vec* $(X_{t_1}, \ldots, X_{t_m})$ and the corresponding finite-dimensional, joint *cdf*

$$F(x_1, t_1; \dots; x_m, t_m) = \boldsymbol{P}\{X_{t_1} \leq x_1; \dots; X_{t_m} \leq x_m\}$$

These finite-dimensional distributions (and their *cdf*'s) of the components X_t are again called ***joint, marginal laws or distributions***. By choosing now in every possible way the points $\{t_1, \dots, t_m\}$ we get then a consistent family of *cdf*'s, and we remember again that, according to the Kolmogorov Theorem 2.37, to have this consistent family is tantamount to define a $\boldsymbol{P}_X$ on the whole $\big(\boldsymbol{R}^T, \mathcal{B}(\boldsymbol{R}^T)\big)$

3.2.2.4 Marginalization

The remarks about joint and marginal distributions discussed in the Sect. 2.3.3 turn out to be instrumental also here. In particular, given the joint *cdf* $F_X(\boldsymbol{x})$ of a *r-vec* $\boldsymbol{X}$, we will always be able to find in a unique way the marginal *cdf*'s by adopting the usual procedure, for example

$$F_{X_k}(x_k) = F_X(+\infty, \dots, x_k, \dots, +\infty)$$

It is easy to see indeed that

$$\begin{aligned} F_{X_k}(x_k) &= \boldsymbol{P}\{X_k \leq x_k\} = \boldsymbol{P}\{X_1 < +\infty, \dots, X_k \leq x_k, \dots, X_n < +\infty\} \\ &= F_X(+\infty, \dots, x_k, \dots, +\infty) \end{aligned}$$

As it has been shown again in the Sect. 2.3.3, in general it is not possible instead to recover *in a unique way* a joint *cdf* F_X from given marginal *cdf*'s F_{X_k}

3.2.2.5 Densities

When the joint *cdf* $F_X(\boldsymbol{x})$ of a *r-vec* $\boldsymbol{X}$ is also *ac* we also have a ***joint density*** (joint *pdf*) $f_X(\boldsymbol{x}) \geq 0$ such that

$$F_X(x_1, \dots, x_n) = \int_{-\infty}^{x_1} \dots \int_{-\infty}^{x_n} f_X(y_1, \dots, y_n)\, dy_1 \dots dy_n$$

Of course it is also

$$f_X(x_1, \dots, x_n) = \frac{\partial^n F_X(x_1, \dots, x_n)}{\partial x_1 \dots \partial x_n}$$

It is possible to show that in this event also the single components X_k have *ac* marginals *cdf*'s $F_{X_k}(x_k)$ with *pdf*'s $f_{X_k}(x_k)$ differentiable from the joint *pdf* $f_X(\boldsymbol{x})$ by means of the usual ***marginalization procedure***

$$f_{X_k}(x_k) = \int_{-\infty}^{+\infty} \dots \int_{-\infty}^{+\infty} f_X(x_1, \dots, x_k, \dots, x_n)\, dx_1 \dots dx_{k-1} dx_{k+1} \dots dx_n$$

with $n-1$ integrations on all the variables except the kth. Apparently it is also

$$f_{X_k}(x_k) = F'_{X_k}(x_k)$$

These $f_{X_k}(x_k)$ are called ***marginal densities*** and here too, while from the joint *pdf* $f_X(\boldsymbol{x})$ we can always deduce—by integration—the marginals $f_{X_k}(x_k)$, a retrieval of the joint *pdf* from the marginals is not in general unique. From the joint *pdf* we can finally calculate $\boldsymbol{P}_X\{B\}$ with $B \in \mathcal{B}(\boldsymbol{R}^n)$: if for instance $B = (a_1, b_1] \times \ldots \times (a_n, b_n]$ in analogy with the univariate case we have

$$\begin{aligned}\boldsymbol{P}_X\{B\} &= \boldsymbol{P}\{X \in B\} = \boldsymbol{P}\{a_1 < X_1 \leq b_1, \ldots, a_n < X_n \leq b_n\} \\ &= \int_{a_1}^{b_1} \ldots \int_{a_n}^{b_n} f_X(x_1, \ldots, x_n)\, dx_1 \ldots dx_n\end{aligned}$$

Example 3.12 Discrete *r-vec's*: Take a ***r-vec*** **with discrete components**: since F_X is now discrete it will be enough to give the the joint discrete distributions to know the law. Consider for example a **multinomial *r-vec*** $\boldsymbol{X} = (X_1, \ldots, X_r)$ with $r = 1, 2, \ldots,$ and law $\mathfrak{B}\,(n;\, p_1, \ldots, p_r)$ defined in the Example 2.6. We remember that this is the case of a random experiment with $r+1$ possible outcomes: the individual components X_j—representing how many times we find the jth outcome among n trials—take values from 0 to n with a joint multinomial law, while it is understood that *rv* X_0 comes from $X_0 + X_1 + \ldots + X_r = n$. From (2.9) we then have

$$\boldsymbol{P}_X\{\boldsymbol{k}\} = \boldsymbol{P}\{X_1 = k_1, \ldots, X_r = k_r\} = \binom{n}{k_1, \ \ldots, \ k_r} p_0^{k_0} p_1^{k_1} \cdot \ \ldots \ \cdot p_r^{k_r} \tag{3.1}$$

with $k_0 + k_1 + \ldots + k_r = n$, and $p_0 + p_1 + \ldots + p_r = 1$. In the case $r = 1$, $\boldsymbol{X}$ reduces to just one component X_1 with binomial law $\mathfrak{B}\,(n;\, p_1)$

Resuming then the discussion of the Sect. 2.1.2, a second example can be the *r-vec* $\boldsymbol{X} = (X_1, \ldots, X_n)$ with the components X_k representing the **outcomes of** n **coin flips**: they take now just the values 0 and 1. As a consequence the $\boldsymbol{X}$ samples are the ordered n-tuples (2.5) that in the Bernoulli model occur with a probability (2.6), so that the joint distribution of the *r-vec* $\boldsymbol{X}$ is now

$$\boldsymbol{P}_X\{a_1, \ldots, a_n\} = \boldsymbol{P}\{X_1 = a_1, \ldots, X_n = a_n\} = p^k q^{n-k} \qquad a_j = 0, 1 \tag{3.2}$$

with $k = \sum_j a_j$ and $q = 1 - p$. A simple calculation would show finally that the marginals of the components X_k are all *id* with a Bernoulli law $\mathfrak{B}\,(1;\, p)$. For the sake of simplicity we will neglect to display the explicit form of the joint *cdf* F_X for these two examples

Example 3.13 Gaussian *r-vec*'s $\boldsymbol{X} \sim \mathfrak{N}(\boldsymbol{b}, \mathbb{A})$: A relevant example of ac *r-vec* is that of the **Gaussian (normal) *r-vec*'s** $\boldsymbol{X} = (X_1, \ldots, X_n)$ with a multivariate, normal *pdf* $\mathfrak{N}(\boldsymbol{b}, \mathbb{A})$ of the type (2.22) presented in the Sect. 2.3.2. The marginal *pdf* of a Gaussian *r-vec* are of course the Gaussian univariate $\mathfrak{N}(b_k, a_k^2)$ of the type (2.31) as it could be seen with a direct calculation by marginalizing $\mathfrak{N}(\boldsymbol{b}, \mathbb{A})$. We must remember however that a *r-vec* with univariate Gaussian marginals is not forcibly also a Gaussian *r-vec*: the joint *pdf*, indeed, is not uniquely defined by such marginals, and hence the joint *pdf* of our *r-vec* could differ from a $\mathfrak{N}(\boldsymbol{b}, \mathbb{A})$, even if its marginals are all $\mathfrak{N}(b_k, a_k^2)$

3.2.3 Independence of rv's

Definition 3.14 Take a probability space $(\Omega, \mathcal{F}, \boldsymbol{P})$ and a family $\boldsymbol{X} = (X_s)_{s \in S}$ of *rv*'s with an arbitrary (finite, countable or uncountable) set of indices S: we say that the components X_s of $\boldsymbol{X}$ are **independent *rv*'s** if, however taken m components of indices $s_1, \dots, s_m$, and however taken the subsets $B_k \in \mathcal{B}(\boldsymbol{R})$ with $k = 1, \dots, m$, it is

$$\boldsymbol{P}\{X_{s_1} \in B_1, \dots, X_{s_m} \in B_m\} = \boldsymbol{P}\{X_{s_1} \in B_1\} \cdot \ldots \cdot \boldsymbol{P}\{X_{s_m} \in B_m\}$$

In short, the independence of the *rv*'s in $\boldsymbol{X}$ boils down to the independence (in the usual sense of the Sect. 1.5) of all the events that we can define by means of an arbitrary, finite collection of *rv*'s of $\boldsymbol{X}$, and in principle this amounts to a very large number of relations. The simplest case is that of $S = \{1, \dots, n\}$ when $\boldsymbol{X}$ is a *r-vec* with n components, but the definition fits also the case of *r-seq*'s and *sp*'s. As already remarked in the Sect. 3.1.2, we finally recall that it is possible to have *rv*'s identically distributed, but different and also independent: in this case we will speak of ***independent and identically distributed rv's (iid)***.

Theorem 3.15 *Take a r-vec* $\boldsymbol{X} = (X_k)_{k=1,\dots,n}$ *on* $(\Omega, \mathcal{F}, \boldsymbol{P})$*: the following two statements are equivalent*

(a) the components X_k *are independent;*
(b) $F_{\boldsymbol{X}}(x_1, \dots, x_n) = F_{X_1}(x_1) \cdot \ldots \cdot F_{X_n}(x_n)$*;*

If moreover $F_{\boldsymbol{X}}(\boldsymbol{x})$ *is ac, then the previous statements are also equivalent to*

(c) $f_{\boldsymbol{X}}(x_1, \dots, x_n) = f_{X_1}(x_1) \cdot \ldots \cdot f_{X_n}(x_n)$.

Proof Omitted: see [1] p. 179 ■

Example 3.16 In a first application of the Theorem 3.15 remark that the *r-vec* $\boldsymbol{X}$ defined in the Example 3.12 with distribution (3.2), and describing the outcome of n coin flips in the Bernoulli model, is apparently composed of *iid rv's*. The joint distribution (3.2) (and hence also the joint *cdf*) turns out indeed to be the product of n identical Bernoulli laws $\mathfrak{B}(1; p)$ representing the marginal laws of the X_k. We can say then that $\boldsymbol{X}$, with the prescribed law (3.2), is a *r-vec* of n ***iid* Bernoulli *rv's***, and we will also symbolically write $\boldsymbol{X} \sim [\mathfrak{B}(1; p)]^n$

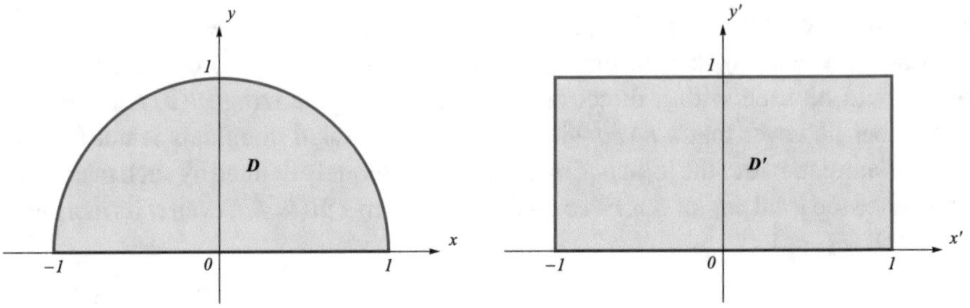

Fig. 3.3 Domains D and D' in the Example 3.16

Resuming then the Example 2.30 with the Fig. 2.12, suppose that f and g respectively are two, different joint *pdf*'s for the bivariate *r-vec's* $\boldsymbol{U}$ and $\boldsymbol{V}$. We have shown that their marginals coincide, but their relations (2.30) with the joint distribution are different: the joint *pdf* of $\boldsymbol{U}$ is the product of its marginals, while this is not true for $\boldsymbol{V}$. We can then conclude from the Theorem 3.15 that $\boldsymbol{U}$ and $\boldsymbol{V}$, with identical marginals, substantially differ because the components of $\boldsymbol{U}$ are independent, and that of $\boldsymbol{V}$ are not

In a third example we will suppose again to take two *r-vec's* $\boldsymbol{U} = (X, Y)$ and $\boldsymbol{V} = (X', Y')$ with constant *pdf* (uniform laws) in two different domains of $\boldsymbol{R}^2$:

$$f_U(x, y) = \begin{cases} \frac{2}{\pi}, & \text{if } (x, y) \in D, \\ 0, & \text{if } (x, y) \notin D, \end{cases} \qquad f_V(x', y') = \begin{cases} \frac{1}{2}, & \text{if } (x', y') \in D', \\ 0, & \text{if } (x', y') \notin D', \end{cases}$$

where the domains D and D' depicted in Fig. 3.3 are

$$D = \{(x, y) \in \boldsymbol{R}^2 \ : \ -1 \leq x \leq +1, \ \ 0 \leq y \leq \sqrt{1 - x^2}\}$$
$$D' = \{(x', y') \in \boldsymbol{R}^2 \ : \ -1 \leq x' \leq +1, \ \ 0 \leq y' \leq 1\}$$

An elementary integration will show that the marginal *pdf*'s of $\boldsymbol{U}$ are

$$f_X(x) = \begin{cases} \frac{2}{\pi}\sqrt{1 - x^2}, & \text{if } (x \in [-1, 1] \\ 0, & \text{if } x \notin [-1, 1] \end{cases} \qquad f_Y(y) = \begin{cases} \frac{4}{\pi}\sqrt{1 - y^2}, & \text{if } y \in [0, 1] \\ 0, & \text{if } y \notin [0, 1] \end{cases}$$

while the marginals of $\boldsymbol{V}$ are

$$f_{X'}(x') = \begin{cases} \frac{1}{2}, & \text{if } x' \in [-1, 1] \\ 0, & \text{if } x' \notin [-1, 1] \end{cases} \qquad f_{Y'}(y') = \begin{cases} 1, & \text{if } y' \in [0, 1] \\ 0, & \text{if } y' \notin [0, 1] \end{cases}$$

so that

$$f_U(x, y) \neq f_X(x) f_Y(y) \qquad f_V(x', y') = f_{X'}(x') f_{Y'}(y')$$

and hence the components of $\boldsymbol{V}$ are independent *rv's*, while that of $\boldsymbol{U}$ are not

3.2.4 Decomposition of Binomial rv's

The sums of *rv*'s and their laws play an extremely important role and will be extensively discussed in the Sect. 3.5.2: we will give here just a short preliminary analysis for particular, discrete *rv*'s that will enable us to get an important result about the binomial laws

Consider a *r-vec* $\boldsymbol{U} = (X, Y)$ whose components take just integer values, and let their joint and marginal distributions be denoted as

$$\boldsymbol{P}_U\{j, k\} = \boldsymbol{P}\{X = j, Y = k\} \qquad \boldsymbol{P}_X\{j\} = \boldsymbol{P}\{X = j\} \qquad \boldsymbol{P}_Y\{k\} = \boldsymbol{P}\{Y = k\}$$

In the following discussion it will be enough to consider X and Y taking just a *finite number* of values

$$X = j \in \{0, 1, \dots, m\} \qquad Y = k \in \{0, 1, \dots, n\} \tag{3.3}$$

but in order to simplify our notations we will suppose that these values are indeed all the relative numbers $\boldsymbol{Z}$ (positive, negative and zero integers), but that only (3.3) have non zero probabilities. Define then the *rv* $W = X + Y$ taking the values

$$W = \ell \in \{0, 1, \dots, n + m\}$$

and look for its distribution $\boldsymbol{P}_W$ based on the available distributions $\boldsymbol{P}_U$, $\boldsymbol{P}_X$ and $\boldsymbol{P}_Y$. In order to do that we will give first a provisional definition of the *discrete convolution* between two discrete distributions with integer values, by remarking also that a more general one will be provided later in the Definition 3.48

Definition 3.17 Take two discrete laws with integer values $\mathfrak{P}$ and $\mathfrak{Q}$ and let $p(j)$ and $q(k)$ be their distributions: we say that the law $\mathfrak{R}$ is their **(discrete) convolution** $\mathfrak{P} * \mathfrak{Q}$ when its distribution $r(\ell)$ is

$$r(\ell) = \sum_{k \in \boldsymbol{Z}} p(\ell - k) q(k) \tag{3.4}$$

Proposition 3.18 *Within the given notations it is*

$$\boldsymbol{P}_W\{\ell\} = \sum_{k \in \boldsymbol{Z}} \boldsymbol{P}_U\{\ell - k, k\}$$

If moreover X and Y are also **independent** *we get*

$$\boldsymbol{P}_W\{\ell\} = \sum_{k \in \boldsymbol{Z}} \boldsymbol{P}_X\{\ell - k\} \boldsymbol{P}_Y\{k\} = (\boldsymbol{P}_X * \boldsymbol{P}_Y)\{\ell\} \tag{3.5}$$

so that $\boldsymbol{P}_W$ *turns out to be the* **(discrete) convolution** *of* $\boldsymbol{P}_X$ *and* $\boldsymbol{P}_Y$

Proof We have indeed

$$\boldsymbol{P}_W\{\ell\} = \boldsymbol{P}\{W = \ell\} = \boldsymbol{P}\{X + Y = \ell\} = \sum_{j+k=\ell} \boldsymbol{P}\{X = j, Y = k\}$$

$$= \sum_{j+k=\ell} \boldsymbol{P}_U\{j, k\} = \sum_{k\in\mathbf{Z}} \boldsymbol{P}_U\{\ell - k, k\}$$

When moreover X and Y are also independent, from the Theorem 3.15 we have $\boldsymbol{P}_U\{j, k\} = \boldsymbol{P}_X\{j\}\boldsymbol{P}_Y\{k\}$ and hence (3.5) is easily deduced ■

In the *Bernoulli model* of the Sect. 2.1.2 the sample space Ω was the set of the n-tuples of results $\omega = (a_1, \dots a_n)$ as in (2.5), where $a_j = 0, 1$ is the outcome of the jth draw, and $k = \sum_j a_j$ is the number of white balls in an n-tuple of draws. We then defined the events

$$A_j = \{\omega \in \Omega \,:\, a_j = 1\} \quad j = 1, \dots, n$$

$$D_k = \left\{\omega \in \Omega \,:\, \sum_{j=0}^{n} a_j = k\right\} \quad k = 0, 1, \dots, n$$

and, taken on Ω the probability (2.6), we have shown that $\boldsymbol{P}\{D_k\}$ are a binomial distribution $\mathfrak{B}(n; p)$, while the A_j are independent with $\boldsymbol{P}\{A_j\} = p$. We are able now to revise this model in terms of *rv*'s: consider first the *r-vec* $\boldsymbol{X} = (X_1, \dots, X_n)$ (as already defined in the Sects. 3.2.2 and 3.2.3) whose independent components are the *rv*'s (indicators)

$$X_j = I_{A_j}$$

namely the outcomes of every draw, and then the (simple) *rv*

$$S_n = \sum_{k=0}^{n} k I_{D_k}$$

namely the number of white balls in n draws. Apparently it is $D_k = \{S_n = k\}$ and $A_j = \{X_j = 1\}$, while among our *rv*'s the following relation holds

$$S_n = \sum_{j=1}^{n} X_j \tag{3.6}$$

because, by definition, for every $\omega \in \Omega$ the value of S_n (number of white balls) is the sum of all the X_j, and hence the *rv*'s in (3.6) are indistinguishable (see Definition 3.4) and also *id*. In the Sect. 2.1.2 we have also seen by direct calculation that

$$X_j \sim \mathfrak{B}(1; p) \qquad S_n \sim \mathfrak{B}(n; p)$$

namely that the sum of n *independent* Bernoulli *rv*'s $X_j \sim \mathfrak{B}(1; p)$ is a binomial *rv* $S_n \sim \mathfrak{B}(n; p)$. We will now revise these results in the light of the Proposition 3.18

Proposition 3.19 *A binomial law $\mathfrak{B}(n; p)$ is the convolution of n Bernoulli laws $\mathfrak{B}(1; p)$ according to*

$$\mathfrak{B}(n; p) = [\mathfrak{B}(1; p)]^{*n} = \underbrace{\mathfrak{B}(1; p) * \ldots * \mathfrak{B}(1; p)}_{n \text{ times}} \tag{3.7}$$

As for the rv's we can then say that:

- *if we take n iid Bernoulli rv's $X_k \sim \mathfrak{B}(1; p)$, their sum $S_n = X_1 + \ldots + X_n$ will follow the binomial law $\mathfrak{B}(n; p)$*
- *if we take a binomial rv $S_n \sim \mathfrak{B}(n; p)$, there exist n iid Bernoulli rv's $X_k \sim \mathfrak{B}(1; p)$ such that S_n is* ***decomposed*** *(in distribution) into their sum, namely*

$$S_n \stackrel{d}{=} X_1 + \ldots + X_n \tag{3.8}$$

Proof The Bernoulli law $\mathfrak{B}(1; p)$ has the distribution

$$p_1(k) = \binom{1}{k} p^k q^{1-k} = \begin{cases} p & \text{if } k = 1 \\ q & \text{if } k = 0 \end{cases}$$

so that from (3.4) the law $\mathfrak{B}(1; p) * \mathfrak{B}(1; p)$ will give only to $0, 1, 2$ the non zero probabilities

$$\begin{aligned} r(0) &= p_1(0)p_1(0) = q^2 \\ r(1) &= p_1(0)p_1(1) + p_1(1)p_1(0) = 2pq \\ r(2) &= p_1(1)p_1(1) = p^2 \end{aligned}$$

which coincide with the distribution

$$p_2(k) = \binom{2}{k} p^k q^{2-k}$$

of the binomial law $\mathfrak{B}(2; p)$. The complete result (3.7) follows by induction, but we will neglect to check it. As for the *rv*'s, if $X_1 \ldots, X_n$ are n *iid* Bernoulli *rv*'s $\mathfrak{B}(1; p)$ *iid*, from the Proposition 3.18 the distribution of their sum S_n will be the convolution of their laws, which according to (3.7) will be the binomial $\mathfrak{B}(n; p)$. If conversely $S_n \sim \mathfrak{B}(n; p)$ we know from (3.7) that its law is convolution of n Bernoulli $\mathfrak{B}(1; p)$. From the Definition 3.5 on the other hand we know that to these n distributions we can always associate n *iid* Bernoulli *rv*'s $X_1, \ldots, X_n$ so that from the Proposition 3.18 results $S_n \stackrel{d}{=} X_1 + \ldots + X_n$ ∎

3.3 Expectation

3.3.1 *Integration and Expectation*

The expectation of a *rv* X is a numerical indicator specifying the location of the barycenter of a distribution $\boldsymbol{P}_X$, and takes origin from the notion of weighed average. For simple *rv*'s

$$X = \sum_{k=1}^{n} x_k I_{D_k}$$

where $D_k = \{X = x_k\} \in \mathcal{F}$ is a decomposition, the expectation is just the *weighed average* of its values

$$\boldsymbol{E}[X] = \sum_{k=1}^{n} x_k \boldsymbol{P}_X\{x_k\} = \sum_{k=1}^{n} x_k \boldsymbol{P}\{X = x_k\} = \sum_{k=1}^{n} x_k \boldsymbol{P}\{D_k\} \tag{3.9}$$

a definition extended in a natural way also to the general discrete *rv*'s (including the case of countably many values) as

$$\boldsymbol{E}[X] = \sum_{k=1}^{\infty} x_k \boldsymbol{P}_X\{x_k\}$$

provided that the series converges. Remark that in particular for every $A \in \mathcal{F}$ we always have

$$\boldsymbol{E}[I_A] = \boldsymbol{P}\{A\} \tag{3.10}$$

a simple result that however highlights a relation between the notions of probability and expectation that will be instrumental for later purposes

When however X is an arbitrary *rv* we can no longer adopt these elementary definitions. In this case we first remark that when X is a *non negative rv* the Theorem 3.10 points to the existence of a non decreasing sequence of non negative, simple *rv*'s $(X_n)_{n\in N}$ such that $X_n(\omega) \uparrow X(\omega)$ for every $\omega \in \Omega$. From the previous elementary definitions we can then define the non negative, monotonic non decreasing numerical sequence $(\boldsymbol{E}[X_n])_{n\in N}$ always admitting a limit (possibly $+\infty$) that we can consider as the definition of the expectation of X. To extend then this procedure to a totally arbitrary *rv* X we remember that this is always representable as a difference between non negative *rv*'s in the form

$$X = X^+ - X^-$$

where $X^+ = \max\{X, 0\}$ and $X^- = -\min\{X, 0\}$ are respectively called *positive and negative parts* of X. As a consequence we can separately take the two non negative *rv*'s X^+ and X^-, define their expectation and finally piece together that of X by difference. Neglecting the technical details, we can then sum up these remarks in the following definition

Definition 3.20 To define the **expectation** of a *rv* X we adopt the following procedure:

- if X is a non negative *rv* we call expectation the limit (possibly $+\infty$)

$$\boldsymbol{E}[X] \equiv \lim_n \boldsymbol{E}[X_n]$$

  where $(X_n)_{n\in\boldsymbol{N}}$ is a monotonic non decreasing sequence of simple *rv*'s such that $X_n \uparrow X$ for every $\omega \in \Omega$, and $\boldsymbol{E}[X_n]$ are defined in an elementary way; the existence of such sequences $(X_n)_{n\in\boldsymbol{N}}$ follows from the Theorem 3.10, and it is possible to prove that the result is independent from the choice of the particular sequence;
- if X is an arbitrary *rv* we will say that its expectation **exists** when at least one of the two non negative numbers $\boldsymbol{E}\left[X^+\right]$, $\boldsymbol{E}\left[X^-\right]$ is finite, and in that case we take

$$\boldsymbol{E}[X] \equiv \boldsymbol{E}\left[X^+\right] - \boldsymbol{E}\left[X^-\right]$$

  if instead both $\boldsymbol{E}\left[X^+\right]$ and $\boldsymbol{E}\left[X^-\right]$ are $+\infty$ we will say that the expectation of X **does not exist**;
- when both the numbers $\boldsymbol{E}\left[X^+\right]$, $\boldsymbol{E}\left[X^-\right]$ are finite, also $\boldsymbol{E}[X]$ is finite and X is said **integrable**, namely to have a finite expectation

$$\boldsymbol{E}[X] < +\infty$$

  since moreover $|X| = X^+ + X^-$, if X is integrable, it turns out to be also **absolutely integrable**, namely

$$\boldsymbol{E}[|X|] < +\infty$$

  and it is apparent that also the reverse holds, so that we can say that a *rv* is integrable iff it is absolutely integrable

Since the procedure outlined in the previous definition closely resembles that used to define the ***Lebesgue integral*** we will also adopt in the following the notation

$$\boldsymbol{E}[X] = \int_\Omega X\, d\boldsymbol{P} = \int_\Omega X(\omega)\, \boldsymbol{P}\{d\omega\}$$

More generally, if g is a measurable function from $(\Omega, \mathcal{F})$ to $(\boldsymbol{R}, \mathcal{B}(\boldsymbol{R}))$, and μ is a measure (possibly not a probability) on $(\Omega, \mathcal{F})$, the Lebesgue integral is defined retracing the procedure of the Definition 3.20 and is denoted as

$$\int_\Omega g\, d\mu = \int_\Omega g(\omega)\, \mu\{d\omega\}$$

When μ is not a probability however such an integral is not an expectation

It is also possible to restrict our integrals to the subsets of Ω: if $A \in \mathcal{F}$ is a subset of Ω, we call ***Lebesgue integral over the set*** A the integrals

$$\int_A X\,d\boldsymbol{P} = \int_\Omega X I_A\,d\boldsymbol{P} = \boldsymbol{E}\left[X I_A\right] \qquad \int_A g\,d\mu = \int_\Omega g I_A\,d\mu$$

respectively for a probability $\boldsymbol{P}$ and for a general measure μ. Remark that (3.10) takes now the more evocative form

$$\boldsymbol{P}\{A\} = \boldsymbol{E}\left[I_A\right] = \int_\Omega I_A\,d\boldsymbol{P} = \int_A d\boldsymbol{P} \tag{3.11}$$

apparently stressing that the probability of an event A is nothing else than the the integral of $\boldsymbol{P}$ on A

Definition 3.21 We will call **moment of order** $\boldsymbol{k}$ of a *rv* X the expectation (if it exists)

$$\boldsymbol{E}\left[X^k\right] = \int_\Omega X^k\,d\boldsymbol{P} \qquad k = 0, 1, 2, \ldots$$

and **absolute moment of order** $\boldsymbol{r}$ the expectation (if it exists)

$$\boldsymbol{E}\left[|X|^r\right] = \int_\Omega |X|^r\,d\boldsymbol{P} \qquad r \geq 0$$

It is important finally to recall the notations usually adopted for the integrals when in particular $(\Omega, \mathcal{F}) = (\boldsymbol{R}^n, \mathcal{B}(\boldsymbol{R}^n))$, g is a Borel function, and $G(\boldsymbol{x})$ is the generalized *cdf* of a Lebesgue-Stieltjes measure μ: in this case we will speak of a ***Lebesgue-Stieltjes integral*** and we will write

$$\int_\Omega g\,d\mu = \int_{\boldsymbol{R}^n} g\,dG = \int_{\boldsymbol{R}^n} g(\boldsymbol{x})\,G(d\boldsymbol{x}) = \int_{\boldsymbol{R}^n} g(x_1, \ldots, x_n)\,G(dx_1, \ldots, dx_n)$$

For $n = 1$ on the other hand we also write

$$\int_{\boldsymbol{R}} g\,dG = \int_{\boldsymbol{R}} g(x)\,G(dx) = \int_{-\infty}^{+\infty} g(x)\,G(dx)$$

while the integral on an interval $(a, b]$ will be

$$\int_{(a,b]} g(x)\,G(dx) = \int_{\boldsymbol{R}} I_{(a,b]}(x)g(x)\,G(dx) = \int_a^b g(x)\,G(dx)$$

If μ is the Lebesgue measure, $G(d\boldsymbol{x})$ is replaced by $d\boldsymbol{x}$ and, for $n = 1$, $G(dx)$ is replaced by dx. When finally μ is a probability $\boldsymbol{P}$, G is replaced by a *cdf* F and the integral on the whole $\boldsymbol{R}^n$ will take the meaning of an expectation

3.3.2 Change of Variables

Theorem 3.22 Change of variables: *Take the r-vec* $\boldsymbol{X} = (X_1, \ldots, X_n)$ *on* $(\Omega, \mathcal{F}, \boldsymbol{P})$ *with joint distribution* $\boldsymbol{P}_X$*, and the Borel function* $g : (\boldsymbol{R}^n, \mathcal{B}(\boldsymbol{R}^n)) \to (\boldsymbol{R}, \mathcal{B}(\boldsymbol{R}))$*; if* $Y = g(\boldsymbol{X})$ *(see Fig. 3.4), we have*

$$\boldsymbol{E}[Y] = \int_\Omega Y(\omega)\, d\boldsymbol{P}(\omega) = \boldsymbol{E}[g(\boldsymbol{X})] = \int_\Omega g(\boldsymbol{X}(\omega))\, d\boldsymbol{P}(\omega)$$
$$= \int_{\boldsymbol{R}^n} g(\boldsymbol{x})\, \boldsymbol{P}_X\{d\boldsymbol{x}\} = \int_{\boldsymbol{R}^n} g(x_1, \ldots, x_n)\, \boldsymbol{P}_X\{dx_1, \ldots, dx_n\}$$

Proof Omitted: see [1] p. 196. Remark from Fig. 3.4 that in general, with $n \geq 2$, $\boldsymbol{X}$ is a *r-vec* and $g(\boldsymbol{x}) = g(x_1, \ldots, x_n)$ an n variables function according to the diagram

$$(\Omega, \mathcal{F}) \xrightarrow{\boldsymbol{X}} \big(\boldsymbol{R}^n, \mathcal{B}(\boldsymbol{R}^n)\big) \xrightarrow{g} (\boldsymbol{R}, \mathcal{B}(\boldsymbol{R}))$$

while $Y = g(\boldsymbol{X})$ is a *rv*. In the simplest case $n = 1$ the *r-vec* $\boldsymbol{X}$ has just one component X, and the diagram of the Theorem 3.22 boils down to

$$(\Omega, \mathcal{F}) \xrightarrow{X} (\boldsymbol{R}, \mathcal{B}(\boldsymbol{R})) \xrightarrow{g} (\boldsymbol{R}, \mathcal{B}(\boldsymbol{R}))$$

with $Y = g(X)$ ■

Since the Definition 3.20 of expectation is rather abstract the previous result and its aftermaths, that basically resort just to the ordinary real integration, are of great practical importance. According to the Theorem 3.22 we can indeed calculate the expectation of $Y = g(\boldsymbol{X})$ (in the sense of the abstract integral of a function from Ω to $\boldsymbol{R}$) as an integral of Borel functions $g(\boldsymbol{x})$ from $\boldsymbol{R}^n$ to $\boldsymbol{R}$. Since moreover the distribution $\boldsymbol{P}_X$ on $(\boldsymbol{R}^n, \mathcal{B}(\boldsymbol{R}^n))$ has a *cdf* F—and possibly a *pdf* f—we can deduce a few familiar rules of integration. To this end remember that if F_X and f_X respectively are the *cdf* and the *pdf* of $\boldsymbol{P}_X$, it is possible to prove (but we will neglect the details) that

$$\int_A g(x)\, F_X(dx) = \int_A g(x) f_X(x)\, dx \qquad \forall A \in \mathcal{B}(\boldsymbol{R}) \tag{3.12}$$

so that, if a *pdf* exists, we can replace $F_X(dx)$ by $f_X(x)\, dx$ in in all the following integrations

Corollary 3.23 *For* $n = 1$*, when the r-vec* $\boldsymbol{X}$ *has just one component* X*, if* $\boldsymbol{P}_X$ *has a discrete cdf* $F_X(x)$ *from Theorem 3.22 we get*

$$\boldsymbol{E}[Y] = \boldsymbol{E}[g(X)] = \int_{-\infty}^{+\infty} g(x) F_X(dx) = \sum_k g(x_k) \boldsymbol{P}_X\{x_k\} \tag{3.13}$$

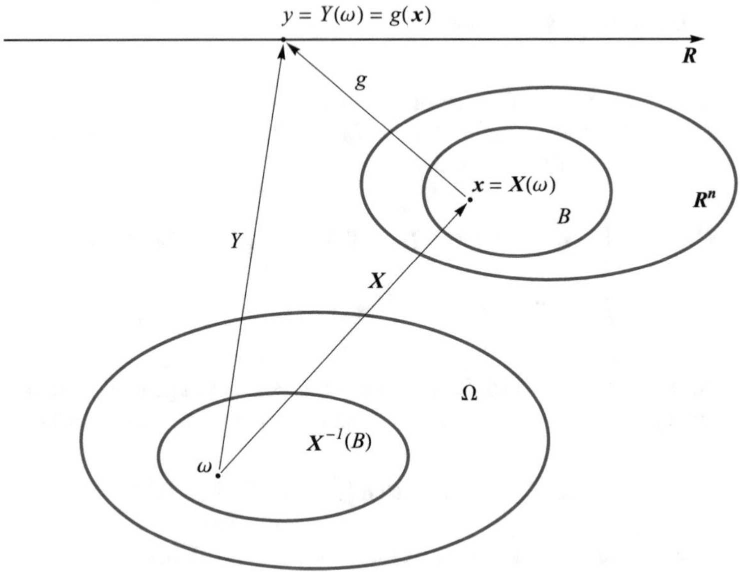

Fig. 3.4 Graphical depiction of the Theorem 3.22

and if in particular $g(x) = x$ (namely $Y = X$) we have

$$\boldsymbol{E}[X] = \sum_k x_k \boldsymbol{P}_X\{x_k\} \tag{3.14}$$

so that we can also write

$$\boldsymbol{E}[Y] = \sum_\ell y_\ell \boldsymbol{P}_Y\{y_\ell\} = \sum_k g(x_k) \boldsymbol{P}_X\{x_k\} \tag{3.15}$$

When instead $F_X(x)$ is ac with pdf $f_X(x)$, it will be

$$\boldsymbol{E}[Y] = \boldsymbol{E}[g(X)] = \int_{-\infty}^{+\infty} g(x) F_X(dx) = \int_{-\infty}^{+\infty} g(x) f_X(x)\, dx \tag{3.16}$$

and if in particular $g(x) = x$ we get the usual formula

$$\boldsymbol{E}[X] = \int_{-\infty}^{+\infty} x F_X(dx) = \int_{-\infty}^{+\infty} x f_X(x)\, dx \tag{3.17}$$

so that we can also write

$$\boldsymbol{E}[Y] = \int_{-\infty}^{+\infty} y f_Y(y)\, dy = \int_{-\infty}^{+\infty} g(x) f_X(x)\, dx \tag{3.18}$$

When $n > 1$, if $F_X(\boldsymbol{x})$ and $f_X(\boldsymbol{x})$ are the joint cdf and pdf of $\boldsymbol{X}$, and $F_Y(y)$, $f_Y(y)$ the cdf and pdf of $Y = g(\boldsymbol{X})$, the Eqs. (3.16) *and* (3.18) *become*

$$\begin{aligned} \boldsymbol{E}[Y] =& \boldsymbol{E}[g(\boldsymbol{X})] = \int_{\boldsymbol{R}} y f_Y(y)\, dy = \int_{\boldsymbol{R}^n} g(\boldsymbol{x}) f_X(\boldsymbol{x})\, d^n\boldsymbol{x} \\ =& \int_{-\infty}^{+\infty} \dots \int_{-\infty}^{+\infty} g(x_1, \dots, x_n) f_X(x_1, \dots, x_n)\, dx_1 \dots dx_n \end{aligned} \tag{3.19}$$

Finally probabilities and cdf 's can be calculated respectively as

$$\boldsymbol{P}_X\big\{[a_1, b_1] \times \dots \times [a_n, b_n]\big\} = \int_{a_1}^{b_1} \dots \int_{a_n}^{b_n} f_X(\boldsymbol{x})\, d^n\boldsymbol{x} \tag{3.20}$$

$$F_X(x_1, \dots, x_n) = \int_{-\infty}^{x_1} \dots \int_{-\infty}^{x_n} f_X(\boldsymbol{y})\, d^n\boldsymbol{y} \tag{3.21}$$

and we also find the marginalization rules

$$f_{X_k}(x_k) = \int_{-\infty}^{+\infty} \dots \int_{-\infty}^{+\infty} f_X(x_1, \dots, x_n)\, dx_1 \dots dx_{k-1} dx_{k+1} \dots dx_n \tag{3.22}$$

Proof Neglecting a complete check of these results, we will just point out that for $n = 1$ with $F_X(x)$ *ac* and *pdf* $f_X(x)$ the Theorem 3.22 becomes

$$\begin{aligned} \boldsymbol{E}[Y] = \boldsymbol{E}[g(X)] &= \int_\Omega Y\, d\boldsymbol{P} = \int_\Omega g(X)\, \boldsymbol{P}\{d\omega\} = \int_{\boldsymbol{R}} g(x)\, \boldsymbol{P}_X\{dx\} \\ &= \int_{\boldsymbol{R}} g(x)\, F_X(dx) = \int_{\boldsymbol{R}} g(x) f_X(x)\, dx \end{aligned}$$

and immediately gives (3.16) and (3.17). Since moreover the (3.17) holds for every *rv*, by means of the *pdf* f_Y of Y, we can also write

$$\boldsymbol{E}[Y] = \int_{-\infty}^{+\infty} y F_Y(dy) = \int_{-\infty}^{+\infty} y f_Y(y)\, dy \tag{3.23}$$

and taking (3.16) along with (3.23) we get (3.18): in short the expectation of $Y = g(X)$ can be calculated in two equivalent ways according to the used *pdf*, either f_X or f_Y, and the Eq. (3.18) is the usual rule for the change of integration variable $y = g(x)$

As for the probability formulas (always with $n = 1$) take in particular as $g(x)$ the following indicator on $(\boldsymbol{R}, \mathcal{B}(\boldsymbol{R}))$

$$g(x) = \chi_B(x) = \begin{cases} 1, & \text{if } x \in B \\ 0, & \text{elsewhere} \end{cases} \qquad B \in \mathcal{B}(\boldsymbol{R})$$

which is a Borel function related to the corresponding indicator on $(\Omega, \mathcal{F})$ by

$$\chi_B(X(\omega)) = I_{X^{-1}(B)}(\omega) \qquad \forall\omega \in \Omega$$

since $\omega \in X^{-1}(B)$ is equivalent to $X(\omega) \in B$. From the Theorem 3.22 we then have

$$\begin{aligned} \boldsymbol{P}\{X \in B\} &= \boldsymbol{P}_X\{B\} = \boldsymbol{P}\{X^{-1}(B)\} = \boldsymbol{E}\left[I_{X^{-1}(B)}\right] = \int_\Omega I_{X^{-1}(B)}\, d\boldsymbol{P} \\ &= \int_\Omega \chi_B(X)\, d\boldsymbol{P} = \int_{\boldsymbol{R}} \chi_B(x)\, \boldsymbol{P}_X\{dx\} = \int_{\boldsymbol{R}} \chi_B(x)\, F_X(dx) \\ &= \int_B F_X(dx) = \int_B f_X(x)\, dx \end{aligned}$$

a result already disclosed at the end of the Sect. 3.1.2. When in particular $B = [a, b]$ we write

$$\boldsymbol{P}_X\{[a, b]\} = \boldsymbol{P}\{a \le X \le b\} = \int_a^b f_X(x)\, dx \tag{3.24}$$

while for $B = (-\infty, x]$ it is

$$\boldsymbol{P}_X\{(-\infty, x]\} = \boldsymbol{P}\{X \le x\} = F_X(x) = \int_{-\infty}^x f_X(t)\, dt \tag{3.25}$$

Remark finally that when a *pdf* f exists the *cdf* apparently is *ac* and hence continuous: as a consequence the probability allotted to the single points is strictly zero, and hence it is immaterial to include or not the endpoints of the intervals. This explains, for instance, why in (3.24) the intervals are closed ■

Example 3.24 If $X \sim \boldsymbol{\delta}_b$ is a **degenerate *rv*** its expectation trivially is

$$\boldsymbol{E}[X] = b \cdot 1 = b \tag{3.26}$$

while if $X \sim \mathfrak{B}(1; p)$ is a **Bernoulli *rv*** we will have

$$\boldsymbol{E}[X] = 1 \cdot p + 0 \cdot (1 - p) = p \tag{3.27}$$

When on the other hand $X \sim \mathfrak{B}(n; p)$ ia a **binomial *rv*** with a simple index rescaling we get

$$\begin{aligned} \boldsymbol{E}[X] &= \sum_{k=1}^n k\binom{n}{k} p^k q^{n-k} = np \sum_{k=1}^n \binom{n-1}{k-1} p^{k-1} q^{n-k} \\ &= np \sum_{k=0}^{n-1} \binom{n-1}{k} p^k q^{n-1-k} = np \end{aligned} \tag{3.28}$$

and finally if $X \sim \mathfrak{P}(\alpha)$ is a **Poisson *rv*** it is

$$\boldsymbol{E}[X] = e^{-\alpha}\sum_{k=0}^{\infty} k\frac{\alpha^k}{k!} = e^{-\alpha}\sum_{k=1}^{\infty}\frac{\alpha^k}{(k-1)!} = \alpha e^{-\alpha}\sum_{k=0}^{\infty}\frac{\alpha^k}{k!} = \alpha \tag{3.29}$$

Example 3.25 Formula (3.17) enables us to calculate the expectation when a *pdf* f_X exists: for a **uniform** ***rv*** $X \sim \mathfrak{U}(a, b)$ with a *pdf* (2.14) we have

$$\boldsymbol{E}[X] = \int_a^b \frac{x}{b-a}\,dx = \frac{1}{b-a}\left[\frac{x^2}{2}\right]_a^b = \frac{a+b}{2} \tag{3.30}$$

while if X is a **Gaussian** ***rv*** $X \sim \mathfrak{N}(b, a^2)$ with *pdf* (2.15), taking $y = (x-b)/a$, and remembering the well known results

$$\int_{-\infty}^{\infty} e^{-y^2/2}\,dy = \sqrt{2\pi} \qquad \int_{-\infty}^{\infty} ye^{-y^2/2}\,dy = 0 \qquad \int_{-\infty}^{\infty} y^2e^{-y^2/2}\,dy = \sqrt{2\pi} \tag{3.31}$$

we have

$$\begin{aligned}\boldsymbol{E}[X] &= \frac{1}{a\sqrt{2\pi}}\int_{-\infty}^{\infty} x\, e^{-(x-b)^2/2a^2}\,dx = \frac{1}{\sqrt{2\pi}}\int_{-\infty}^{\infty}(ay+b)e^{-y^2/2}\,dy \\ &= \frac{b}{\sqrt{2\pi}}\int_{-\infty}^{\infty} e^{-y^2/2}\,dy = b \end{aligned} \tag{3.32}$$

If then X is an **exponential** ***rv*** $X \sim \mathfrak{E}(a)$ with *pdf* (2.17) the expectation is

$$\boldsymbol{E}[X] = \int_0^{+\infty} ax\, e^{-ax} = \frac{1}{a}\left[-(1+ax)e^{-ax}\right]_0^{+\infty} = \frac{1}{a} \tag{3.33}$$

and if it is a **Laplace** ***rv*** $X \sim \mathfrak{L}(a)$ with *pdf* (2.18) the expectation is

$$\boldsymbol{E}[X] = \int_{-\infty}^{+\infty} \frac{a}{2} x\, e^{-a|x|} = 0 \tag{3.34}$$

It is important to remark, instead, that if X is a **Cauchy** ***rv*** $X \sim \mathfrak{C}(a, b)$ with *pdf* (2.19) the expectation does not exist according to the Definition 3.20, namely it does neither converge, nor diverge. By using indeed the Heaviside function (2.13), it is easy to see that $X^+ = X\vartheta(X)$ and $X^- = -X\vartheta(-X)$: taking then for simplicity the Cauchy law $\mathfrak{C}(a, 0)$ with $b = 0$ (so that its *pdf* (2.19) turns out to be symmetric with $f_X(-x) = f_X(x)$) we find

$$\boldsymbol{E}\left[X^+\right] = \boldsymbol{E}\left[X^-\right] = \frac{a}{\pi}\int_0^{+\infty}\frac{x}{a^2+x^2}\,dx = \frac{a}{\pi}\left[\frac{1}{2}\ln(a^2+x^2)\right]_0^{+\infty} = +\infty$$

and hence $\boldsymbol{E}[X]$ takes the form $\infty - \infty$, namely the expectation is not defined as a Lebesgue integral. That its principal value

$$\lim_{M\to+\infty}\frac{a}{\pi}\int_{-M}^{+M}\frac{x}{a^2+x^2}\,dx=0$$

does instead exist is in fact immaterial: the expectation is a Lebesgue integral, not the principal value of a Riemann integral. We will see later in the Sect. 4.6 that this difference is not just a mathematical nicety and has instead a few far reaching implications

Proposition 3.26 *Take the integrable rv's X and Y defined on* $(\Omega, \mathcal{F}, \boldsymbol{P})$*:*

1. $\boldsymbol{E}[aX+bY]=a\,\boldsymbol{E}[X]+b\,\boldsymbol{E}[Y]$ *with* $a, b \in \boldsymbol{R}$
2. $\big|\boldsymbol{E}[X]\big| \le \boldsymbol{E}[|X|]$
3. *if* $X=0$*,* $\boldsymbol{P}$*-a.s., then* $\boldsymbol{E}[X]=0$*; if moreover X is an arbitrary rv and A an event*

$$\boldsymbol{E}[XI_A]=\int_A X\,d\boldsymbol{P}=0 \qquad \text{if } \boldsymbol{P}\{A\}=0$$

4. *if* $X \le Y$*,* $\boldsymbol{P}$*-a.s. then* $\boldsymbol{E}[X]\le\boldsymbol{E}[Y]$*, and if* $X=Y$ $\boldsymbol{P}$*-a.s., then* $\boldsymbol{E}[X]=\boldsymbol{E}[Y]$
5. *if* $X\ge 0$ *and* $\boldsymbol{E}[X]=0$*, then* $X=0$*,* $\boldsymbol{P}$*-a.s., namely X is degenerate* δ_0
6. *if* $\boldsymbol{E}[XI_A]\le\boldsymbol{E}[YI_A]$*,* $\forall A\in\mathcal{F}$*, then* $X\le Y$*,* $\boldsymbol{P}$*-a.s., and in particular if* $\boldsymbol{E}[XI_A]=\boldsymbol{E}[YI_A]$*,* $\forall A\in\mathcal{F}$*, then* $X=Y$*,* $\boldsymbol{P}$*-a.s.*
7. *if X and Y are independent, then also XY is integrable and*

$$\boldsymbol{E}[XY]=\boldsymbol{E}[X]\cdot\boldsymbol{E}[Y]$$

Proof Omitted: see [1] p. 183-5 and 191. These results, in a probabilistic notation, are indeed well known properties of the Lebesgue integral ■

Example 3.27 As a particular application of these results, remark that from the property 1. The expectation $\boldsymbol{E}[\,\cdot\,]$ turns out to be a linear functional giving rise to a few simplifications: remember for instance that to calculate the expectation (3.28) of a binomial *rv* $S_n \sim \mathfrak{B}(n;p)$ in the Example 3.24 we adopted the elementary definition (3.9): this result however comes even faster by recalling that according to the equation (3.8) every binomial *rv* coincides in distribution with the sum of n iid Bernoulli $\mathfrak{B}(1;p)$ *rv*'s $X_1,\dots,X_n$. Since the expectation depends only on the distribution of a *rv*, and since from (3.27) we know that $\boldsymbol{E}\left[X_j\right]=p$, from the expectation linearity we then immediately have that

$$\boldsymbol{E}[S_n]=\boldsymbol{E}\left[\sum_{j=1}^n X_j\right]=\sum_{j=1}^n\boldsymbol{E}\left[X_j\right]=np \tag{3.35}$$

in an apparent agreement with (3.28)

Of course the expectations comply also with a few important *inequalities* that are typical of the integrals: they are summarized in their probabilistic setting in the Appendix B. In the next Sect. 3.3.3 we will however separately introduce the *Chebyshev Inequality* (Proposition 3.41) because of its relevance in the subsequent discussion of the *Law of Large Numbers*

3.3.3 Variance and Covariance

The expectation of a *rv* X is a number specifying its distribution *barycenter*. As a centrality indicator, however, it is not unique, and on the other hand it does not convey all the information carried by the law of X. The expectation, for instance, could even fail to be one of the possible values of X: the expectation p of a Bernoulli *rv* $\mathfrak{B}(1; p)$ (taking just the values 0 and 1) is in general neither 0, nor 1. There are on the other hand several other indicators able to represent the *center* of a distribution: one is the ***mode*** which, when a *pdf* exists, is just the value (or the values) of x where $f_X(x)$ has a local maximum. It is also easy to see that in general the expectation does not coincide with the mode: when in particular the distribution is not symmetric the mode and the expectation are different

Beside these centrality indicators, however, it would be important to find also other numerical parameters showing other features of the law of a *rv* X. It would be relevant in particular to be able to give a measure of the *dispersion* of the *rv* values around its expectation. The *deviations* of the X values w.r.t. $\boldsymbol{E}[X]$ can be given as $X - \boldsymbol{E}[X]$ and we could imagine to concoct a meaningful parameter just by calculating its expectation $\boldsymbol{E}[X - \boldsymbol{E}[X]]$. But we immediately see that this number identically vanishes

$$\boldsymbol{E}[X - \boldsymbol{E}[X]] = \boldsymbol{E}[X] - \boldsymbol{E}[X] = 0$$

and hence that it can not be a measure of the X dispersion. Since however it is apparent that this vanishing essentially results from $X - \boldsymbol{E}[X]$ taking positive and negative values, it is customary to consider rather the quadratic deviations $(X - \boldsymbol{E}[X])^2$ that are never negative and that as a consequence will have in general a non zero expectation

Definition 3.28 If X, Y are *rv*'s and $\boldsymbol{X} = (X_1, \ldots, X_n)$ is a *r-vec*, and all the items are square integrable, then

- we call **variance** of X the non negative number (finite or infinite)

$$\boldsymbol{V}[X] = \boldsymbol{E}\left[(X - \boldsymbol{E}[X])^2\right] = \sigma_X^2$$

  and **standard deviation** its positive square root $\sigma_X = +\sqrt{\boldsymbol{V}[X]}$
- we call **covariance** of X and Y the number

$$\boldsymbol{cov}[X, Y] = \boldsymbol{E}[(X - \boldsymbol{E}[X])(Y - \boldsymbol{E}[Y])] \qquad (\boldsymbol{cov}[X, X] = \boldsymbol{V}[X])$$

and (if $\boldsymbol{V}[X] \neq 0,\ \boldsymbol{V}[Y] \neq 0$) **correlation coefficient** the number

$$\rho[X, Y] = \frac{\boldsymbol{cov}[X, Y]}{\sqrt{\boldsymbol{V}[X]}\sqrt{\boldsymbol{V}[Y]}}$$

- when $\boldsymbol{cov}[X, Y] = 0$, namely $\rho[X, Y] = 0$, we will say that X and Y are **uncorrelated *rv*'s**
- we finally call **covariance matrix** (respectively **correlation matrix**) of the *r-vec* $\boldsymbol{X}$ the $n \times n$ matrix $\mathbb{R} = \|r_{ij}\|$ (respectively $\mathbb{P} = \|\rho_{ij}\|$) with elements $r_{ij} = \boldsymbol{cov}\left[X_i, X_j\right]$ (respectively $\rho_{ij} = \rho\left[X_i, X_j\right]$)

Proposition 3.29 *If X, Y are square integrable rv's, and a, b are numbers, then*

1. $\boldsymbol{cov}[X, Y] = \boldsymbol{E}[XY] - \boldsymbol{E}[X]\,\boldsymbol{E}[Y]$; *in particular* $\boldsymbol{V}[X] = \boldsymbol{E}\left[X^2\right] - \boldsymbol{E}[X]^2$
2. *if* $\boldsymbol{V}[X] = 0$, *then* $X = \boldsymbol{E}[X]$, $\boldsymbol{P}$*-a.s.*
3. $\boldsymbol{V}[a + b\,X] = b^2\,\boldsymbol{V}[X]$
4. $\boldsymbol{V}[X + Y] = \boldsymbol{V}[X] + \boldsymbol{V}[Y] + 2\,\boldsymbol{cov}[X, Y]$
5. *if X, Y are independent, then they are also uncorrelated*

Proof

1. The result follows by calculating the product in the definition

$$\boldsymbol{cov}[X, Y] = \boldsymbol{E}[XY] - 2\boldsymbol{E}[X]\,\boldsymbol{E}[Y] + \boldsymbol{E}[X]\,\boldsymbol{E}[Y] = \boldsymbol{E}[XY] - \boldsymbol{E}[X]\,\boldsymbol{E}[Y]$$

   Taking then $X = Y$ we also get the result about the variance
2. Since $(X - \boldsymbol{E}[X])^2 \geq 0$, the outcome results from 5 in Proposition 3.26: if indeed $\boldsymbol{V}[X] = 0$, then $X - \boldsymbol{E}[X] = 0$, $\boldsymbol{P}$-a.s. (namely $X - \boldsymbol{E}[X]$ is a degenerate $\boldsymbol{\delta}_0$) and hence $X = \boldsymbol{E}[X]$, $\boldsymbol{P}$-a.s.
3. From the expectation linearity we have $\boldsymbol{E}[a + bX] = a + b\boldsymbol{E}[X]$, and hence

$$\boldsymbol{V}[a + bX] = \boldsymbol{E}\left[(a + bX - a - b\boldsymbol{E}[X])^2\right] = b^2\boldsymbol{E}\left[(X - \boldsymbol{E}[X])^2\right] = b^2\boldsymbol{V}[X]$$

4. We indeed have

$$\begin{aligned}\boldsymbol{V}[X + Y] &= \boldsymbol{E}\left[(X + Y - \boldsymbol{E}[X + Y])^2\right] = \boldsymbol{E}\left[(X - \boldsymbol{E}[X] + Y - \boldsymbol{E}[Y])^2\right]\\ &= \boldsymbol{V}[X] + \boldsymbol{V}[Y] + 2\,\boldsymbol{E}[(X - \boldsymbol{E}[X])\,(Y - \boldsymbol{E}[Y])]\\ &= \boldsymbol{V}[X] + \boldsymbol{V}[Y] + 2\,\boldsymbol{cov}[X, Y]\end{aligned}$$

   Remark that (at variance with the expectation $\boldsymbol{E}$) $\boldsymbol{V}$ is not a *linear* functional. In particular we have just shown that in general $\boldsymbol{V}[X + Y]$ is not the sum $\boldsymbol{V}[X] + \boldsymbol{V}[Y]$; this happens only when the *rv*'s X, Y are ***uncorrelated*** because in this event we have

$$\boldsymbol{V}[X + Y] = \boldsymbol{V}[X] + \boldsymbol{V}[Y] \qquad \text{if} \quad \boldsymbol{cov}[X, Y] = 0$$

5. Because of the X, Y independence, we have from the definition

$$\boldsymbol{cov}[X, Y] = \boldsymbol{E}[X - \boldsymbol{E}[X]]\,\boldsymbol{E}[Y - \boldsymbol{E}[Y]] = 0$$

namely X and Y are also uncorrelated ■

Proposition 3.30 *Given two rv's X and Y, we always have*

$$|\rho[X, Y]| \leq 1$$

Moreover it is $|\rho[X, Y]| = 1$ iff there are two numbers $a \neq 0$ and b such that $Y = aX + b$, $\boldsymbol{P}$-a.s.; in particular $a > 0$ if $\rho[X, Y] = +1$, and $a < 0$ if $\rho[X, Y] = -1$

Proof Omitted: see [2] pp. 145–6. These properties of the correlation coefficient ρ do not hold for the covariance which instead takes arbitrary positive and negative real values. The present proposition shows that ρ is a measure of the *linear dependence* between X and Y: when indeed ρ takes its extremal values ± 1, Y is a linear function of X ■

Example 3.31 Independence vs non-correlation: Point 5 in Proposition 3.29 states that two independent *rv*'s are also uncorrelated. In general however the reverse does not hold: two uncorrelated *rv*'s can be dependent (not independent). Consider for instance a *rv* α taking the three values $0, \frac{\pi}{2}, \pi$ with equal probabilities $\frac{1}{3}$, and define then the *rv*'s $X = \sin\alpha$ and $Y = \cos\alpha$. It is easy to see that X and Y are uncorrelated because

$$\boldsymbol{E}[X] = \frac{1}{3}\cdot 0 + \frac{1}{3}\cdot 1 + \frac{1}{3}\cdot 0 = \frac{1}{3} \qquad \boldsymbol{E}[Y] = \frac{1}{3}\cdot 1 + \frac{1}{3}\cdot 0 + \frac{1}{3}\cdot(-1) = 0$$

$$\boldsymbol{E}[XY] = \frac{1}{3}\cdot(1\cdot 0) + \frac{1}{3}\cdot(0\cdot 1) + \frac{1}{3}\cdot(-1\cdot 0) = 0$$

so that

$$\boldsymbol{E}[X]\,\boldsymbol{E}[Y] = 0 = \boldsymbol{E}[XY]$$

On the other hand they are also not independent because

$$\boldsymbol{P}\{X = 1\} = \frac{1}{3} \qquad \boldsymbol{P}\{Y = 1\} = \frac{1}{3} \qquad \boldsymbol{P}\{X = 1,\ Y = 1\} = 0$$

so that

$$\boldsymbol{P}\{X = 1,\ Y = 1\} = 0 \neq \frac{1}{9} = \boldsymbol{P}\{X = 1\}\,\boldsymbol{P}\{Y = 1\}$$

Proposition 3.32 *Necessary and sufficient condition for a matrix $n \times n$, $\mathbb{R}$ to be covariance matrix of a r-vec $\boldsymbol{X} = (X_1, \ldots, X_n)$, is to be symmetric and non negative definite; or equivalently, that it exists a matrix $n \times n$, $\mathbb{C}$ such that $\mathbb{R} = \mathbb{C}\mathbb{C}^T$, where $\mathbb{C}^T$ is the transposed matrix of $\mathbb{C}$*

Proof It follows from the definition that the covariance matrix $\mathbb{R}$ of a *r-vec* $\boldsymbol{X}$ is always symmetric ($r_{ij} = r_{ji}$), and it is easy to check that it is also non negative definite: if indeed we take n real numbers $\lambda_1, \ldots, \lambda_n$ we find

$$\sum_{i,j=1}^{n} r_{ij}\lambda_i\lambda_j = \sum_{i,j=1}^{n} \lambda_i\lambda_j \boldsymbol{E}\left[(X_i - \boldsymbol{E}[X_i])(X_j - E[X_j])\right]$$
$$= \boldsymbol{E}\left[\left(\sum_{i=1}^{n} \lambda_i(X_i - \boldsymbol{E}[X_i])\right)^2\right] \geq 0$$

The present proposition states in fact that these properties are characteristic of the covariance matrices, in the sense that also the reverse holds: every symmetric and non negative definite matrix is a legitimate covariance matrix of some *r-vec* $\boldsymbol{X}$. We neglect the proof (see [1] p. 235) of this statement and of the remainder of the proposition ■

Example 3.33 We consider first some discrete laws: if $X \sim \boldsymbol{\delta}_b$ is a **degenerate *rv*** we apparently have $\boldsymbol{V}[X] = 0$; if on the other hand $X \sim \mathfrak{B}(1; p)$ is a **Bernoulli *rv*** since $\boldsymbol{E}[X] = p$ we have at once

$$\boldsymbol{V}[X] = \boldsymbol{E}\left[(X - \boldsymbol{E}[X])^2\right] = (1-p)^2\, p + (0-p)^2\,(1-p) = p\,(1-p)$$

If then $S_n \sim \mathfrak{B}(n; p)$ is a **binomial *rv***, which according to (3.8) is in distribution the sum of n iid Bernoulli $\mathfrak{B}(1; p)$ *rv*'s $X_1, \ldots, X_n$, from the X_j independence we have

$$\boldsymbol{V}[S_n] = \boldsymbol{V}\left[\sum_{j=1}^{n} X_j\right] = \sum_{j=1}^{n} \boldsymbol{V}\left[X_j\right] = \sum_{j=1}^{n} p(1-p) = np(1-p) \qquad (3.36)$$

If finally $X \sim \mathfrak{P}(\alpha)$ is a **Poisson *rv***, with $\boldsymbol{E}[X] = \alpha$ according to (3.29), it is expedient to start by calculating

$$\boldsymbol{E}[X(X-1)] = \sum_{k=0}^{\infty} k(k-1)\frac{e^{-\alpha}\alpha^k}{k!} = e^{-\alpha}\sum_{k=2}^{\infty}\frac{\alpha^k}{(k-2)!} = \alpha^2 e^{-\alpha}\sum_{k=0}^{\infty}\frac{\alpha^k}{k!} = \alpha^2$$

so that

$$\boldsymbol{E}\left[X^2\right] = \alpha^2 + \boldsymbol{E}[X] = \alpha^2 + \alpha$$

and hence

$$\boldsymbol{V}[X] = \boldsymbol{E}\left[X^2\right] - \boldsymbol{E}[X]^2 = \alpha^2 + \alpha - \alpha^2 = \alpha$$

Remark that for the Poisson laws it is $\boldsymbol{E}[X] = \boldsymbol{V}[X] = \alpha$. Going then to the laws with a *pdf*, for a **uniform *rv*** $X \sim \mathfrak{U}(a, b)$ it is

$$\boldsymbol{E}\left[X^2\right] = \int_a^b \frac{x^2}{b-a}\,dx = \frac{1}{b-a}\left[\frac{x^3}{3}\right]_a^b = \frac{b^3-a^3}{3(b-a)}$$

so that from (3.30) we get

$$\boldsymbol{V}[X] = \boldsymbol{E}\left[X^2\right] - \boldsymbol{E}[X]^2 = \frac{b^3-a^3}{3(b-a)} - \frac{(a+b)^2}{4} = \frac{(b-a)^2}{12} \tag{3.37}$$

For an **exponential** ***rv*** $X \sim \mathfrak{E}(a)$ with the change of variable $y = ax$ we find

$$\boldsymbol{E}\left[X^2\right] = \int_0^{+\infty} x^2 a e^{-ax}\,dx = \frac{1}{a^2}\int_0^{+\infty} y^2 e^{-y}\,dy = \frac{1}{a^2}\left[-(2+2y+y^2)e^{-y}\right]_0^{+\infty} = \frac{2}{a^2}$$

and hence from (3.33)

$$\boldsymbol{V}[X] = \boldsymbol{E}\left[X^2\right] - \boldsymbol{E}[X]^2 = \frac{2}{a^2} - \frac{1}{a^2} = \frac{1}{a^2} \tag{3.38}$$

In a similar way for a **Laplace** ***rv*** $X \sim \mathfrak{L}(a)$ we get

$$\boldsymbol{V}[X] = \frac{2}{a^2} \tag{3.39}$$

If finally $X \sim \mathfrak{N}(b, a^2)$ is a **Gaussian** ***rv***, by taking into account the Gaussian integrals (3.31), with the change of variable $y = (x-b)/a$ we have

$$\boldsymbol{E}\left[X^2\right] = \int_{-\infty}^{+\infty} x^2 \frac{e^{-(x-b)^2/2a^2}}{a\sqrt{2\pi}}\,dx = \frac{1}{\sqrt{2\pi}}\int_{-\infty}^{+\infty} (ay+b)^2 e^{-y^2/2}\,dy = a^2 + b^2$$

and hence from (3.32)

$$\boldsymbol{V}[X] = \boldsymbol{E}\left[X^2\right] - \boldsymbol{E}[X]^2 = (a^2+b^2) - b^2 = a^2 \tag{3.40}$$

For a **Cauchy** ***rv*** $X \sim \mathfrak{C}(a, b)$, however, a variance can not be defined first because its expectation does not exist as remarked at the end of the Sect. 3.3.2, and then because in any case its second moment diverges as can be seen for example for $b = 0$ with the change of variable $y = x/a$

$$\boldsymbol{E}\left[X^2\right] = \frac{a}{\pi}\int_{-\infty}^{+\infty} \frac{x^2}{a^2+x^2}\,dx = \frac{a^2}{\pi}\int_{-\infty}^{+\infty} \frac{y^2}{1+y^2}\,dy = \frac{a^2}{\pi}\left[y - \arctan y\right]_{-\infty}^{+\infty} = +\infty$$

Example 3.34 Bivariate Gaussian vectors: If $\boldsymbol{X} = (X, Y) \sim \mathfrak{N}(\boldsymbol{b}, \mathbb{A})$ is a bivariate Gaussian (normal) *r-vec* we know that in its joint *pdf* (2.24) there are five free parameters: the two components of $\boldsymbol{b} = (b_1, b_2) \in \boldsymbol{R}^2$, and the three numbers $a_1 > 0$, $a_2 > 0$, $|r| \leq 1$ derived from the elements of the symmetric, positive defined matrix $\mathbb{A}$ as

$$a_k = \sqrt{a_{kk}} \qquad r = \frac{a_{12}}{\sqrt{a_{11}a_{22}}} = \frac{a_{21}}{\sqrt{a_{11}a_{22}}}$$

We have pointed out in the Exemple 2.31 that also the univariate marginals are normal $\mathfrak{N}(b_k, a_k^2)$ with *pdf*

$$f_X(x) = \int_{-\infty}^{\infty} f_X(x,y)\,dy = \frac{1}{a_1\sqrt{2\pi}}\,e^{-(x-b_1)^2/2a_1^2}$$
$$f_Y(y) = \int_{-\infty}^{\infty} f_X(x,y)\,dx = \frac{1}{a_2\sqrt{2\pi}}\,e^{-(y-b_2)^2/2a_2^2}$$

A direct calculus would show that the probabilistic meaning of the five parameters appearing in a bivariate $\mathfrak{N}(\boldsymbol{b}, \mathbb{A})$ is

$$b_1 = \boldsymbol{E}[X] \qquad b_2 = \boldsymbol{E}[Y]$$
$$a_1^2 = a_{11} = \boldsymbol{V}[X] \qquad a_2^2 = a_{22} = \boldsymbol{V}[Y] \qquad r = \frac{a_{12}}{\sqrt{a_{11}a_{22}}} = \frac{a_{21}}{\sqrt{a_{11}a_{22}}} = \rho[X,Y]$$

so that the vector of the means $\boldsymbol{b}$ and the covariance matrix $\mathbb{A}$ are

$$\boldsymbol{b} = \begin{pmatrix} b_1 \\ b_2 \end{pmatrix} = \begin{pmatrix} \boldsymbol{E}[X] \\ \boldsymbol{E}[Y] \end{pmatrix}$$
$$\mathbb{A} = \begin{pmatrix} a_{11} & a_{12} \\ a_{21} & a_{22} \end{pmatrix} = \begin{pmatrix} \boldsymbol{V}[X] & \boldsymbol{cov}[X,Y] \\ \boldsymbol{cov}[X,Y] & \boldsymbol{V}[Y] \end{pmatrix} = \begin{pmatrix} a_1^2 & a_1a_2r \\ a_1a_2r & a_2^2 \end{pmatrix}$$

We emphasized above that two independent *rv*'s X, Y are also uncorrelated, but that in general the reverse does not hold. It is then relevant to remark that in a joint Gaussian *r-vec* $\boldsymbol{X}$ the uncorrelated components X_k are also independent. In other words: **the components of a multivariate Gaussian *r-vec* $\boldsymbol{X} \sim \mathfrak{N}(\boldsymbol{b}, \mathbb{A})$ are independent *iff* they are uncorrelated**. This follows—for instance in the bivariate case—from the fact that if the components X, Y are uncorrelates we have $r = \rho[X,Y] = 0$, and hence the joint *pdf* (2.24) boils down to

$$f_X(x,y) = \frac{1}{2\pi\,a_1a_2}\,e^{-(x-b_1)^2/a_1^2}\,e^{-(y-b_2)^2/a_2^2}$$

so that $f_X(x,y) = f_X(x)\cdot f_Y(y)$, and hence according to the Theorem 3.15 X, Y are also independent

Proposition 3.35 Chebyshev inequality: *If X is a non-negative, integrable rv we have*

$$\boldsymbol{P}\{X \geq \epsilon\} \leq \frac{\boldsymbol{E}[X]}{\epsilon} \qquad \forall\epsilon > 0 \tag{3.41}$$

Proof The result follows immediately from

$$\boldsymbol{E}[X] \geq \boldsymbol{E}\left[X I_{\{X\geq\epsilon\}}\right] \geq \epsilon \boldsymbol{E}\left[I_{\{X\geq\epsilon\}}\right] = \epsilon \boldsymbol{P}\{X \geq \epsilon\}$$

where $\epsilon > 0$ is of course arbitrary ■

Corollary 3.36 *If X is a square integrable rv, then for every $\epsilon > 0$ it is*

$$\boldsymbol{P}\{|X| \geq \epsilon\} = \boldsymbol{P}\left\{X^2 \geq \epsilon^2\right\} \leq \frac{\boldsymbol{E}\left[X^2\right]}{\epsilon^2}$$

$$\boldsymbol{P}\left\{\left|X - \boldsymbol{E}[X]\right| \geq \epsilon\right\} \leq \frac{\boldsymbol{E}\left[(X - \boldsymbol{E}[X])^2\right]}{\epsilon^2} = \frac{\boldsymbol{V}[X]}{\epsilon^2} \tag{3.42}$$

Proof Just apply the inequality (3.41) to the proposed *rv*'s ■

3.4 Conditioning

3.4.1 Conditional Distributions

In the Sect. 1.4 we defined the conditional probability for two events only when the probability of the conditioning event does not vanish. This restriction, however, is not satisfied for instance by the negligible events as $\{Y = y\}$ when Y is an *ac rv*: we know indeed that in this case $\boldsymbol{P}\{Y = y\} = 0$. We need then to extend our definitions and notations in order to account even for these, not irrelevant cases

Definition 3.37 If X, Y are two *rv*'s with a joint *cdf* $F_{XY}(x, y)$ which is y-differentiable, and if Y is *ac* with *pdf* $f_Y(y)$, we will call ***cdf* of X conditioned by the event** $\{Y = y\}$ the function

$$F_{X|Y}(x|y) = F_X(x|Y = y) \equiv \frac{\partial_y F_{XY}(x, y)}{f_Y(y)} \tag{3.43}$$

for every y such that $f_Y(y) \neq 0$ (namely $\boldsymbol{P}_Y$-a.s.), while for the y such that $f_Y(y) = 0$ (a $\boldsymbol{P}_Y$-negligible set) $F_X(x|Y = y)$ takes arbitrary values, possibly zero. If moreover also X is *ac* and the joint *pdf* is $f_{XY}(x, y)$, then **the *pdf* of X conditioned by the event** $\{Y = y\}$ is

$$f_{X|Y}(x|y) = f_X(x|Y = y) \equiv \frac{f_{XY}(x, y)}{f_Y(y)} \tag{3.44}$$

for the y such that $f_Y(y) \neq 0$, and zero for the y such that $f_Y(y) = 0$

In order to intuitively account for these definitions consider the joint and marginal *cdf*'s $F_{XY}(x, y)$, $F_X(x)$ and $F_Y(y)$ of X, Y, and the *pdf* $f_Y(y) = F'_Y(y)$ of Y. By supposing then that $F_{XY}(x, y)$ is y-derivabile, take first the *modified* conditioning event $\{y < Y \le y + \Delta y\}$ which presumably has a non vanishing probability: the Definition 3.37 is then recovered in the limit for $\Delta y \to 0$. From the elementary definition of conditioning we have indeed that

$$\begin{aligned} F_X(x|y < Y \le y + \Delta y) &= \boldsymbol{P}\{X \le x | y < Y \le y + \Delta y\} \\ &= \frac{\boldsymbol{P}\{X \le x,\ y < Y \le y + \Delta y\}}{\boldsymbol{P}\{y < Y \le y + \Delta y\}} \\ &= \frac{F_{XY}(x, y + \Delta y) - F_{XY}(x, y)}{F_Y(y + \Delta y) - F_Y(y)} = \frac{\frac{F_{XY}(x, y+\Delta y) - F_{XY}(x, y)}{\Delta y}}{\frac{F_Y(y+\Delta y) - F_Y(y)}{\Delta y}} \end{aligned}$$

so that in the limit for $\Delta y \to 0$ we find (3.43)

$$F_X(x \mid Y = y) \equiv \lim_{\Delta y \to 0} F_X(x \mid y < Y \le y + \Delta y) = \frac{\partial_y F_{XY}(x, y)}{F'_Y(y)} = \frac{\partial_y F_{XY}(x, y)}{f_Y(y)}$$

If we finally suppose that also X has a *pdf* $f_X(x)$, and that $f_{XY}(x, y)$ is the joint *pdf*, a further x-derivation of (3.43) gives rise to the conditional *pdf* (3.44). The formulas (3.43) and (3.44) define the conditional distributions for every y such that $f_Y(y) > 0$, namely $\boldsymbol{P}_Y$-*as*. Where instead $f_Y(y) = 0$ the value of $F_X(x \mid Y = y)$ (or of the corresponding *pdf*) is arbitrary (for instance zero) since this choice does not affect the results of the calculations. Remark moreover that if X, Y are ***independent***, from the Theorem 3.15 it follows at once that

$$f_{X|Y}(x|y) = f_X(x) \tag{3.45}$$

Similar results hold apparently also for discrete *rv*'s, but for the fact that in this case the conditional *pdf*'s are replaced by the conditional probabilities according to the definitions of the Sect. 1.4.

Proposition 3.38 *If X, Y are two rv's with joint pdf $f_{XY}(x, y)$, then*

$$\boldsymbol{P}_X\{A|Y = y\} = \int_A f_{X|Y}(x|y)\,dx = \frac{1}{f_Y(y)} \int_A f_{XY}(x, y)\,dx \tag{3.46}$$

$$\boldsymbol{P}_{XY}\{A \times B\} = \int_B \boldsymbol{P}_X\{A|Y = y\} f_Y(y)\,dy \tag{3.47}$$

$$\boldsymbol{P}_X\{A\} = \int_{-\infty}^{+\infty} \boldsymbol{P}_X\{A|Y = y\} f_Y(y)\,dy \tag{3.48}$$

Proof From (3.44) we have first (3.46)

$$\boldsymbol{P}_X\{A|Y = y\} = \boldsymbol{P}\{X \in A|Y = y\} = \int_A f_{X|Y}(x|y)\,dx = \frac{1}{f_Y(y)} \int_A f_{XY}(x, y)\,dx$$

and thence also (3.47) and (3.48) result:

$$
\begin{aligned}
\boldsymbol{P}_{XY}\{A \times B\} = \boldsymbol{P}\{X \in A, Y \in B\} &= \int_{A\times B} f_{XY}(x, y)\, dxdy \\
&= \int_A dx \int_B dy\, f_{X|Y}(x|y) f_Y(y) \\
&= \int_B \boldsymbol{P}_X\{A|Y = y\} f_Y(y)\, dy \\
\boldsymbol{P}_X\{A\} = \boldsymbol{P}\{X \in A\} &= \int_{-\infty}^{+\infty} \boldsymbol{P}_X\{A|Y = y\} f_Y(y)\, dy
\end{aligned}
$$

In particular the (3.48) shows how to calculate $\boldsymbol{P}_X$ from the conditional distribution (3.46) ■

Proposition 3.39 *If X, Y are two rv's with joint pdf $f_{XY}(x, y)$, then*

$$f_{XY}(x, y|a \le Y \le b) = \frac{f_{XY}(x, y)}{\int_a^b f_Y(y')\, dy'}\, \chi_{[a,b]}(y) \tag{3.49}$$

$$f_X(x|a \le Y \le b) = \frac{\int_a^b f_{XY}(x, y')\, dy'}{\int_a^b f_Y(y')\, dy'} \tag{3.50}$$

$$f_Y(y|a \le Y \le b) = \frac{f_Y(y)}{\int_a^b f_Y(y')\, dy'}\, \chi_{[a,b]}(y) \tag{3.51}$$

where the indicator of the subset B in $(\boldsymbol{R}, \mathcal{B}(\boldsymbol{R}))$ is

$$\chi_B(x) = \begin{cases} 1, & \text{if } x \in B \\ 0, & \text{else} \end{cases} \qquad B \in \mathcal{B}(\boldsymbol{R})$$

Proof From the definitions we have first

$$
\begin{aligned}
F_{XY}(x, y|a \le Y \le b) &= \frac{\boldsymbol{P}\{X \le x,\ Y \le y,\ a \le Y \le b\}}{\boldsymbol{P}\{a \le Y \le b\}} \\
&= \begin{cases} 0 & \text{if } y \le a \\ \frac{\boldsymbol{P}\{X \le x,\ a \le Y \le y\}}{\boldsymbol{P}\{a \le Y \le b\}} = \frac{F_{XY}(x,y) - F_{XY}(x,a)}{F_Y(b) - F_Y(a)} & \text{if } a \le y \le b \\ \frac{\boldsymbol{P}\{X \le x,\ a \le Y \le b\}}{\boldsymbol{P}\{a \le Y \le b\}} = \frac{F_{XY}(x,b) - F_{XY}(x,a)}{F_Y(b) - F_Y(a)} & \text{if } b \le y \end{cases}
\end{aligned}
$$

and then (3.49) follows by remembering that

$$
\begin{aligned}
f_{XY}(x, y|a \le Y \le b) &= \partial_x \partial_y F_{XY}(x, y|a \le Y \le b) \\
F_Y(b) - F_Y(a) &= \int_a^b f_Y(y')\, dy'
\end{aligned}
$$

From (3.49) we then derive (3.50) and (3.51) by marginalization ■

Proposition 3.40 *If $\boldsymbol{X} = (X_1, X_2) \sim \mathfrak{N}(\boldsymbol{b}, \mathbb{A})$ is a bivariate, Gaussian r-vec with pdf (2.24), the conditional law of X_2 w.r.t. $X_1 = x_1$ is again Gaussian with parameters*

$$\mathfrak{N}\left(b_2 + r(x_1 - b_1)\frac{a_2}{a_1},\ (1 - r^2)a_2^2\right) \tag{3.52}$$

Proof We know that the bivariate *pdf* of $\boldsymbol{X}$ is (2.24), and that its two marginals are $\mathfrak{N}(b_k, a_k^2)$ with *pdf* (2.31). A direct application of (3.44) brings then to the following conditional *pdf*

$$f_{X_2|X_1}(x_2|x_1) = \frac{e^{-\frac{1}{2(1-r^2)}\left[\frac{(x_1-b_1)^2}{a_1^2} - 2r\frac{(x_1-b_1)(x_2-b_2)}{a_1a_2} + \frac{(x_2-b_2)^2}{a_2^2}\right]}}{2\pi a_1 a_2\sqrt{1-r^2}}\, a_1\sqrt{2\pi}\, e^{\frac{(x_1-b_1)^2}{2a_1^2}}$$

$$= \frac{e^{-\frac{1}{2(1-r^2)}\left[r^2\frac{(x_1-b_1)^2}{a_1^2} - 2r\frac{(x_1-b_1)(x_2-b_2)}{a_1a_2} + \frac{(x_2-b_2)^2}{a_2^2}\right]}}{\sqrt{2\pi a_2^2(1-r^2)}}$$

$$= \frac{e^{-\frac{1}{2a_2^2(1-r^2)}\left[(x_2-b_2) - r(x_1-b_1)\frac{a_2}{a_1}\right]^2}}{\sqrt{2\pi a_2^2(1-r^2)}}$$

and hence to the result (3.52) ■

So far we have considered just the reciprocal conditioning between two *rv*'s, but this was required only to simplify the notation. We will remember then that given two *r-vec*'s $\boldsymbol{X} = (X_1, \dots, X_n)$ and $\boldsymbol{Y} = (Y_1, \dots, Y_m)$ with a joint *pdf*

$$f_{\boldsymbol{XY}}(x_1, \dots, x_n, y_1, \dots, y_m)$$

a procedure identical to that adopted in the case of two *rv*'s gives rise to the definition of the conditional *pdf* of $\boldsymbol{X}$ w.r.t.the event $\{Y_1 = y_1, \dots, Y_m = y_m\}$

$$f_{\boldsymbol{X}|\boldsymbol{Y}}(x_1, \dots, x_n \mid y_1, \dots, y_m) \equiv \frac{f_{\boldsymbol{XY}}(x_1, \dots, x_n, y_1, \dots, y_m)}{f_{\boldsymbol{Y}}(y_1, \dots, y_m)} \tag{3.53}$$

3.4.2 Conditional Expectation

We already know that if B is a non-zero probability event, then the conditional probability $\boldsymbol{P}\{\cdot \mid B\}$ can be defined in an elementary way (Sect. 1.4) and constitutes a new probability on $(\Omega, \mathcal{F})$. As a consequence a *rv* X in a natural way will have a conditional distribution $\boldsymbol{P}_X\{\cdot \mid B\}$, the conditional *cdf* $F_X(\cdot \mid B)$ and *pdf* $f_X(\cdot \mid B)$ and a conditional expectation $\boldsymbol{E}[X|B]$. And even when the conditioning event is negligible as $\{Y = y\}$ with Y an *ac rv*, we have shown in the previous section how to define $\boldsymbol{P}_X\{\cdot \mid Y = y\}$, $F_X(\cdot \mid Y = y)$ and $f_X(\cdot \mid Y = y)$. We can then follow the same procedure presented in the Sect. 3.3.1 to define the conditional expectations by means of these new conditional measures. We will suppose in the following that our *rv*'s are always endowed with a *pdf*

Definition 3.41 Given the *rv's* X, Y and a Borel function $g(x)$, we will call **conditional expectation of** $g(X)$ **w.r.t.** $\{Y = y\}$ the y-function

$$m(y) \equiv \boldsymbol{E}\left[g(X)|Y = y\right] = \int_{-\infty}^{+\infty} g(x)\, f_{X|Y}(x|y)\, dx \tag{3.54}$$

We will call instead **conditional expectation of** $g(X)$ **w.r.t. the *rv*** Y the *rv*

$$\boldsymbol{E}\left[g(X)|Y\right] \equiv m\,(Y) \tag{3.55}$$

It is important to stress that to define the *rv* (3.55) we must first notice that the expectation (3.54) is a function $m(y)$ of the value y of the conditioning *rv* Y, and only then we can introduce—based on the Proposition 3.6—the *rv* $m(Y)$ usually denoted by the new symbol $\boldsymbol{E}\left[g(X)|Y\right]$. Remark again that the expectation $m(Y(\omega)) = \boldsymbol{E}\left[g(X)|Y\right](\omega)$ is a *rv*, and no longer a number or a function as usually an expectation is: this kind of *rv*'s will play a relevant role in the following sections

Proposition 3.42 *Given two rv's* X, Y *on* $(\Omega, \mathcal{F}, \boldsymbol{P})$, *the following properties of the conditional expectations always hold:*

1. $\boldsymbol{E}\left[\boldsymbol{E}\left[X|Y\right]\right] = \boldsymbol{E}\left[X\right]$
2. $\boldsymbol{E}\left[X|Y\right] = \boldsymbol{E}\left[X\right]$ $\boldsymbol{P}$-*a.s.* *if X and Y are independent*
3. $\boldsymbol{E}\left[\varphi(X, Y)|Y = y\right] = \boldsymbol{E}\left[\varphi(X, y)|Y = y\right]$ $\boldsymbol{P}_Y$-*a.s.*
4. $\boldsymbol{E}\left[\varphi(X, Y)|Y = y\right] = \boldsymbol{E}\left[\varphi(X, y)\right]$ $\boldsymbol{P}_Y$-*a.s.* *if X and Y are independent*
5. $\boldsymbol{E}\left[X\, g(Y)|Y\right] = g(Y)\, \boldsymbol{E}\left[X|Y\right]$ $\boldsymbol{P}$-*a.s.*

Proof

1. From (3.54) and (3.44) it is

$$\begin{aligned}
\boldsymbol{E}\left[\boldsymbol{E}\left[X|Y\right]\right] &= \boldsymbol{E}\left[m(Y)\right] = \int_{\boldsymbol{R}} m(y) f_Y(y)\, dy = \int_{\boldsymbol{R}} \boldsymbol{E}\left[X|Y = y\right] f_Y(y)\, dy \\
&= \int_{\boldsymbol{R}} \left[\int_{\boldsymbol{R}} x f_{X|Y}(x \mid y)\, dx\right] f_Y(y)\, dy \\
&= \int_{\boldsymbol{R}} \left[\int_{\boldsymbol{R}} x \frac{f_{XY}(x, y)}{f_Y(y)}\, dx\right] f_Y(y)\, dy \\
&= \int_{\boldsymbol{R}} x \left[\int_{\boldsymbol{R}} f_{XY}(x, y)\, dy\right] dx = \int_{\boldsymbol{R}} x f_X(x)\, dx = \boldsymbol{E}\left[X\right]
\end{aligned}$$

2. From the independence and from (3.45) it follows

$$m(y) = \boldsymbol{E}[X|Y = y] = \int_{\boldsymbol{R}} x\, f_{X|Y}(x|y)\, dx = \int_{\boldsymbol{R}} x\, f_X(x)\, dx = \boldsymbol{E}[X]$$

so that $\boldsymbol{E}[X|Y] = m(Y) = \boldsymbol{E}[X]$

3. From (3.49) we can write

$$\begin{aligned}
&\boldsymbol{E}[\varphi(X,Y)|y \leq Y \leq y + \Delta y] \\
&\quad = \int_{\boldsymbol{R}} dx \int_{\boldsymbol{R}} dz\, \varphi(x,z) f_{XY}(x,z|y \leq Y \leq y + \Delta y) \\
&\quad = \int_{-\infty}^{+\infty} dx \int_{y}^{y+\Delta y} dz\, \varphi(x,z)\, \frac{f_{XY}(x,z)}{F_Y(y+\Delta y) - F_Y(y)}
\end{aligned}$$

Since on the other hand

$$F_Y(y+\Delta y) - F_Y(y) = F_Y'(y)\Delta y + \mathrm{o}(\Delta y) = f_Y(y)\Delta y + \mathrm{o}(\Delta y)$$

we also have

$$\begin{aligned}
\lim_{\Delta y \to 0} \int_{y}^{y+\Delta y} dz\, \varphi(x,z)\, \frac{f_{XY}(x,z)}{F_Y(y+\Delta y) - F_Y(y)} &= \varphi(x,y)\, \frac{f_{XY}(x,y)}{f_Y(y)} \\
&= \varphi(x,y) f_{X|Y}(x \mid y)
\end{aligned}$$

and finally

$$\begin{aligned}
\boldsymbol{E}[\varphi(X,Y)|Y = y] &= \lim_{\Delta y \to 0} \boldsymbol{E}[\varphi(X,Y)|y \leq Y \leq y + \Delta y] \\
&= \int_{-\infty}^{+\infty} \varphi(x,y) f_{X|Y}(x \mid y)\, dx = \boldsymbol{E}[\varphi(X,y)|Y = y]
\end{aligned}$$

4. Since X and Y are independent, the result follows from the previous one and from (3.45)
5. From 3. Of the present Proposition it follows in particular that

$$\boldsymbol{E}[X\, g(Y)|Y = y] = \boldsymbol{E}[X\, g(y)|Y = y] = g(y)\, \boldsymbol{E}[X|Y = y]$$

and the last statement ensues by plugging Y as argument in this function ■

By using the conditional *pdf*'s (3.53) we can also define the conditional expectations w.r.t. negligible events of the type $\{Y_1 = y_1, \ldots, Y_m = y_m\}$, namely

$$\begin{aligned}
m(y_1, \ldots, y_m) &= \boldsymbol{E}[X|Y_1 = y_1, \ldots, Y_m = y_m] \\
&= \int_{\boldsymbol{R}} x f_{X|Y}(x|y_1, \ldots, y_m)\, dx,
\end{aligned}$$

and hence the ***conditional expectations w.r.t. a r-vec***

$$\boldsymbol{E}\left[X|\boldsymbol{Y}\right] = \boldsymbol{E}\left[X|Y_1,\ldots,Y_m\right] = m(Y_1,\ldots,Y_m) = m(\boldsymbol{Y}) \tag{3.56}$$

The properties of these *rv*'s are similar to those of the conditional expectations w.r.t. a single *rv* introduced earlier in the present section: we will not list them here

Example 3.43 Lifetime: Let us suppose that the operating time without failures (**lifetime**) of the components in a device is a *rv* Y with *pdf* $f_Y(y)$: if the device starts working at the time $y = 0$, $f_Y(y)$ will vanish for $y < 0$. We want to calculate

$$f_{Y-y_0}(y|Y \geq y_0) \qquad \text{e} \qquad \boldsymbol{E}\left[Y - y_0|Y \geq y_0\right]$$

namely the *pdf* and the expectation (**mean lifetime**) of the residual lifetime of a component, supposing that it is still working at the time $y_0 > 0$. Taking then $\boldsymbol{P}\{Y \geq y_0\} > 0$, from (3.51) with $a = y_0$ and $b = +\infty$ we have first

$$\boldsymbol{E}\left[Y - y_0|Y \geq y_0\right] = \int_{\boldsymbol{R}} (y - y_0) f_Y(y|Y \geq y_0)\, dy = \frac{\int_{y_0}^{+\infty} (y - y_0) f_Y(y)\, dy}{\int_{y_0}^{+\infty} f_Y(y)\, dy} \tag{3.57}$$

On the other hand, to find the *pdf* of the residual lifetime $Y - y_0$, we remark that

$$\begin{aligned} F_{Y-y_0}(y|Y \geq y_0) &= \boldsymbol{P}\{Y - y_0 \leq y|Y \geq y_0\} \\ &= \boldsymbol{P}\{Y \leq y + y_0|Y \geq y_0\} = F_Y(y + y_0|Y \geq y_0) \end{aligned}$$

so that from (3.51) we have

$$\begin{aligned} f_{Y-y_0}(y|Y \geq y_0) &= \partial_y F_{Y-y_0}(y|Y \geq y_0) = \partial_y F_Y(y_0 + y|Y \geq y_0) \\ &= f_Y(y_0 + y|Y \geq y_0) = \frac{f_Y(y_0 + y)\, \chi_{(y_0,+\infty)}(y_0 + y)}{\int_{y_0}^{+\infty} f_Y(y')\, dy'} \\ &= \frac{f_Y(y_0 + y)\, \chi_{(0,+\infty)}(y)}{\int_{y_0}^{+\infty} f_Y(y')\, dy'} \end{aligned} \tag{3.58}$$

Apparently the result (3.57) could also be deduced from (3.58) by direct calculation:

$$\begin{aligned} \boldsymbol{E}\left[Y - y_0|Y \geq y_0\right] &= \int_{-\infty}^{+\infty} y f_{Y-y_0}(y|Y \geq y_0)\, dy = \frac{\int_0^{+\infty} y f_Y(y_0 + y)\, dy}{\int_{y_0}^{+\infty} f_Y(y')\, dy'} \\ &= \frac{\int_{y_0}^{+\infty} (y - y_0) f_Y(y)\, dy}{\int_{y_0}^{+\infty} f_Y(y)\, dy} \end{aligned}$$

It is interesting now to see what happens when the lifetime $Y \sim \mathfrak{E}(a)$ is an exponential *rv*. In this case we know from (2.17) and (3.33) that

$$f_Y(y) = ae^{-ay}\vartheta(y) \qquad\qquad \boldsymbol{E}[Y] = \frac{1}{a}$$

and since for $z = y - y_0$ we have

$$\int_{y_0}^{+\infty} f_Y(y)\,dy = \int_{y_0}^{+\infty} ae^{-ay}\,dy = e^{-ay_0}$$

$$\int_{y_0}^{+\infty} (y - y_0) f_Y(y)\,dy = \int_0^{+\infty} z f_Y(z + y_0)\,dz = \int_0^{+\infty} za\,e^{-a(z+y_0)}\,dz = \frac{e^{-ay_0}}{a}$$

$$f_Y(y_0 + y)\,\chi_{(0,+\infty)}(y) = a\,e^{-a(y_0+y)}\,\chi_{(0,+\infty)}(y) = e^{-ay_0} f_Y(y)$$

we also see from (3.57) and (3.58) that

$$\boldsymbol{E}[Y - y_0 | Y \geq y_0] = \frac{1}{a} = \boldsymbol{E}[Y] \qquad\qquad f_{Y-y_0}(y|Y \geq y_0) = f_Y(y)$$

In other words: not only the mean lifetime of a component (under the condition that it worked properly up to the time $y = y_0$) does not depend on y_0 and always coincides with $\boldsymbol{E}[Y]$, but also the *pdf* of $Y - y_0$ (conditioned by $Y \geq y_0$) does not depend on y_0 and coincides with the un-conditional *pdf*. This behavior is characteristic of the exponential *rv's* (we also say that they are **memoryless**, or that they show no aging) in the sense that there are no other distributions enjoying this property

Example 3.44 Buffon's needle: A needle of unit length is thrown at random on a table where a few parallel lines are drawn at a unit distance: what is the probability that the needle will lie across one of these lines? Since the lines are drawn periodically on the table, it will be enough to study the problem with only two lines by supposing that the needle center does fall between them. The position of the said center along the direction of the parallel lines is also immaterial: to keep this into account we could also add another independent *rv* to our problem, but in the end we would simply marginalize it without changing the result. The needle position will then be given just by two *rv's*: the distance X of its center from the left line, and the angle Θ between the needle and a perpendicular to the parallel lines (see Fig. 3.5). That the needle is **thrown at random** here means that the pair of *rv's* X, Θ is uniform in $[0, 1] \times [-\frac{\pi}{2}, \frac{\pi}{2}]$, namely that

$$f_{X\Theta}(x, \theta) = \frac{1}{\pi}\,\chi_{[0,1]}(x)\,\chi_{[-\frac{\pi}{2},\frac{\pi}{2}]}(\theta)$$

It is easy to see then that the marginal *pdf*'s are

$$f_X(x) = \chi_{[0,1]}(x) \qquad\qquad f_\Theta(\theta) = \frac{1}{\pi}\,\chi_{[-\frac{\pi}{2},\frac{\pi}{2}]}(\theta)$$

and hence that X and Θ are independent. Take now

$$B = \left\{ (x, \theta) \;:\; either \;\; x \leq \frac{1}{2}\cos\theta, \;\; \text{or } x \geq 1 - \frac{1}{2}\cos\theta, \;\; \text{with} \; -\frac{\pi}{2} \leq \theta \leq \frac{\pi}{2} \right\}$$

so that our event will be

$$A = \{\text{the needle lies across a line}\} = \big\{\omega \in \Omega \;:\; (X, \Theta) \in B\big\}$$

while $I_A = \chi_B(X, \Theta)$, where $\chi_B(x, \theta)$ is the indicator of B in $\boldsymbol{R}^2$. The result can now be found in several equivalent ways: we will use in sequence (3.10), the point 1 of the Proposition 3.42, the uniformity of Θ, and finally the point 4 of the Proposition 3.42, namely

$$\begin{aligned} \boldsymbol{P}\{A\} &= \boldsymbol{E}\,[I_A] = \boldsymbol{E}\,[\chi_B(X, \Theta)] = \boldsymbol{E}\,[\,\boldsymbol{E}\,[\chi_B(X, \Theta)|\,\Theta]\,] \\ &= \int_{-\pi/2}^{\pi/2} \boldsymbol{E}\,[\chi_B(X, \Theta)|\Theta = \theta]\,\frac{d\theta}{\pi} = \frac{1}{\pi}\int_{-\pi/2}^{\pi/2} \boldsymbol{E}\,[\chi_B(X, \theta)]\,d\theta \end{aligned}$$

Now we should just recall that X is uniform to get

$$\boldsymbol{E}\,[\chi_B(X, \theta)] = \boldsymbol{P}\Big\{\big\{X \leq \tfrac{1}{2}\cos\theta\big\} \cup \big\{X \geq 1 - \tfrac{1}{2}\cos\theta\big\}\Big\} = \frac{1}{2}\cos\theta + \frac{1}{2}\cos\theta = \cos\theta$$

and hence

$$\boldsymbol{P}\{A\} = \frac{1}{\pi}\int_{-\pi/2}^{\pi/2} \cos\theta\, d\theta = \frac{2}{\pi}$$

This result has been used to give an **empirical estimate** of the number π: throw the needle n times and define n iid Bernoulli *rv*'s Y_k (with $k = 1, \ldots, n$), such that $Y_k = 1$ if the needle lies across a line in the kth toss, and $Y_k = 0$ if not: if p is the probability that the needle will fall across a line in every single toss, and if $\nu_n = Y_1 + \cdots + Y_n$ is the *rv* counting the number of times the needle does that in n tosses, then it is spontaneous (and the **Law of Large Numbers** that will be discussed in the subsequent Sect. 4.3 will make this a precise statement) to think that, with a sufficiently large n, the value of the relative frequency ν_n/n will be a good estimate of the probability p. Since then from the previous discussion we know that $p = 2/\pi$, a good estimate of the value of π will be given by an empirical value of the *rv* $2n/\nu_n$ with an n large enough. This procedure to approximate π has been used several times in history[1] and constitutes the first known instance of the application of the statistical regularities to numerical calculus problems: a method subsequently called **Monte Carlo** that we will discuss again in the following chapters

[1]See [3] p. 37: the estimate has been done first in 1850 by the Swiss astronomer R. Wolf (1816–1893): by tossing the needle 5 000 times he got 3.1596 as the approximation for π.

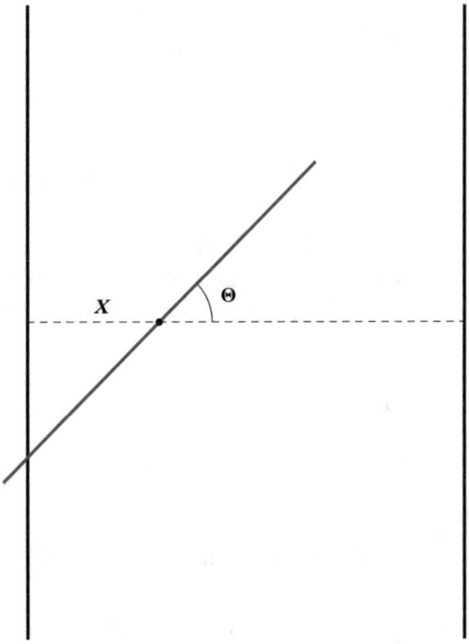

Fig. 3.5 Buffon's needle

3.4.3 Optimal Mean Square Estimation

In order to stress the relations among the three *rv*'s X, Y and $\boldsymbol{E}[X|Y]$ as defined in (3.54) and (3.55), we must recall first that in general the statistical dependence of X and Y does not necessarily require the existence of a Borel function $h(y)$ such that $X = h(Y)$: in other words, the *statistical dependence* does not imply a *functional dependence* (see also Sect. 3.1.3). On the other hand, given two statistically dependent *rv*'s X, Y, we could wish to use some Y measurement to get information on (to have an *estimate* of) the values of X. In statistical practice this is achieved by using the *rv* $h(Y)$ (for some suitable function h) as an *estimator* of X. Since however, as previously remarked, we can not hope in general to find an h such that $X = h(Y)$, our estimate will always be affected by errors so that we will need a criterion to choose an optimal estimator of X: the most known is to search for the *best estimator in mean square* (*ms*). We first define the *mean square error* (*mse*) made by estimating X by means of an estimator $h(Y)$

$$\boldsymbol{E}\left[(X - h(Y))^2\right]$$

and then we choose as the best estimator $h^*(Y)$ that which minimizes the *mse*:

$$\boldsymbol{E}\left[(X - h^*(Y))^2\right] = \inf_h \boldsymbol{E}\left[(X - h(Y))^2\right]$$

To find such an optimal estimator, namely the Borel function h^* that minimizes the *mse*, is a typical variational problem which in any case admit an exact formal solution (as stated in the next proposition): the Borel function $h^*(y)$ restituting the best estimator in mean square coincides with the $m(y) = \boldsymbol{E}[X|Y = y]$ defined in (3.54)

Proposition 3.45 *The best estimator in ms of X through Y is* $\boldsymbol{E}[X|Y]$, *namely it is given by the Borel function*

$$h^*(y) = m(y) = \boldsymbol{E}[X|Y = y]$$

as defined in (3.54)

Proof If $h(Y)$ is an arbitrary estimator and $h^*(Y) = \boldsymbol{E}[X|Y]$, we have

$$\begin{aligned}\boldsymbol{E}\left[(X - h(Y))^2\right] &= \boldsymbol{E}\left[(X - h^*(Y) + h^*(Y) - h(Y))^2\right] \\ &= \boldsymbol{E}\left[(X - h^*(Y))^2\right] + \boldsymbol{E}\left[(h^*(Y) - h(Y))^2\right] \\ &\qquad\qquad +2\boldsymbol{E}\left[(X - h^*(Y))(h^*(Y) - h(Y))\right]\end{aligned}$$

On the other hand from the points 1 and 5 of the Proposition 3.42 it is

$$\begin{aligned}\boldsymbol{E}\left[(X - h^*(Y))(h^*(Y) - h(Y))\right] &= \boldsymbol{E}\Big[\boldsymbol{E}\left[(X - h^*(Y))(h^*(Y) - h(Y)) \,\middle|\, Y\right]\Big] \\ &= \boldsymbol{E}\Big[(h^*(Y) - h(Y))\,\boldsymbol{E}\left[(X - h^*(Y)) \,\middle|\, Y\right]\Big]\end{aligned}$$

and since

$$\boldsymbol{E}\left[(X - h^*(Y)) \,\middle|\, Y\right] = \boldsymbol{E}[X|Y] - \boldsymbol{E}\left[h^*(Y)|Y\right] = \boldsymbol{E}[X|Y] - h^*(Y) = 0$$

we finally have

$$\boldsymbol{E}\left[(X - h(Y))^2\right] = \boldsymbol{E}\left[(X - h^*(Y))^2\right] + \boldsymbol{E}\left[(h^*(Y) - h(Y))^2\right]$$

But apparently it is $\boldsymbol{E}\left[(h^*(Y) - h(Y))^2\right] \geq 0$, and hence

$$\boldsymbol{E}\left[(X - h(Y))^2\right] \geq \boldsymbol{E}\left[(X - h^*(Y))^2\right]$$

for every Borel function h ■

The function $m(y) = \boldsymbol{E}[X|Y = y]$ is also known as *regression curve* of X on Y: this name comes from the studies of sir F. Galton (1822–1911) about the heights of the human generations (parents and children) in a given population. In terms of conditional expectations his results indicated that when the parents are taller than the average population, then the children tend to be shorter than the parents; when instead the parents are shorter than the mean, then the children tend to be taller than them. In both cases the children height is said to *regress* toward the mean value.

3.5 Combinations of *rv*'s

3.5.1 Functions of *rv*'s

Proposition 3.46 *Take a rv X with pdf $f_X(x)$: if $y = \varphi(x)$ is a continuous, regular function whose definition interval includes the values of X, and can be decomposed in n disjoint intervals $[a_k, b_k]$ where φ is differentiable and strictly monotonic, with nowhere vanishing derivative; then the pdf $f_Y(y)$ of the rv $Y = \varphi(X)$ is*

$$f_Y(y) = \sum_{k=1}^{n} \frac{f_X\big(x_k(y)\big)}{\big|\varphi'\big(x_k(y)\big)\big|}\,\chi_{[\alpha_k,\beta_k]}(y) \tag{3.59}$$

where $[\alpha_k, \beta_k]$ are the intervals of the values taken by φ for $x \in [a_k, b_k]$, and for every given y the $x_k(y)$ are the (at most n) solutions of the equation $\varphi(x) = y$

Proof Let us suppose first that X takes values in $[a, b]$ (namely that $f_X(x)$ vanishes outside this interval), and that $\varphi(x)$ is defined, differentiable and strictly increasing ($\varphi'(x) > 0$) in $[a, b]$. If $[\alpha, \beta]$ is the interval of the values taken by $\varphi(x)$, let us denote with $x_1(y) = \varphi^{-1}(y)$ the unique solution of the equation $\varphi(x) = y$ which exists (and is monotonic as a function of y) when $y \in [\alpha, \beta]$. It is then apparent that

$$F_Y(y) = \boldsymbol{P}\{Y \le y\} = \begin{cases} 0 \text{ for } y < \alpha \\ 1 \text{ for } y > \beta \end{cases}$$

while for $y \in [\alpha, \beta]$, by taking as integration variable $z = \varphi(x)$, $x = \varphi^{-1}(z) = x_1(z)$, we get

$$\begin{aligned} F_Y(y) &= \boldsymbol{P}\{Y \le y\} = \boldsymbol{P}\{\varphi(X) \le y\} = \boldsymbol{P}\big\{X \le \varphi^{-1}(y)\big\} = \boldsymbol{P}\{X \le x_1(y)\} \\ &= \int_a^{x_1(y)} f_X(x)\,dx = \int_\alpha^y f_X\big(x_1(z)\big)x_1'(z)\,dz = \int_\alpha^y f_Y(z)\,dz \end{aligned}$$

As a consequence we will have

$$f_Y(y) = f_X\big(x_1(y)\big)\,x_1'(y)\chi_{[\alpha,\beta]}(y) = \begin{cases} f_X\big(x_1(y)\big)\,x_1'(y) & \text{for } \alpha \le y \le \beta \\ 0 & \text{elsewhere} \end{cases}$$

If instead φ is strictly decreasing a similar calculation would lead to

$$f_Y(y) = -f_X\big(x_1(y)\big)\,x_1'(y)\chi_{[\alpha,\beta]}(y)$$

so that on every case, when φ is strictly monotonic on $[a, b]$, we can write

$$f_Y(y) = f_X\big(x_1(y)\big)\,|x_1'(y)|\,\chi_{[\alpha,\beta]}(y) \tag{3.60}$$

Since on the other hand from a well known result of the elementary analysis

$$x_1'(y) = \frac{1}{\varphi'\big(x_1(y)\big)}$$

our transformation for a monotonic function φ will be

$$f_Y(y) = \frac{f_X\big(x_1(y)\big)}{\big|\varphi'\big(x_1(y)\big)\big|}\,\chi_{[\alpha,\beta]}(y) \tag{3.61}$$

namely (3.59) when the sum is reduced to one term. When instead φ is not strictly monotonic on the set of the X values, in many cases of interest its definition interval can be decomposed in the union of n disjoint intervals $[a_k, b_k]$ in whose interior φ is differentiable and strictly monotonic, with nowhere vanishing derivatives. If now $[\alpha_k, \beta_k]$ are the intervals of the values taken by φ for $x \in [a_k, b_k]$, and if for a given y we denote as $x_k(y)$ the (at most n) solutions of the equation $\varphi(x) = y$, the result (3.59) is deduced as in the monotonic case.[2] ■

It is important to stress that the number of terms of the sum (3.59) depends on y, because for every y we will find only the $m \le n$ summands corresponding to the solutions of $\varphi(x) = y$ such that $\chi_{[\alpha_k,\beta_k]}(y) = 1$. When on the other hand $\varphi(x) = y$ has no solution there are no summands at all and $f_Y(y) = 0$.

In a more general setting $\boldsymbol{Y} = \varphi(\boldsymbol{X})$ transforms a *r-vec* $\boldsymbol{X}$ with n components X_j into a *r-vec* $\boldsymbol{Y}$ with the $m \ne n$ components

$$Y_k = \varphi_k(X_1, \ldots, X_n), \qquad k = 1, \ldots, m$$

Without a loss of generality we can however always suppose $n = m$ because:

- if $m < n$, we can always add to $\boldsymbol{Y}$ $n - m$ auxiliary components coincident with $X_{m+1}, \ldots, X_n$; after solving the problem in this form by determining the joint *pdf* $f_Y(y_1, \ldots, y_n)$, we will eliminate the excess variables $y_{m+1}, \ldots, y_n$ by marginalization;
- if $m > n$, $m - n$ among the Y_k apparently will turn out to be functions of the other n components; we then solve the problem for the first n *rv*'s Y_k, and then the

[2]Remark that the result (3.59) can be reformulated as

$$f_Y(y) = \int_{-\infty}^{+\infty} f_X(x)\,\delta\big[y - \varphi(x)\big]\,dx \tag{3.62}$$

by using the Dirac distribution $\delta(x)$ that in our notations satisfies the relation

$$\delta\big[y - \varphi(x)\big] = \sum_{k=1}^{n} \frac{\delta\big[x - x_k(y)\big]}{\big|\varphi'\big(x_k(y)\big)\big|}\,\chi_{[\alpha_k,\beta_k]}(y)$$

See for instance [4] p. 22.

distribution of the remaining $m - n$ *rv*'s is deduced as functions of the previous ones

Taking then $n = m$, we will just state without proof the main result. For a given $\boldsymbol{y}$ let $\boldsymbol{x}^j(\boldsymbol{y})$ be the (at most n) solutions of the n equations system $y_k = \varphi_k(x_1, \dots, x_n)$: then the joint *pdf* of the *r-vec* $\boldsymbol{Y}$ is

$$f_{\boldsymbol{Y}}(\boldsymbol{y}) = \sum_{j=1}^{n} \frac{f_{\boldsymbol{X}}(\boldsymbol{x}^j(\boldsymbol{y}))}{|J(\boldsymbol{x}^j(\boldsymbol{y}))|} \chi_j(\boldsymbol{y}) \tag{3.63}$$

where $J(\boldsymbol{x})$ is the Jacobian determinant of the transformation with elements $\partial\varphi_k/\partial x_l$, while the $\chi_j(\boldsymbol{y})$ take value 1 if the jth solution exists in $\boldsymbol{y}$, and 0 otherwise. This apparently generalizes (3.59) with the same provisions about the number (possibly vanishing) of the terms in the sum.

Example 3.47 Linear functions: When $Y = aX + b$, namely $\varphi(x) = ax + b$, with $a \neq 0$, the equation $y = \varphi(x)$ always has a unique solution $x_1(y) = (y - b)/a$. As a consequence

$$f_{aX+b}(y) = \frac{1}{|a|} f_X\left(\frac{y-b}{a}\right)$$

In particular, if $X \sim \mathfrak{N}(0, 1)$ is a standard normal *rv*, then $Y = aX + b \sim \mathfrak{N}(b, a^2)$; and conversely, if $X \sim \mathfrak{N}(b, a^2)$, then $Y = (X - b)/a \sim \mathfrak{N}(0, 1)$ is a standard normal.
Quadratic functions: If $Y = X^2$, namely $\varphi(x) = x^2$, the equation $y = \varphi(x)$ has two solutions $x_1(y) = -\sqrt{y}$ and $x_2(y) = +\sqrt{y}$ for $y > 0$ (they coincide for $y = 0$). Taking then $\vartheta(y)$ the Heaviside funcrtion (2.13) we will have

$$f_{X^2}(y) = \frac{1}{2\sqrt{y}} \left[f_X(\sqrt{y}) + f_X(-\sqrt{y})\right] \vartheta(y)$$

When in particular $X \sim \mathfrak{N}(0, 1)$ we get

$$f_{X^2}(y) = \frac{e^{-y/2}}{\sqrt{2\pi y}} \vartheta(y) \tag{3.64}$$

and we will see in the next section that this is called a χ_1^2 law with 1 degree of freedom
Exponential functions: When $Y = e^X$ and $X \sim \mathfrak{N}(b, a^2)$ from (3.59) we find

$$f_{e^X}(y) = \frac{e^{-(\ln y - b)^2/2a^2}}{ay\sqrt{2\pi}} \vartheta(y)$$

a law called **log-normal** and denoted by $\ln\mathfrak{N}(b, a^2)$. To show that the expectation and variance of $Y \sim \ln\mathfrak{N}(b, a^2)$ are

$$\boldsymbol{E}[Y] = e^{b+a^2/2} \qquad \boldsymbol{V}[Y] = e^{2b+a^2}(e^{a^2} - 1) \tag{3.65}$$

remark that by taking $z = \frac{x-a^2-b}{a}$ we get

$$\boldsymbol{E}[Y] = \boldsymbol{E}\left[e^X\right] = \int_{-\infty}^{+\infty} e^x \frac{e^{-\frac{(x-b)^2}{2a^2}}}{a\sqrt{2\pi}}\, dx = e^{b+a^2/2} \int_{-\infty}^{+\infty} \frac{e^{-\frac{z^2}{2}}}{\sqrt{2\pi}}\, dz = e^{b+a^2/2}$$

and since $2X \sim \mathfrak{N}(2b, 4a^2)$, from the previous result it also follows that

$$\boldsymbol{E}\left[Y^2\right] = \boldsymbol{E}\left[e^{2X}\right] = e^{2b+2a^2}$$

and hence

$$\boldsymbol{V}[Y] = \boldsymbol{E}\left[Y^2\right] - \boldsymbol{E}[Y]^2 = e^{2b+2a^2} - e^{2b+a^2} = e^{2b+a^2}(e^{a^2} - 1)$$

A last example of application of (3.59) can be found in a celebrated problem known as **Bertrand's paradox** and discussed in the Appendix C.

3.5.2 *Sums of Independent rv's*

Definition 3.48 We call **convolution** of two *pdf*'s f and g the function

$$\begin{aligned}(f * g)(x) &= (g * f)(x) \\ &= \int_{-\infty}^{\infty} f(x-y)g(y)\, dy = \int_{-\infty}^{\infty} g(x-y)f(y)\, dy\end{aligned}$$

It is easy to see that the convolution of two *pdf*'s again is a *pdf*

Proposition 3.49 *Given two independent rv's X and Y with pdf's $f_X(x)$ and $f_Y(y)$, the pdf of their sum $Z = X + Y$ is*

$$f_Z(x) = (f_X * f_Y)(x) = (f_Y * f_X)(x)$$

namely is the convolution of the respective pdf's

Proof If two *rv*'s X and Y have the joint *pdf* $f_{XY}(x, y)$ and we take $Z = \varphi(X, Y)$ with $z = \varphi(x, y)$ a Borel function, by adopting the shorthand notation

$$\{\varphi \le z\} = \{(x, y) \in \boldsymbol{R}^2 : \varphi(x, y) \le z\}$$

it is easy to see that the *cdf* of Z is

$$F_Z(z) = \boldsymbol{P}\{Z \leq z\} = \boldsymbol{P}\{\varphi(X, Y) \leq z\} = \int_{\{\varphi \leq z\}} f_{XY}(x, y)\, dxdy$$

When in particular $\varphi(x, y) = x + y$, and X, Y are independent, namely $f_{XY}(x, y) = f_X(x) f_Y(y)$, with the change of variable $u = x + y$ we get

$$\begin{aligned} F_Z(z) &= \int_{\{x+y \leq z\}} f_X(x) f_Y(y)\, dx\, dy = \int_{-\infty}^{\infty} \left[\int_{-\infty}^{z-x} f_Y(y)\, dy \right] f_X(x)\, dx \\ &= \int_{-\infty}^{\infty} \left[\int_{-\infty}^{z} f_Y(u - x)\, du \right] f_X(x)\, dx = \int_{-\infty}^{z} \left[\int_{-\infty}^{\infty} f_Y(u - x) f_X(x)\, dx \right] du \end{aligned}$$

or also, by inverting the integration order,

$$F_Z(z) = \int_{-\infty}^{z} \left[\int_{-\infty}^{\infty} f_X(u - y) f_Y(y)\, dy \right] du$$

We can then say that the *pdf* of $Z = X + Y$ is

$$f_Z(z) = \int_{-\infty}^{\infty} f_Y(z - x) f_X(x)\, dx = \int_{-\infty}^{\infty} f_X(z - y) f_Y(y)\, dy$$

namely that $f_Z = f_X * f_Y = f_Y * f_X$ ■

The previous results can also be extended to more than two *rv*'s: given n independent *rv*'s $X_1, \ldots, X_n$ admitting *pdf*'s, then the *pdf* of their sum $Z = X_1 + \cdots + X_n$ is the n-convolution

$$f_Z(x) = (f_{X_1} * \ldots * f_{X_n})(x) \tag{3.66}$$

Example 3.50 Sums of uniform *rv*'s: When $X_1, \ldots, X_n$ are iid $\mathfrak{U}(-1, 1)$ *rv*'s their *pdf* can be given for instance by means of the Heaviside $\vartheta(x)$ (2.13)

$$f_{X_k}(x) = f(x) = \frac{1}{2}\, \vartheta(1 - |x|) \qquad k = 1, \ldots, n$$

A direct calculation then shows that

$$\begin{aligned} f_{X_1+X_2}(x) &= \frac{2 - |x|}{4}\, \vartheta(2 - |x|), \\ f_{X_1+X_2+X_3}(x) &= \left[\vartheta(1 - |x|) \frac{3 - x^2}{8} + \vartheta(|x| - 1) \frac{(3 - |x|)^2}{16} \right] \vartheta(3 - |x|) \end{aligned}$$

while for $Y = X_1 + \cdots + X_n$ we inductively get

$$f_Y(x) = \frac{\vartheta(n-|x|)}{2^n(n-1)!}\sum_{k=0}^{\lfloor (n+x)/2 \rfloor}(-1)^k\binom{n}{k}(n+x-2k)^{n-1}$$

where $\lfloor\alpha\rfloor$ is the integer part (floor) of the real number α. As a consequence we find that sums of iid uniform *rv*'s are not at all uniform: for instance $f_{X_1+X_2}$ is triangular on $[-2, 2]$, while $f_{X_1+X_2+X_3}$ consists of three parabolic segments continuously connected on $[-3, 3]$

Sums of Gaussian *rv's*: The previous example shows that not every law convolute with a law of the same type produces a law in the same family. It is interesting then to remark that, if $X \sim \mathfrak{N}(b_1, a_1^2)$ and $Y \sim \mathfrak{N}(b_2, a_2^2)$ are independent Gaussian *rv*'s, a direct calculation would show that $X + Y \sim \mathfrak{N}(b_1 + b_2,\ a_1^2 + a_2^2)$, and symbolically

$$\mathfrak{N}(b_1, a_1^2) * \mathfrak{N}(b_2, a_2^2) = \mathfrak{N}(b_1 + b_2,\ a_1^2 + a_2^2) \tag{3.67}$$

This important result, that can be extended to an arbitrary number ov *rv*'s, is known as **reproductive property** of the Gaussian family of laws: we will prove it later (together with similar results for other families of laws) in the Sect. 4.2.3 by means of the characteristic functions.

$\boldsymbol{\chi_n^2}$ **distributions**: If $X_1, \ldots, X_n$ are *iid* $\mathfrak{N}(0, 1)$ *rv*'s, from (3.64) and by iterated convolutions of $f_{X_k^2}(x)$, we get for $Z = X_1^2 + \ldots + X_n^2$ the *pdf*

$$f_Z(x) = \frac{x^{n/2-1}e^{-x/2}}{2^{n/2}\Gamma(n/2)}\,\vartheta(x)$$

which is known as χ^2 distribution with n degrees of freedom and denoted with the symbol $\boldsymbol{\chi_n^2}$. Here $\Gamma(x)$ is the gamma function defined as

$$\Gamma(x) = \int_0^{+\infty} z^{x-1}e^{-z}\,dz \tag{3.68}$$

with the well known properties

$$\Gamma(x) = (x-1)\Gamma(x-1) \qquad \Gamma(1) = 1 \qquad \Gamma\left(\frac{1}{2}\right) = \sqrt{\pi}$$

so that in particular

$$\Gamma\left(\frac{n}{2}\right) = \begin{cases} (n-2)!!\,2^{-n/2} & \text{for even } n \\ (n-2)!!\,2^{-(n-1)/2} & \text{for odd } n \end{cases}$$

It is possible to prove that the expectation and the variance of a χ_n^2 *rv* Z are

$$\boldsymbol{E}[Z] = n \qquad \boldsymbol{V}[Z] = 2n$$

Student $\mathfrak{T}_n$ distributions: With $X_0, X_1, \ldots, X_n$ *iid* $\mathfrak{N}(0, a^2)$ *rv's*, take

$$T = \frac{X_0}{\sqrt{(X_1^2 + \cdots + X_n^2)/n}} = \frac{X_0/a}{\sqrt{(X_1^2 + \cdots + X_n^2)/na^2}} = \frac{X_0/a}{\sqrt{Z/n}}$$

From the previous examples we know that the X_k/a are $\mathfrak{N}(0, 1)$, while $Z = (X_1^2 + \cdots + X_n^2)/a^2$ is $\boldsymbol{\chi}_n^2$. It is possible then to prove that the *pdf* of T is

$$f_T(t) = \frac{1}{\sqrt{\pi n}} \frac{\Gamma\left(\frac{n+1}{2}\right)}{\Gamma\left(\frac{n}{2}\right)} \left(1 + \frac{t^2}{n}\right)^{-(n+1)/2}$$

which is called Student-T distribution and is denoted with the symbol $\mathfrak{T}_n$. With $n = 1$ the Student distribution coincides with the Cauchy $\mathfrak{C}(1, 0)$. It i possible to prove that expectation and variance of a Student T with law $\mathfrak{T}_n$ are

$$\boldsymbol{E}[T] = 0 \quad \text{for } n \geq 2 \qquad \boldsymbol{V}[T] = \frac{n}{n-2} \quad \text{for } n \geq 3$$

and do not exist for different values of n.

References

1. Shiryaev, A.N.: Probability. Springer, New York (1996)
2. Meyer, P.L.: Introductory Probability and Statistical Applications. Addison-Wesley, Reading (1970)
3. Gnedenko, B.V.: The Theory of Probability. MIR, Moscow (1978)
4. Vladimirov, V.S.: Methods of the Theory of Generalized Functions. Taylor & Francis, London (2002)

Chapter 4
Limit Theorems

4.1 Convergence

The Limit Theorems are statements about limits of sequences of sums of *rv*'s when the number of addenda grows to infinity. The convergence of a sequence of *rv*'s $(X_n)_{n\in N}$ can however have many non equivalent meanings, and hence we must first of all list the more usual kinds of convergence and their mutual relations

Definition 4.1 Given a sequence of *rv*'s $(X_n)_{n\in N}$ on $(\Omega, \mathcal{F}, \boldsymbol{P})$, we say that

- ***it converges in probability*** to the *rv* X, and we will write $X_n \xrightarrow{P} X$, when

$$\boldsymbol{P}\{|X_n - X| > \epsilon\} \xrightarrow{n} 0 , \qquad \forall \epsilon > 0$$

- ***it converges almost surely (*** $\boldsymbol{P}$-a.s.***), or with probability 1*** to the *rv* X, and we will write $X_n \xrightarrow{as} X$, or even $X_n \xrightarrow{n} X$ $\boldsymbol{P}$-a.s., when either

$$\boldsymbol{P}\{X_n \to X\} = 1 \qquad or \qquad \boldsymbol{P}\{X_n \nrightarrow X\} = 0$$

  where $\{X_n \nrightarrow X\}$ is the set of ω such that $(X_n)_{n\in N}$ does not converge to X
- ***it converges in*** L^p (with $0 < p < +\infty$) to the *rv* X and we will write $X_n \xrightarrow{L^p} X$, when

$$\boldsymbol{E}\left[|X_n - X|^p\right] \xrightarrow{n} 0$$

  If in particular $p = 2$ we also say that $(X_n)_{n\in N}$ ***converges in mean square (ms)*** and we adopt the notation $X_n \xrightarrow{ms} X$. The exact meaning of the symbols $L^p = L^p(\Omega, \mathcal{F}, \boldsymbol{P})$ is discussed in the Appendix D
- ***it converges in distribution***, and we will write $X_n \xrightarrow{d} X$, when

$$\boldsymbol{E}\left[f(X_n)\right] \xrightarrow{n} \boldsymbol{E}\left[f(X)\right] , \qquad \forall f \in C(\boldsymbol{R})$$

N. Cufaro Petroni, *Probability and Stochastic Processes for Physicists*, UNITEXT for Physics, https://doi.org/10.1007/978-3-030-48408-8_4

where $\mathcal{C}(\boldsymbol{R})$ is the set of the functions that are bounded and continuous on $\boldsymbol{R}$

Definition 4.2 Given a sequence of *cdf*'s $\big(F_n(x)\big)_{n\in N}$

- ***it converges weakly*** to the *cdf* $F(x)$, and we will write $F_n \xrightarrow{w} F$, when

$$\int_R f(x)F_n(dx) \xrightarrow{n} \int_R f(x)F(dx)\,, \qquad \forall\, f \in \mathcal{C}(\mathrm{R})$$

  where $\mathcal{C}(\boldsymbol{R})$ is the set of the bounded and continuous functions
- ***it converges in general*** to the *cdf* $F(x)$, and we will write $F_n \xrightarrow{g} F$, when

$$F_n(x) \xrightarrow{n} F(x)\,, \qquad \forall x \in P_C(F)$$

  where $P_C(F)$ is the set of points $x \in \boldsymbol{R}$ where $F(x)$ is continuous

Proposition 4.3 *A sequence of cdf's* $\big(F_n(x)\big)_{n\in N}$ *converges weakly to the cdf* $F(x)$ *iff it converges in general*

Proof Omitted: see [1] p. 314 ■

Given now a sequence of *rv*'s $(X_n)_{n\in N}$ with their *cdf*'s F_{X_n}, it is apparent that $(X_n)_{n\in N}$ converges in distribution to the *rv* X with *cdf* F_X *iff* $F_{X_n} \xrightarrow{w} F_X$, or equivalently *iff* $F_{X_n} \xrightarrow{g} F_X$. Practically—with a few clarifications about their meaning—the convergences in distribution, weak and in general are equivalent. This entails that the convergence in distribution of a sequence of *rv*'s can be proved by looking just to their *cdf*'s, namely to their laws: in particular it will be enough to prove that $F_{X_n}(x) \xrightarrow{n} F_X(x)$ wherever the limit *cdf* $F_X(x)$ is continuous.

The four types of convergence of the Definition 4.1, however, are not equivalent and their mutual relationships are listed in the following theorem and are graphically summarized in the Fig. 4.1

Theorem 4.4 *Given a sequence of rv's* $(X_n)_{n\in N}$ *and the rv* X*, we have*

1. $X_n \xrightarrow{as} X \implies X_n \xrightarrow{P} X$
2. $X_n \xrightarrow{L^p} X \implies X_n \xrightarrow{P} X\,, \qquad p > 0$
3. $X_n \xrightarrow{P} X \implies X_n \xrightarrow{d} X$
4. $X_n \xrightarrow{d} c \implies X_n \xrightarrow{P} c\,,$ *if c is a number (degenerate convergence)*

Proof Omitted: see [1] p. 256 and 262 ■

Inferences different form the previous ones are not instead generally guaranteed as could be seen from a few simple counterexamples. That notwithstanding it is possible to find supplementary hypotheses to have other inferences beyond those of the Theorem 4.4: a few well known supplementary conditions are collected in the subsequent theorem.

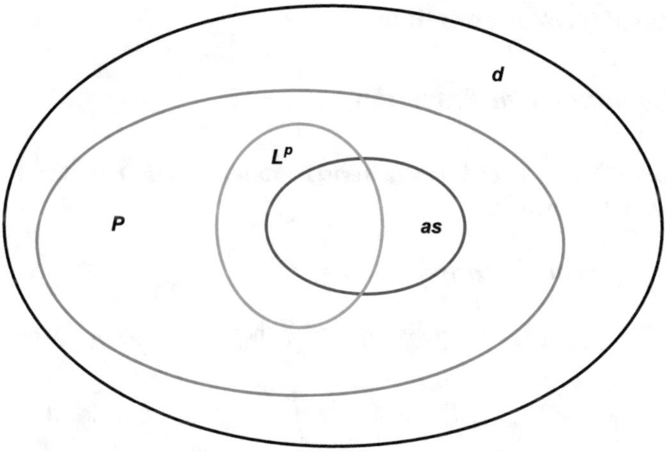

Fig. 4.1 Relations among the four types of convergence according to the Theorem 4.4

Theorem 4.5 *Given a sequence of rv's $(X_n)_{n\in N}$ and a rv X*

1. *if $X_n \xrightarrow{P} X$, then it exists a subsequence $(X_{n_k})_{k\in N}$ such that $X_{n_k} \xrightarrow{as} X$;*
2. *if $X_n \xrightarrow{L^p} X$, then it exists a subsequence $(X_{n_k})_{k\in N}$ such that $X_{n_k} \xrightarrow{as} X$;*
3. *if $X_n \xrightarrow{as} X$, and if it exists a rv $Y \geq 0$ with $\boldsymbol{E}[|Y|] < +\infty$ and such that $|X_n - X| < Y$, then we also have $X_n \xrightarrow{L^p} X$.*

Proof Omitted: see [1] p. 258 and [2] p. 164 ■

Theorem 4.6 Degenerate convergence in *ms*: *A sequence of rv's $(X_n)_{n\in N}$ converges in ms to the number m (degenerate convergence) iff*

$$\boldsymbol{E}[X_n] \xrightarrow{n} m \qquad \boldsymbol{V}[X_n] \xrightarrow{n} 0 \tag{4.1}$$

Proof We have indeed

$$\begin{aligned}(X_n - m)^2 &= [(X_n - \boldsymbol{E}[X_n]) + (\boldsymbol{E}[X_n] - m)]^2 \\ &= (X_n - \boldsymbol{E}[X_n])^2 + (\boldsymbol{E}[X_n] - m)^2 + 2(X_n - \boldsymbol{E}[X_n])(\boldsymbol{E}[X_n] - m)\end{aligned}$$

and since apparently

$$\boldsymbol{E}[(X_n - \boldsymbol{E}[X_n])(\boldsymbol{E}[X_n] - m)] = (\boldsymbol{E}[X_n] - m)\boldsymbol{E}[X_n - \boldsymbol{E}[X_n]] = 0$$

we also get

$$\boldsymbol{E}\left[(X_n - m)^2\right] = \boldsymbol{V}[X_n] + (\boldsymbol{E}[X_n] - m)^2$$

so that the degenerate convergence in *ms* is equivalent to the conditions (4.1) ■

4.2 Characteristic Functions

4.2.1 Definition and Properties

Definition 4.7 We will call **characteristic function (*chf*)** of the *r-vec* $\boldsymbol{X} = (X_1, \dots, X_n)$ the function

$$\varphi_{\boldsymbol{X}}(\boldsymbol{u}) = \varphi_{\boldsymbol{X}}(u_1, \dots, u_n) = \boldsymbol{E}\left[e^{i\,\boldsymbol{u}\cdot\boldsymbol{X}}\right] \qquad \boldsymbol{u} \in \boldsymbol{R}^n \tag{4.2}$$

where $\boldsymbol{u} \cdot \boldsymbol{X} = \sum_k u_k X_k$. If there is a *pdf* of $\boldsymbol{X}$ then the *chf* $\varphi_{\boldsymbol{X}}(\boldsymbol{u})$ takes the form

$$\varphi_{\boldsymbol{X}}(\boldsymbol{u}) = \int_{\boldsymbol{R}^n} e^{i\,\boldsymbol{u}\cdot\boldsymbol{x}} f_{\boldsymbol{X}}(\boldsymbol{x})\, d^n\boldsymbol{x} = \int_{-\infty}^{+\infty} \dots \int_{-\infty}^{+\infty} e^{i\,\boldsymbol{u}\cdot\boldsymbol{x}} f_{\boldsymbol{X}}(x_1 \dots x_n)\, dx_1 \dots dx_n$$

and if the *r-vec* $\boldsymbol{X}$ has just one component X the *chf* becomes

$$\varphi_X(u) = \int_{-\infty}^{+\infty} e^{iux} f_X(x)\, dx \qquad u \in \boldsymbol{R}$$

namely the ***Fourier transform*** of the *pdf*

Proposition 4.8 *If $\varphi_X(u)$ is the chf of the rv X, for every $u \in \boldsymbol{R}$ we have*

$$\varphi_X(u) = \overline{\varphi_X(-u)} \qquad |\varphi_X(u)| \le \varphi_X(0) = 1$$

where $\overline{z}$ is the complex conjugate of the complex number z. Moreover $\varphi_X(u)$ is uniformly continuous on $\boldsymbol{R}$, and is even and real iff $f_X(x)$ is even

Proof The first result immediately ensues from the definition, while for the second it is enough to remark that

$$|\varphi_X(u)| = \left|\int_{-\infty}^{+\infty} e^{iux} f(x)\, dx\right| \le \int_{-\infty}^{+\infty} f(x)\, dx = \varphi_X(0) = 1$$

If moreover $f_X(x)$ is even the imaginary part of $\varphi_X(u)$ vanishes for symmetry, while the real part is apparently even. We omit the proof of the uniform continuity: for further details see [1] p. 278 ■

Proposition 4.9 *If $\varphi_X(u)$ is the chf of a rv X, and if $Y = aX + b$ with a, b two numbers, then*

$$\varphi_Y(u) = e^{ibu}\varphi_X(au) \tag{4.3}$$

If $\boldsymbol{X} = (X_1, \dots, X_n)$ is a r-vec, denoted respectively as $\varphi_{\boldsymbol{X}}(u_1, \dots, u_n)$ and $\varphi_{X_k}(u_k)$ the joint and marginal chf's, then

$$\varphi_{X_k}(u_k) = \varphi_X(0, \dots, u_k, \dots, 0) \tag{4.4}$$

If finally the components X_k are independent and $S_n = X_1 + \ldots + X_n$, then

$$\varphi_{S_n}(u) = \varphi_{X_1}(u) \cdot \ldots \cdot \varphi_{X_n}(u) \tag{4.5}$$

Proof If $Y = aX + b$, the Eq. (4.3) results form the Definition 4.2 because

$$\varphi_Y(u) = \boldsymbol{E}\left[e^{iuY}\right] = e^{iub}\boldsymbol{E}\left[e^{i(au)X}\right] = e^{ibu}\varphi_X(au)$$

Also the Eq. (4.4) immediately results form the Definition 4.2; finally, if the X_k are also independent, we find (4.5) because

$$\begin{aligned}\varphi_{S_n}(u) &= \boldsymbol{E}\left[e^{iuS_n}\right] = \boldsymbol{E}\left[e^{iuX_1} \cdot \ldots \cdot e^{iuX_n}\right] \\ &= \boldsymbol{E}\left[e^{iuX_1}\right] \cdot \ldots \cdot \boldsymbol{E}\left[e^{iuX_n}\right] = \varphi_{X_1}(u) \cdot \ldots \cdot \varphi_{X_n}(u)\end{aligned}$$

This last property will be particularly relevant in the discussion of the Limit Theorems and of the reproductive properties: while indeed from (3.66) the *pdf* $f_{S_n}(x)$ of the sum of n independent *rv*'s is the convolution product $(f_{X_1} * \ldots * f_{X_n})(x)$ of the *pdf*'s, its *chf* $\varphi_{S_n}(u)$ is instead the ordinary product $\varphi_{X_1}(u) \cdot \ldots \cdot \varphi_{X_n}(u)$ of their *chf*'s ■

Example 4.10 To find the *chf* of the discrete laws we just calculate the expectation (4.2) as a sum: first the *chf* of a ***degenerate*** *rv* $X \sim \boldsymbol{\delta}_b$ is

$$\varphi_X(u) = e^{ibu} \tag{4.6}$$

then for a ***Bernoulli*** *rv* $X \sim \mathfrak{B}(1; p)$ we have

$$\varphi_X(u) = p\,e^{iu} + q \tag{4.7}$$

For a ***binomial*** *rv* $S_n \sim \mathfrak{B}(n; p)$ it is expedient to recall that from (3.8) we have $S_n \stackrel{d}{=} X_1 + \ldots + X_n$ with X_k Bernoulli *iid* and hence from (4.5) and (4.7) we get

$$\varphi_X(u) = (p\,e^{iu} + q)^n \tag{4.8}$$

Finally for a ***Poisson*** *rv* $X \sim \mathfrak{P}(\alpha)$ the *chf* is

$$\varphi_X(u) = \sum_{k=0}^{\infty} e^{iuk} e^{-\alpha}\frac{\alpha^k}{k!} = e^{-\alpha}\sum_{k=0}^{\infty}\frac{\left(\alpha e^{iu}\right)^k}{k!} = e^{\alpha(e^{iu}-1)} \tag{4.9}$$

When instead there is a *pdf*, the *chf* is found by performing an appropriate integration: for a ***uniform*** *rv* $X \sim \mathfrak{U}(a, b)$ we have

$$\varphi_X(u) = \int_a^b \frac{e^{iux}}{b-a}\,dx = \frac{e^{ibu} - e^{iau}}{i(b-a)u} \tag{4.10}$$

and in particular for $X \sim \mathfrak{U}(-1, 1)$ it is

$$\varphi_X(u) = \frac{\sin u}{u} \tag{4.11}$$

For a ***Gaussian*** *rv* $X \sim \mathfrak{N}(b, a^2)$ we recall from the Example 3.47 that $Y = (X - b)/a \sim \mathfrak{N}(0, 1)$ while from (4.3) we have

$$\varphi_X(u) = e^{ibu}\varphi_Y(au)$$

so that it will be enough to calculate the *chf* φ_Y of a standard Gaussian $\mathfrak{N}(0, 1)$. From the convergence properties of the power expansion of exponentials we then have

$$\begin{aligned}\varphi_Y(u) = \boldsymbol{E}\left[e^{iuY}\right] &= \frac{1}{\sqrt{2\pi}}\int_{-\infty}^{+\infty} e^{iux}e^{-x^2/2}\,dx \\ &= \frac{1}{\sqrt{2\pi}}\int_{-\infty}^{+\infty} e^{-x^2/2}\sum_{n=0}^{\infty}\frac{(iux)^n}{n!}\,dx = \sum_{n=0}^{\infty}\frac{(iu)^n}{n!}\frac{1}{\sqrt{2\pi}}\int_{-\infty}^{+\infty} x^n e^{-x^2/2}\,dx\end{aligned}$$

and since

$$\frac{1}{\sqrt{2\pi}}\int_{-\infty}^{+\infty} x^n e^{-x^2/2}\,dx = \begin{cases} 0 & \text{for } n = 2k+1 \\ (2k-1)!! & \text{for } n = 2k \end{cases}$$

we get

$$\varphi_Y(u) = \sum_{k=0}^{\infty}\frac{(iu)^{2k}}{(2k)!}(2k-1)!! = \sum_{k=0}^{\infty}\left(-\frac{u^2}{2}\right)^k\frac{1}{k!} = e^{-u^2/2} \tag{4.12}$$

and finally

$$\varphi_X(u) = e^{ibu - a^2u^2/2} \tag{4.13}$$

In particular when $X \sim \mathfrak{N}(0, a^2)$ the *pdf* and the *chf* respectively are

$$f_X(x) = \frac{1}{a\sqrt{2\pi}}e^{-x^2/2a^2} \qquad \varphi_X(u) = e^{-a^2u^2/2}$$

and hence we plainly see first that the *chf* of a Gaussian *pdf* is again a Gaussian function, and second that there is an inverse relation between the width (variance) a^2 of the *pdf* and the width $1/a^2$ of the *chf*. Some elementary integration shows then that the *chf* of an ***exponential*** *rv* $X \sim \mathfrak{E}(a)$ is

$$\varphi_X(u) = \int_0^{+\infty} ae^{-ax}e^{ixu}\,dx = \frac{a}{a-iu} = \frac{a^2+iau}{a^2+u^2} \tag{4.14}$$

while that of a ***Laplace*** *rv* $X \sim \mathfrak{L}(a)$ is

$$\varphi_X(u) = \int_{-\infty}^{+\infty} \frac{a}{2}\,e^{-a|x|}e^{ixu}\,dx = \frac{a^2}{a^2+u^2} \tag{4.15}$$

For a ***Cauchy*** *rv* $X \sim \mathfrak{C}(a, b)$ the *chf*

$$\varphi_X(u) = \int_{-\infty}^{+\infty} \frac{a}{\pi}\,\frac{e^{ixu}}{a^2+(x-b)^2}\,dx = e^{-a|u|+ibu} \tag{4.16}$$

is finally derived in the complex field from the residue theorem

Theorem 4.11 *If X is a rv with* chf $\varphi(u)$*, if* $\boldsymbol{E}\left[|X|^n\right] < +\infty, \forall n \in \boldsymbol{N}$*, and if*

$$\overline{\lim_n}\,\frac{\boldsymbol{E}\left[|X|^n\right]^{1/n}}{n} = \frac{1}{R} < +\infty$$

then $\varphi(u)$ is derivable at every order $n \in \boldsymbol{N}$ with

$$\varphi^{(n)}(u) = \boldsymbol{E}\left[(iX)^n e^{iuX}\right] \qquad \varphi^{(n)}(0) = i^n\,\boldsymbol{E}\left[X^n\right] \tag{4.17}$$

Moreover for $|u| < R/3$ the Taylor expansion holds

$$\varphi(u) = \sum_{n=0}^{\infty} \frac{(iu)^n}{n!}\,\boldsymbol{E}\left[X^n\right] = \sum_{n=0}^{\infty} \frac{u^n}{n!}\,\varphi^{(n)}(0) \tag{4.18}$$

If instead $\boldsymbol{E}\left[|X|^k\right] < +\infty$ only for a finite number n of exponents $k = 1, \ldots, n$, then $\varphi(u)$ is derivable only up to the order n, and the Taylor formula holds

$$\varphi(u) = \sum_{k=0}^{n} \frac{(iu)^k}{k!}\,\boldsymbol{E}\left[X^k\right] + o(u^n) = \sum_{k=0}^{n} \frac{u^k}{k!}\,\varphi^{(k)}(0) + o(u^n) \tag{4.19}$$

with an infinitesimal (for $u \to 0$) remainder of order larger than n

Proof Omitted: see [1] p. 278. Remark that—after checking the conditions to perform the limit under the integral—the Eq. (4.17) is nothing else than a derivation under the integral. As for the expansion (4.18), this too heuristically derives from the Taylor series expansion of an exponential according to

$$\varphi(u) = \boldsymbol{E}\left[e^{iuX}\right] = \boldsymbol{E}\left[\sum_{n=0}^{\infty} \frac{(iu)^n}{n!}\,X^n\right] = \sum_{n=0}^{\infty} \frac{(iu)^n}{n!}\,\boldsymbol{E}\left[X^n\right] = \sum_{n=0}^{\infty} \frac{u^n}{n!}\,\varphi^{(n)}(0)$$

These results elucidate the important relation between a *chf* $\varphi(u)$ and the moments $\boldsymbol{E}[X^n]$ of a *rv* X: further details about the so-called ***problem of moments*** and about the ***cumulants*** can be found in the Appendix E ■

With a similar proof the previous theorem can be extended to expansions around a point $u = v$, instead of $u = 0$, and in this case—with a suitable convergence radius—we will find the formula

$$\varphi(u) = \sum_{n=0}^{\infty} \frac{i^n (u-v)^n}{n!} \boldsymbol{E}\left[X^n e^{ivX}\right]$$

Theorem 4.12 Uniqueness theorem: *If $f(x)$ and $g(x)$ are two pdf's with the same*chf, *namely if*

$$\int_{-\infty}^{+\infty} e^{iux} f(x)\, dx = \int_{-\infty}^{+\infty} e^{iux} g(x)\, dx\,, \qquad \forall\, u \in \boldsymbol{R}$$

then it is $f(x) = g(x)$ for every $x \in \boldsymbol{R}$, with the possible exception of a set of points of vanishing Lebesgue measure

Proof Omitted: see [1] p. 282 ■

Theorem 4.13 Inversion formula: *If $\varphi(u)$ is the chf of an ac law, then the corresponding pdf is*

$$f(x) = \frac{1}{2\pi} \lim_{T\to+\infty} \int_{-T}^{T} e^{-iux}\varphi(u)\, du = \frac{1}{2\pi}\, VP \int_{-\infty}^{+\infty} e^{-iux}\varphi(u)\, du \tag{4.20}$$

Proof Omitted: see [1] p. 283 ■

Theorem 4.14 *Necessary and sufficient condition for the independence of the components of a* r-vec $\boldsymbol{X} = (X_1, \ldots, X_n)$ *is the relation*

$$\varphi_{\boldsymbol{X}}(u_1, \ldots, u_n) = \varphi_{X_1}(u_1) \cdot \ldots \cdot \varphi_{X_n}(u_n)$$

namely that the joint chf $\varphi_{\boldsymbol{X}}(\boldsymbol{u})$ is the product of the marginal chf's $\varphi_{X_k}(u_k)$ of the individual components

Proof Omitted: see [1] p. 286 ■

All these results point out that the law of a *rv* can equivalently be represented either by its *pdf* (or by its *cdf* when this is not *ac*), or by its *chf*: the knowledge of one allows, indeed, to get the other in an unique way, and vice-versa. Furthermore all the relevant information (expectation and other moments) can be independently calculated either from the *pdf*, of from the *chf*. Before accepting the idea that the law of a *rv* can be well represented by its *chf*, we must however highlight a rather subtle point: it is not easy sometimes to find if a given function $\varphi(u)$ is an acceptable *chf* of some law. That a function $f(x)$ is a possible *pdf* it is rather straightforward to check: it is enough to be a real, non negative normalized function. For a *chf* instead it is not enough, for instance, that $\varphi(u)$ admit an inverse Fourier transform according to the formula (4.20): we need to know (without performing a direct, often difficult calculation) that the inverse transform is a good *pdf*. In short we need an intrinsic profiling of $\varphi(u)$ allowing us to be sure that it is a good *chf*.

Theorem 4.15 Bochner theorem: *A continuous function $\varphi(u)$ is a chf iff it is non-negative definite,*[1] *and $\varphi(0) = 1$*

Proof Omitted: see [1] p. 287 and [3] vol. II, p. 622. Here we will remark only that if $\varphi(u)$ is the *chf* of a *rv* X we already know that it is (uniformly) continuous and that $\varphi(0) = 1$. It is easy moreover to check that, for every $u_1, \dots, u_n$, and however taken n complex numbers $z_1, \dots, z_n$, it is

$$\sum_{j,k=1}^{n} z_j\overline{z_k}\,\varphi(u_j - u_k) = \sum_{j,k=1}^{n} z_j\overline{z_k}\,\boldsymbol{E}\left[e^{i(u_j-u_k)X}\right] = \boldsymbol{E}\left[\sum_{j,k=1}^{n} z_j\overline{z_k}\,e^{iu_jX}e^{-iu_kX}\right]$$
$$= \boldsymbol{E}\left[\sum_{j=1}^{n} z_j e^{iu_jX}\,\overline{\sum_{k=1}^{n} z_k e^{iu_kX}}\right] = \boldsymbol{E}\left[\left|\sum_{j=1}^{n} z_j e^{iu_jX}\right|^2\right] \geq 0$$

namely $\varphi(u)$ is non negative definite. The Bochner theorem states that also the reverse holds: every function of a real variable with complex values $\varphi(u)$, and with the said three properties is a good *chf* ■

The close relationship between the *cdf* $F(x)$ (or its *pdf* $f(x)$) and the *chf* $\varphi(t)$ of a law suggests that the weak convergence of a sequence of *cdf*'s $\left(F_n(x)\right)_{n\in\boldsymbol{N}}$ can be assessed by looking at the pointwise convergence of the corresponding sequence of *chf*'s $\left(\varphi_n(u)\right)_{n\in\boldsymbol{N}}$. The following theorem then itemizes under what conditions the weak convergence $F_n \xrightarrow{w} F$ is equivalent to the pointwise convergence $\varphi_n(u) \to \varphi(u)$ of the corresponding *chf*'s

Theorem 4.16 Paul Lévy continuity theorem: *Given a sequence of cdf's $\left(F_n(x)\right)_{n\in\boldsymbol{N}}$ and the corresponding sequence of chf's $\left(\varphi_n(u)\right)_{n\in\boldsymbol{N}}$*

---

[1] A function $\varphi(u)$ is non-negative definite when, however chosen n points $u_1, \dots, u_n$, the matrix $\|\varphi(u_j - u_k)\|$ turns out to be non-negative definite, namely when, however chose n complex numbers $z_1, \dots, z_n$, we always have

$$\sum_{j,k=1}^{n} z_j\overline{z}_k\,\varphi(u_j - u_k) \geq 0. \tag{4.21}$$

1. *if* $F_n \xrightarrow{w} F$ *and if* $F(x)$ *turns out to be a cdf, then also* $\varphi_n(u) \xrightarrow{n} \varphi(u)$ *for every* $u \in \boldsymbol{R}$, *and* $\varphi(u)$ *turns out to be the chf of* $F(x)$;
2. *if the limit* $\varphi(u) = \lim_n \varphi_n(u)$ *exists for every* $u \in \boldsymbol{R}$, *and if* $\varphi(u)$ *is continuous in* $u = 0$, *then* $\varphi(u)$ *is the chf of a cdf* $F(x)$ *and it results that* $F_n \xrightarrow{w} F$
3. *if in particular we a priori know that* $F(x)$ *is a cdf and* $\varphi(u)$ *is its chf, then* $F_n \xrightarrow{w} F$ iff $\varphi_n(u) \xrightarrow{n} \varphi(u)$ *for every* $u \in \boldsymbol{R}$

Proof Omitted: see [1] p. 322. ■

4.2.2 Gaussian Laws

The *r-vec*'s with joint Gaussian law $\mathfrak{N}(\boldsymbol{b}, \mathbb{A})$ play a very prominet role in probability and statistics. First, as we will see in the Sect. 4.4, this follows from the so-called Central Limit Theorem stating that sums of a large number of independent *rv*'s, with arbitrary laws under rather broad conditions, tend to become Gaussian. This is of course the conceptual basis for the error law stating that random errors in the empirical measurements—errors resulting precisely from the sum of a large number of small, independent and uncontrollable disturbances—are approximately Gaussian. Second, the Gaussian *rv*'s enjoy a few relevant properties, for instance:

- their laws $\mathfrak{N}(\boldsymbol{b}, \mathbb{A})$ are completely qualified by a small number of parameters
- they exhibit a total equivalence of independence and non correlation, a property not shared with other *rv*'s (see Sect. 3.3.3)
- they have finite momenta of every order and can then be analyzed with the functional analysis tools discussed in the Appendix D.

As a consequence it will be very useful to find an effective way to completely represent the family $\mathfrak{N}(\boldsymbol{b}, \mathbb{A})$ of Gaussian laws, and here we will look at this problem from the standpoint of their *chf*'s.

If there is only one component $X \sim \mathfrak{N}(b, a^2)$ we know that for $a > 0$ the *pdf* is

$$f_X(x) = \frac{1}{a\sqrt{2\pi}} e^{(x-b)^2/2a^2}$$

Since $a^2 = \boldsymbol{V}[X]$, when $a \downarrow 0$ the law of X intuitively converges to that of a degenerate *rv* taking only the value $X = b$, $\boldsymbol{P}$-a.s.. We know on the other hand that a *rv* degenerate in b follows a typically not continuous law $\boldsymbol{\delta}_b$ that admit no *pdf*. As a consequence—to the extent that we represent a law only with its *pdf*–we are obliged to set apart the case $a > 0$ (when X has a proper Gaussian *pdf*) from the case $a = 0$ (when X degenerates in b and no longer has a *pdf*), and to accept that the two description do not go smoothly one into the other when $a \downarrow 0$. To bypass this awkwardness let us recall therefore that a *rv* can be effectively described also through its *chf*, and that for our *rv*'s we find from (4.6) and (4.13)

$$\varphi_X(u) = \begin{cases} e^{ibu} & \text{if } a = 0, \text{ law } \boldsymbol{\delta}_b \\ e^{ibu-u^2a^2/2} & \text{if } a > 0, \text{ law } \mathfrak{N}(b, a^2) \end{cases}$$

It is apparent then that—at variance with its *pdf*–the *chf* with $a = 0$ smoothly results form that with $a > 0$ in the limit $a \downarrow 0$, so that we can now speak of a unified family of laws $\mathfrak{N}(b, a^2)$ for $a \geq 0$, with $\mathfrak{N}(b, 0) = \boldsymbol{\delta}_b$, in the sense that all these distributions are represented by the *chf*'s

$$\varphi_X(u) = e^{ibu-u^2a^2/2} \qquad a \geq 0$$

where the degenerate case is nothing else (as intuitively expected) than the limit $a \downarrow 0$ of the non degenerate case.

These remarks can now be extended also to Gaussian *r-vec*'s $\boldsymbol{X} \sim \mathfrak{N}(\boldsymbol{b}, \mathbb{A})$: in terms of *pdf*'s we would be obliged to discriminate between singular ($|\mathbb{A}| = 0$ non negative definite), and non singular ($|\mathbb{A}| > 0$, positive definite) covariance matrices. For *r-vec*'s with more than one component the difficulty is compounded by the possible dissimilar behavior of the individual components: it is not ruled out, indeed, the circumstance that only some components turn out to be degenerate giving rise to a distribution which is neither discrete nor *ac*. The usage of the *chf*'s allows instead to give again a coherent, unified description.

Definition 4.17 We will say that $\boldsymbol{X} = (X_1, \dots, X_n) \sim \mathfrak{N}(\boldsymbol{b}, \mathbb{A})$ is a ***Gaussian (normal)*** *r-vec* with average vector $\boldsymbol{b} = (b_1, \dots, b_n) \in \boldsymbol{R}^n$ and symmetric, non negative definite covariance matrix $\mathbb{A} = \|a_{kl}\|$, if its *chf* is

$$\varphi_X(\boldsymbol{u}) = \varphi_X(u_1, \dots, u_n) = e^{i\,\boldsymbol{b}\cdot\boldsymbol{u}}\, e^{-\boldsymbol{u}\cdot\mathbb{A}\boldsymbol{u}/2} \qquad \boldsymbol{u} \in \boldsymbol{R}^n \tag{4.22}$$

where $\boldsymbol{b} \cdot \boldsymbol{u} = \sum_k b_k u_k$ is the Euclidean scalar product between vectors in $\boldsymbol{R}^n$

The *chf* (4.22) is a generalization of the *chf* (4.13) that is recovered when b is a number and the covariance matrix is reduced to a unique element a^2. Remark that—at variance with the *pdf* (2.22)—only the matrix $\mathbb{A}$, and not its inverse $\mathbb{A}^{-1}$, appears in the *chf* (4.22), that accordingly is not affected by a possible singularity. Since however the singular case has been treated as an extension of the non singular Gaussian *r-vec*, it is be expedient to check that the Definition 4.17 is indeed acceptable and coherent.

Proposition 4.18 *In the non singular case ($|\mathbb{A}| \neq 0$) the (4.22) is the chf of the Gaussian pdf (2.22); in the singular case ($|\mathbb{A}| = 0$) the same (4.22) turns out to be the chf of a law $\mathfrak{N}(\boldsymbol{b}, \mathbb{A})$ that we will still call Gaussian in spite of the fact that there is no pdf*

Proof If $\mathbb{A}$ is non singular ($|\mathbb{A}| \neq 0$) its inverse $\mathbb{A}^{-1}$ exists and it is possible to show by a direct calculation of the inverse Fourier transform (here omitted) that the $\varphi_X(\boldsymbol{u})$ of Definition 4.17 is precisely the *chf* of a *r-vec* $\boldsymbol{X} \sim \mathfrak{N}(\boldsymbol{b}, \mathbb{A})$ with a normal, multivariate *pdf* (2.22)

$$f_X(\boldsymbol{x}) = \sqrt{\frac{|\mathbb{A}^{-1}|}{(2\pi)^n}}\, e^{-\frac{1}{2}(\boldsymbol{x}-\boldsymbol{b})\cdot\mathbb{A}^{-1}(\boldsymbol{x}-\boldsymbol{b})}$$

When instead $\mathbb{A}$ is singular ($|\mathbb{A}| = 0$), $\mathbb{A}^{-1}$ does not exist and (4.22) can no longer be considered as the Fourier transform of some *pdf*. That notwithstanding it is possible to show that (4.22) continues to be the *chf* of some *r-vec*, albeit lacking a *pdf*. For $n \in \boldsymbol{N}$ take indeed the matrix $\mathbb{A}_n = \mathbb{A} + \frac{1}{n}\,\mathbb{I}$ ($\mathbb{I}$ is the identity matrix) that turns out to be symmetric, non negative definite and—at variance with $\mathbb{A}$ – non singular for every $n \in \boldsymbol{N}$. Then $\mathbb{A}_n^{-1}$ exists for every $n \in \boldsymbol{N}$ and the function

$$\varphi_n(\boldsymbol{u}) = e^{i\,\boldsymbol{b}\cdot\boldsymbol{u}}\,e^{-\boldsymbol{u}\cdot\mathbb{A}_n\boldsymbol{u}/2}$$

is the *chf* of a *r-vec* distributed as $\mathfrak{N}(\boldsymbol{b}, \mathbb{A}_n)$ with a suitable Gaussian *pdf*. Since moreover for every $\boldsymbol{u}$ we of course find

$$\lim_n \varphi_n(\boldsymbol{u}) = e^{i\,\boldsymbol{b}\cdot\boldsymbol{u}}\,e^{-\boldsymbol{u}\cdot\mathbb{A}\boldsymbol{u}/2} = \varphi_{\boldsymbol{X}}(\boldsymbol{u})$$

and the limit function (4.22) is continuous in $\boldsymbol{u} = (0, \dots, 0)$, the Continuity Theorem 4.16 entails that $\varphi_{\boldsymbol{X}}(\boldsymbol{u})$ is the *chf* of a law, even if it does not admit a *pdf*. The *r-vec*'s resulting from this limit procedure can then legitimately be considered as Gaussian *r-vec*'s $\mathfrak{N}(\boldsymbol{b}, \mathbb{A})$ for the singular case $|\mathbb{A}| = 0$ ■

Proposition 4.19 *Given a Gaussian* r-vec $\boldsymbol{X} = (X_1, \dots, X_n) \sim \mathfrak{N}(\boldsymbol{b}, \mathbb{A})$ *it is*

$$b_k = \boldsymbol{E}\,[X_k]\,; \qquad a_{kl} = \boldsymbol{cov}\,[X_k, X_l] \qquad a_{kk} = a_k^2 = \boldsymbol{V}\,[X_k]$$

and its components $X_k \sim \mathfrak{N}\,(b_k, a_k^2)$ *are independent iff they are uncorrelated*

Proof The probabilistic meaning of $\boldsymbol{b}$ and $\mathbb{A} = \|a_{kl}\|$ (already discussed in the Example 3.34 for the bivariate, non degenerate case) are derivable from the *chf* (4.22) with a direct calculation here omitted. It is easy instead to show that the individual components X_k are Gaussian $\mathfrak{N}\,(b_k, a_k^2)$ (as previously stated without a proof in the Example 3.13): from (4.4) we immediately get that the marginal *chf*'s of our Gaussian *r-vec* are in fact

$$\varphi_{X_k}(u_k) = e^{ib_k u_k}\,e^{-u_k^2 a_k^2/2}$$

and hence they too are Gaussian $\mathfrak{N}\,(b_k, a_k^2)$. The equivalence between independence and non correlation of the components (already discussed in the Example 3.34 for the non degenerate, bivariate case) can now be proved in general: first it is a foregone conclusion that if the X_k are independent they also are uncorrelated. Viceversa, if the component of a Gaussian *r-vec* $\mathfrak{N}(\boldsymbol{b}, \mathbb{A})$ are uncorrelated the covariance matrix $\mathbb{A}$ turns out to be diagonal with $a_{kl} = \delta_{kl} a_k^2$ and hence its *chf* is

$$\varphi_{\boldsymbol{X}}(u_1, \dots, u_n) = e^{i\,\boldsymbol{b}\cdot\boldsymbol{u}}\,e^{-\sum_k a_k^2 u_k^2/2} = \prod_{k=1}^{n}\left(e^{ib_k u_k}\,e^{-a_k^2 u_k^2/2}\right) = \prod_{k=1}^{n}\varphi_{X_k}(u_k)$$

where $\varphi_{X_k}(u_k)$ are the *chf* of the individual components. As a consequence, from the Theorem 4.14, the components of $\boldsymbol{X}$ are independent ■

Proposition 4.20 *Given the r-vec* $\boldsymbol{X} = (X_1, \ldots, X_n)$*, the following statements are equivalent.*

1. $\boldsymbol{X} \sim \mathfrak{N}(\boldsymbol{b}, \mathbb{A})$
2. $\boldsymbol{c} \cdot \boldsymbol{X} \sim \mathfrak{N}(\boldsymbol{c} \cdot \boldsymbol{b}\,,\, \boldsymbol{c} \cdot \mathbb{A}\boldsymbol{c})$ *for every* $\boldsymbol{c} \in \boldsymbol{R}^n$
3. $\boldsymbol{X} = \mathbb{C}\boldsymbol{Y} + \boldsymbol{b}$ *where* $\boldsymbol{Y} \sim \mathfrak{N}(0, \mathbb{I})$*,* $\mathbb{C}$ *is non singular, and* $\mathbb{A} = \mathbb{C}\mathbb{C}^T$.

Proof Omitted: see [1] p. 301-2. Remark in the point 3 that the *r-vec* $\boldsymbol{Y}$ is Gaussian with components Y_k that are standard $\mathfrak{N}(0, 1)$ and independent because its covariance matrix is δ_{jk}. As a consequence the components of an arbitrary, Gaussian *r-vec* $\boldsymbol{X} \sim \mathfrak{N}(\boldsymbol{b}, \mathbb{A})$ always are linear combinations of the independent, standard normal components of the *r-vec* $\boldsymbol{Y} \sim \mathfrak{N}(0, \mathbb{I})$; and viceversa, the components of an arbitrary Gaussian *r-vec* $\boldsymbol{X} \sim \mathfrak{N}(\boldsymbol{b}, \mathbb{A})$ can always be made standard and independent by means of suitable linear combinations ■

4.2.3 Composition and Decomposition of Laws

The locution ***reproductive properties of a family of laws*** usually refers to a family which is closed under convolution, in the sense that the ***composition*** through convolution of the *pdf*'s of two or more laws of the said family again produces a law of the same family. We already met the reproductive properties (3.67) of the normal *rv*'s $\mathfrak{N}(b, a^2)$ in the Sect. 3.5.2, but we postponed the proof in order to shirk lengthy and uneasy integrations. The *chf*'s allow instead even here a remarkable simplification because, as we know from the Proposition 4.9, the convolution of the *pdf*'s is replaced by the simple product of the *chf*'s.

Example 4.21 The reproductive properties (3.67) of the ***Gaussian laws*** $\mathfrak{N}(b, a^2)$

$$\mathfrak{N}(b_1, a_1^2) * \mathfrak{N}(b_2, a_2^2) = \mathfrak{N}(b_1 + b_2\,,\, a_1^2 + a_2^2)$$

are simply proved by recalling Proposition 4.9 and (4.13), and remarking that the product of the *chf*'s $\varphi_1(u)$ and $\varphi_2(u)$ of the laws $\mathfrak{N}(b_1, a_1^2)$ and $\mathfrak{N}(b_2, a_2^2)$ is

$$\varphi(u) = \varphi_1(u)\varphi_2(u) = e^{ib_1u - a_1^2u^2/2}e^{ib_2u - a_2^2u^2/2} = e^{i(b_1+b_2)u - (a_1^2+a_2^2)u^2/2}$$

namely the *chf* of the law $\mathfrak{N}(b_1 + b_2\,,\, a_1^2 + a_2^2)$. As a consequence the family of laws $\mathfrak{N}(b, a^2)$ with parameters a and b is closed under convolution. As a particular case, for $a_1 = a_2 = 0$, we also retrieve the reproductive properties of the ***degenerate laws*** $\boldsymbol{\delta}_b$

$$\boldsymbol{\delta}_{b_1} * \boldsymbol{\delta}_{b_2} = \boldsymbol{\delta}_{b_1+b_2} \tag{4.23}$$

By the same token we can also prove that the ***Poisson laws*** $\mathfrak{P}(\alpha)$ enjoy the same property in the sense that

$$\mathfrak{P}(\alpha_1) * \mathfrak{P}(\alpha_2) = \mathfrak{P}(\alpha_1 + \alpha_2) \tag{4.24}$$

From (4.9) we indeed see that the product of the *chf*'s of $\mathfrak{P}(\alpha_1)$ and $\mathfrak{P}(\alpha_2)$ is

$$\varphi(u) = e^{\alpha_1(e^{iu}-1)}e^{\alpha_2(e^{iu}-1)} = e^{(\alpha_1+\alpha_2)(e^{iu}-1)}$$

namely the *chf* of $\mathfrak{P}(\alpha_1 + \alpha_2)$.

The parametric families δ_b, $\mathfrak{N}(b, a^2)$ e $\mathfrak{P}(\alpha)$—whose relevance will be emphasized in the subsequent discussion about the limit theorems—enjoy a further important property: they are closed also under convolution ***decomposition***. If for instance we decompose a Gaussian law $\mathfrak{N}(b, a^2)$ into the convolution of two other laws, the latter must also be Gaussian laws from the family $\mathfrak{N}(b, a^2)$. In other words, not only the convolution of two Gaussians always produces a Gaussian, but only by composing two Gaussians we can get a Gaussian law. In this sense we say that the family $\mathfrak{N}(b, a^2)$ is closed under convolution composition and decomposition. A similar result holds for the families δ_b and $\mathfrak{P}(\alpha)$, but, at variance with the compositions, the theorems about decompositions are rather difficult to prove: for details see [4] vol. I, Sect. 20.2, p. 283.

Example 4.22 There are more parametric families of laws that are closed under convolution: it is easy for instance to prove from (4.16) the reproductive properties of the ***Cauchy laws*** $\mathfrak{C}(a, b)$

$$\mathfrak{C}(a_1, b_1) * \mathfrak{C}(a_2, b_2) = \mathfrak{C}(a_1 + a_2, b_1 + b_2) \tag{4.25}$$

We should however refrain from supposing a too wide generalization of this property: for instance a convolution of exponential laws $\mathfrak{E}(a)$ does not produce an exponential law: it is easy to see from 4.14 that if $\varphi_1(u)$ and $\varphi_2(u)$ are *chf*'s of $\mathfrak{E}(a_1)$ and $\mathfrak{E}(a_2)$, their product

$$\varphi(u) = \varphi_1(u)\varphi_2(u) = \frac{a_1}{a_1 - iu}\,\frac{a_2}{a_2 - iu}$$

is not the *chf* of an exponential. If instead we combine exponential laws with the same parameter a we find a new family of laws that will be useful in the following: the product of the *chf*'s $\varphi_a(u)$ of n exponentials $\mathfrak{E}(a)$ with the same a is indeed

$$\varphi(u) = \varphi_a^n(u) = \left(\frac{a}{a - iu}\right)^n \tag{4.26}$$

and it is possible to show with a direct calculation that the corresponding *pdf* is

$$f_Z(x) = \frac{(ax)^{n-1}}{(n-1)!}\,ae^{-ax}\vartheta(x) \qquad n = 1, 2, \ldots \tag{4.27}$$

where $\vartheta(x)$ is the Heaviside function. These are known as ***Erlang laws*** $\mathfrak{E}_n(a) = \mathfrak{E}^{*n}(a)$, and we have just shown indeed that an Erlang $\mathfrak{E}_n(a)$ *rv* always is decomposable in the sum of n independent exponential $\mathfrak{E}(a)$ *rv's*. It is also easy to check from (3.33) and (3.38) that if $X \sim \mathfrak{E}_n(a)$, then

$$\boldsymbol{E}[X] = \frac{n}{a} \qquad \boldsymbol{V}[X] = \frac{n}{a^2} \tag{4.28}$$

Remark the formal reciprocity between the Erlang (4.27) and the Poisson distributions: the expression

$$\frac{x^k e^{-x}}{k!}\,\vartheta(x) \qquad k = 0, 1, 2 \ldots$$

represents indeed at the same time both a discrete Poisson distribution $\mathfrak{P}(x)$ with values k (and parameter $x > 0$), and an Erlang *pdf* $\mathfrak{E}_{k+1}(1)$ of order $k+1$, with values x (and parameter $a = 1$): this reciprocity will be further elucidated later on by the discussion of the Poisson process in the Sect. 6.1.

4.3 Laws of Large Numbers

The classical *limit theorems* are statements about limits (for $n \to \infty$) of sums $S_n = X_1 + \cdots + X_n$ of sequences $(X_n)_{n\in N}$ of *rv*'s, where a prominent role is played by the families of laws δ_b, $\mathfrak{N}(b, a^2)$ and $\mathfrak{P}(\alpha)$. We should at once remark, however, that the limit theorems are not an aftermath of the composition and decomposition properties of the Sect. 4.2.3. They are instead deep results that go beyond the boundaries of the previous discussion. First of all, while the composition and decomposition properties pertain to *finite sums* of *rv*'s, the limit theorems touch to *limits of sequences of sums* of *rv*'s. Second, while for instance the composition and decomposition properties of a Gaussian *rv* states that this law always is the sum of a finite number of independent *rv*'s the are again Gaussians, in the Central Limit Theorem the normal laws comes out as the limit in distribution of sums of independent *rv*'s with *arbitrary laws*, within rather broad conditions. We will begin our treatment with the Law of Large Numbers that, at variance with the Gaussian and Poisson theorems that will be discussed in the subsequent sections, are a case of *degenerate convergence*: the sequence S_n does indeed converge toward a *number*, namely toward a *rv* taking just one value $\boldsymbol{P}$-a.s.. The oldest version of this important result, the Bernoulli Theorem (1713), is briefly recalled in the Appendix F.

Theorem 4.23 Weak Law of Large Numbers: *Given a sequence $(X_n)_{n\in N}$ of rv's iid with $\boldsymbol{E}[|X_n|] < +\infty$, and taken $S_n = X_1 + \cdots + X_n$ and $\boldsymbol{E}[X_n] = m$, it turns out that*

$$\frac{S_n}{n} \xrightarrow{\boldsymbol{P}} m$$

Proof In the present formulation the X_k are not in general Bernoulli *rv*'s as in the original Bernoulli's proof, and hence the S_n are not binomial, so that the proof can not be given along the lines of Appendix F where the binomial laws play the central role. To bypass the problem remark first that, from the point 4 of the Theorem 4.4, the *degenerate convergence* in probability (for us toward m) is *equivalent* to the convergence in distribution to the same constant and hence we can legitimately utilize the Lévy Theorem 4.16. If $\varphi(u)$ is the *chf* of the X_n, the *chf*'s of the S_n/n will be

$$\varphi_n(u) = \boldsymbol{E}\left[e^{iuS_n/n}\right] = \prod_{k=1}^{n} \boldsymbol{E}\left[e^{iuX_k/n}\right] = \left[\varphi\left(\frac{u}{n}\right)\right]^n$$

Our *rv*'s are integrable by hypothesis, and hence from (4.19) we get

$$\varphi(u) = 1 + ium + o(u) \qquad u \to 0$$

so that, with fixed, arbitrary u,

$$\varphi\left(\frac{u}{n}\right) = 1 + i\frac{u}{n}m + o\left(\frac{1}{n}\right) \qquad n \to \infty$$

For every $u \in \boldsymbol{R}$ we then have

$$\varphi_n(u) = \left[1 + i\frac{u}{n}m + o\left(\frac{1}{n}\right)\right]^n \xrightarrow{n} e^{imu}$$

and since e^{imu} is the *chf* of a *rv* degenerate in m, the result follows from the Theorem 4.16. ■

There is a variant of this weak Law of Large Numbers that is fit also for sequences of *rv*'s that are independent, but *not identically distributed*. To this end it is expedient to remark that the Theorem 4.23 can also be put in the form

$$\frac{S_n - \boldsymbol{E}[S_n]}{n} \xrightarrow{P} 0 \tag{4.29}$$

that no longer refers to a common expectation value, and hence is suitable for independent, but not identically distributed X_n. The next theorem shows that the identical distribution hypothesis can be replaced by another about the variance $\boldsymbol{V}[X_n]$ that of course must be now supposed to be finite.

Theorem 4.24 *If the rv's in $(X_n)_{n\in N}$ are independent with $\boldsymbol{E}\left[|X_n|^2\right] < +\infty$, taken $S_n = X_1 + \cdots + X_n$, if we can find a number $C > 0$ such that*

$$\boldsymbol{V}[X_n] < C \qquad \forall n \in \boldsymbol{N}$$

then it is

$$\frac{S_n - \boldsymbol{E}[S_n]}{n} \xrightarrow{\boldsymbol{P}} 0$$

Proof From the Chebyshev inequality (3.42), and however chosen $\epsilon > 0$

$$\begin{aligned}\boldsymbol{P}\left\{\left|\frac{S_n - \boldsymbol{E}[S_n]}{n}\right| \geq \epsilon\right\} &\leq \frac{1}{\epsilon^2}\boldsymbol{V}\left[\frac{S_n - \boldsymbol{E}[S_n]}{n}\right] = \frac{1}{n^2\epsilon^2}\boldsymbol{V}\left[\sum_{k=1}^{n}(X_k - \boldsymbol{E}[X_k])\right] \\ &= \frac{1}{n^2\epsilon^2}\sum_{k=1}^{n}\boldsymbol{V}[X_k - \boldsymbol{E}[X_k]] = \frac{1}{n^2\epsilon^2}\sum_{k=1}^{n}\boldsymbol{V}[X_k] \leq \frac{nC}{n^2\epsilon^2} \\ &= \frac{C}{n\epsilon^2} \xrightarrow{n} 0\end{aligned}$$

and the theorem if proved by definition of convergence in probability ■

The Law of Large Numbers plays an extremely important role in Probability because it allows to confidently estimate the expectation of a *rv* X by averaging on a large number of independent observations. To this end we consider a sequence $(X_n)_{n\in N}$ of independent measurements of X (so that the X_n are *iid*) and we calculate their average S_n/n. According to the Theorem 4.23 we can then *confidently* say that the difference between the empirical value of S_n/n and the theoretical $\boldsymbol{E}[X]$ is infinitesimal with n. For the time being, however, the locution *confidently* is problematic: our previous formulations of the Law of Large Numbers guarantees indeed the convergence of S_n/n to $\boldsymbol{E}[X]$ only in probability, and not $\boldsymbol{P}$-a.s.. As a consequence, strictly speaking, the probability that S_n/n *does not converge* to $\boldsymbol{E}[X]$ can be different from zero. If we had not other results stronger than the Theorems 4.23 and 4.24, we could suspect that, with non zero probability, the average of a sequence of measurements does not actually converge to $\boldsymbol{E}[X]$, and this would be particularly alarming in all the empirical applications. For this reason great efforts have been devoted to find a strong Law of Large Numbers (namely in force $\boldsymbol{P}$-a.s.) in order to guarantee the correctness of all the empirical procedures with probability 1.

Theorem 4.25 Strong Law of Large Numbers: *Given a sequence $(X_n)_{n\in N}$ of rv's iid with $\boldsymbol{E}[|X_n|] < +\infty$, and taken $S_n = X_1 + \cdots + X_n$ and $\boldsymbol{E}[X_n] = m$, it turns out that*

$$\frac{S_n}{n} \xrightarrow{as} m$$

Proof Omitted: see [1] p. 391. Remark that the hypotheses of the present theorem coincide with that of the *weak* Theorem 4.23: the different result ($\boldsymbol{P}$-a.s. convergence instead of convergence in probability) is a produce only of the more advanced techniques of demonstration. It is also possible to show that here too we can dismiss the assumption of identical distribution of the X_n replacing it with some constraint on the variances (see [1] p. 389), and that the result still holds if the expectation exists but it is not necessarily finite (see [1] p. 393). We will refrain however to enter into

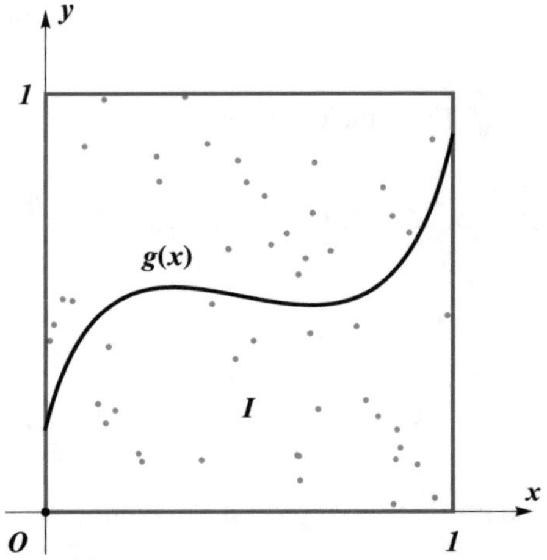

Fig. 4.2 Calculation of the integral (4.30) with the Monte Carlo method

the technical details of these important advances, and we will show instead a few examples of practical application of the Law of Large Numbers ■

Example 4.26 Consider a continuous function $g(x) : [0, 1] \to [0, 1]$ and suppose you want to calculate in a numerical way (namely without finding a primitive) the integral

$$I = \int_0^1 g(x)\, dx \tag{4.30}$$

We will show here that this is possible by taking advantage of the statistical regularities highlighted by the Law of Large Numbers: a method known as ***Monte Carlo*** that we will present in two possible variants.

Take first the *r-vec* $\boldsymbol{U} = (X, Y)$ with values in $\big(\boldsymbol{R}^2, \mathcal{B}(\boldsymbol{R}^2)\big)$ and independent components uniformly distributed in $[0, 1]$ so that

$$f_U(x, y) = \begin{cases} 1 & (x, y) \in [0, 1] \times [0, 1] \\ 0 & \text{else} \end{cases}$$

and $\boldsymbol{U}$ is uniformly distributed in $[0, 1] \times [0, 1]$. If then

$$A = \big\{(x, y) \in [0, 1] \times [0, 1] \,:\, g(x) \geq y\big\} \in \mathcal{B}(\boldsymbol{R}^2) \qquad B = \big\{(X, Y) \in A\big\} \in \mathcal{F}$$

$$\chi_A(x, y) = \begin{cases} 1 & (x, y) \in A \\ 0 & \text{else} \end{cases} \qquad I_B(\omega) = \begin{cases} 1 & \omega \in B \\ 0 & \text{else} \end{cases}$$

the *rv* $Z = I_B = \chi_A(X, Y)$ is a Bernoulli $\mathfrak{B}(1; p)$ with

$$p = \boldsymbol{E}[Z] = \boldsymbol{P}\{B\} = \boldsymbol{P}\{(X,Y) \in A\} = \boldsymbol{P}\{Y \le g(X)\}$$
$$= \int_A f_U(x,y)\,dx\,dy = \int_0^1 \left[\int_0^{g(x)} dy\right] dx = \int_0^1 g(x)\,dx = I$$

In short the value of the integral I is the probability of the event $Y \le g(X)$ for a point of coordinates X, Y taken at random in $[0,1] \times [0,1]$, and this also coincides with the expectation of Z. Hence I can be calculated by estimating $\boldsymbol{E}[Z]$ with the strong Law of Large Numbers: take n points (X_k, Y_k), $k = 1, \dots, n$ uniform in $[0,1] \times [0,1]$, let $Z_k = \chi_A(X_k, Y_k)$ be the corresponding sequence of iid *rv's*, and define $S_n = Z_1 + \dots + Z_n$; the value $I = \boldsymbol{E}[Z]$ is then well approximated by S_n/n for large values of n. In practice this amounts to calculate $I = p = \boldsymbol{P}\{Y \le g(X)\}$ by first enumerating the random points uniform in $[0,1] \times [0,1]$ that fall under the curve $y = g(x)$ in the Fig. 4.2, and then dividing the result by the total number of drawn points.

The numerical calculation of I can also be performed with an alternative procedure by remarking that if X is a uniform $\mathfrak{U}(0,1)$ *rv*, and if $Y = g(X)$, it turns out that

$$\boldsymbol{E}\,Y = \boldsymbol{E}\big[g(X)\big] = \int_0^1 g(x)\,dx = I$$

As a consequence we can calculate I by estimating the expectation of $Y = g(X)$ as an average of trials: if $(X_n)_{n\in N}$ is a sequence of uniform $\mathfrak{U}(0,1)$ *iid rv's*, the Law of Large Numbers states that with probability 1 we will have

$$\frac{1}{n}\sum_{k=1}^{n} g(X_k) \xrightarrow{\;n\;} \boldsymbol{E}[g(X)] = I$$

and hence with a fair number of measurements we can always approximate the value of I with the required precision.

4.4 Gaussian Theorems

The Gaussian theorems are statements about *convergence in distribution* of the sums S_n toward the standard normal law $\mathfrak{N}(0,1)$, and since we are interested rather in *the form* of the limit distribution than in its expectation or its variance, it will be expedient to preliminarily standardize the sequences at issue. Recalling that a *rv* is *standardized* when $\boldsymbol{E}[X] = 0$ and $\boldsymbol{V}[X] = 1$ we will study in the following the standardized sums

$$S_n^* = \frac{S_n - \boldsymbol{E}[S_n]}{\sqrt{\boldsymbol{V}[S_n]}} \tag{4.31}$$

In the oldest versions of these theorems the sums S_n were binomial *rv*'s (see Appendix F), but the modern formulations are much more general and can be proved under a wide selection of hypotheses.

Theorem 4.27 Central Limit Theorem for *iid rv's*: *Take a sequence $(X_n)_{n\in N}$ of iid rv's with $\boldsymbol{E}\left[X_n^2\right] < +\infty$ and $\boldsymbol{V}[X_n] > 0$, and define $S_n = X_1 + \ldots + X_n$ and S_n^* as in (4.31): then it is*

$$S_n^* \xrightarrow{d} \mathfrak{N}(0, 1)$$

Proof Since the convergence in distribution of a sequence of *rv*'s is equivalent to the convergence in general of the corresponding sequence of *cdf*'s (see Sect. 4.1), our theorem states that

$$\boldsymbol{P}\{S_n^* \le x\} \xrightarrow{n} \Phi(x)\,, \qquad \forall x \in \boldsymbol{R}$$

where

$$\Phi(x) = \frac{1}{\sqrt{2\pi}} \int_{-\infty}^{x} e^{-z^2/2}\, dz$$

is the standard error function (2.16) that it is continuous for every x. To prove the statement we will then take advantage of the P. Lévy Theorem 4.16: since the X_n are *iid*, take

$$m = \boldsymbol{E}[X_n] \qquad \sigma^2 = \boldsymbol{V}[X_n] \qquad \varphi(u) = \boldsymbol{E}\left[e^{iu(X_n - m)}\right]$$

and remark first that

$$\boldsymbol{E}[S_n] = nm \qquad \boldsymbol{V}[S_n] = n\sigma^2 \qquad S_n^* = \frac{1}{\sigma\sqrt{n}} \sum_{k=1}^{n} (X_k - m)$$

From the independence of the X_n's we have then

$$\varphi_n(u) = \boldsymbol{E}\left[e^{iuS_n^*}\right] = \boldsymbol{E}\left[\prod_{k=1}^{n} e^{iu(X_k - m)/\sigma\sqrt{n}}\right] = \prod_{k=1}^{n} \boldsymbol{E}\left[e^{iu(X_k - m)/\sigma\sqrt{n}}\right] = \left[\varphi\left(\frac{u}{\sigma\sqrt{n}}\right)\right]^n$$

where $\varphi(u)$ is the *chf* of $X_n - m$ with finite moments at least up to the second order and

$$\boldsymbol{E}[X_n - m] = 0 \qquad \boldsymbol{E}\left[(X_n - m)^2\right] = \sigma^2$$

From the (4.19) we know that

$$\varphi(u) = 1 - \frac{\sigma^2 u^2}{2} + o(u^2) \qquad u \to 0$$

so that, with a fixed arbitrary u, and $n \to \infty$, we have

$$\varphi_n(u) = \left[1 - \frac{u^2}{2n} + o\left(\frac{1}{n}\right)\right]^n \xrightarrow{n} e^{-u^2/2}$$

Since we know from (4.13) that $e^{-u^2/2}$ is the *chf* of $\mathfrak{N}(0, 1)$, the theorem is proved according to the P. Lévy Theorem 4.16 ■

Remark that from $\boldsymbol{V}[S_n] = n\sigma^2$, and taking advantage of the equivalence between the degenerate convergences in probability and in distribution, we could reformulate the result (4.29) of the Law of Large numbers as

$$\frac{S_n - \boldsymbol{E}[S_n]}{\boldsymbol{V}[S_n]} \xrightarrow{d} \delta_0 = \mathfrak{N}(0, 0)$$

where the denominator grows as n, while the Central Limit Theorem 4.27 states that

$$S_n^* = \frac{S_n - \boldsymbol{E}[S_n]}{\sqrt{\boldsymbol{V}[S_n]}} \xrightarrow{d} \mathfrak{N}(0, 1)$$

A juxtaposition of these two assertions highlights analogies and differences between the two results: in the Central Limit Theorem there is the square root of the variance, so that the denominator grows only as $\sqrt{n}$, and this intuitively explains why in this case the convergence is no longer *degenerate*. We will finally recall another variant of the Central Limit Theorem that, by imposing further technical conditions, allows one to jettison the hypothesis of identical distribution of the X_n's.

Theorem 4.28 Central Limit Theorem for independent *rv's*: *Take a sequence $(X_n)_{n\in N}$ of independent rv's with $\boldsymbol{E}\left[X_n^2\right] < +\infty$ and $\boldsymbol{V}[X_n] > 0$, define $S_n = X_1 + \ldots + X_n$ and S_n^* as in (4.31), and posit*

$$m_n = \boldsymbol{E}[X_n] \qquad V_n = \sqrt{\sigma_1^2 + \cdots + \sigma_n^2}$$

*If it exists a $\delta > 0$ such that (**Lyapunov conditions**)*

$$\frac{1}{V_n^{2+\delta}} \sum_{k=1}^{n} \boldsymbol{E}\left[|X_k - m_k|^{2+\delta}\right] \xrightarrow{n} 0$$

then it is

$$S_n^* \xrightarrow{d} \mathfrak{N}(0, 1)$$

Proof Omitted: see [1] p. 332 ■

4.5 Poisson Theorems

In the old *binomial* formulations of the Gaussian Theorems (see for instance the Local Limit Theorem in the Appendix F) the proof of the convergence toward a normal law resulted from the approximation of the values of a binomial distribution by means of

Gaussian functions, but—because of the structural differences between the discrete and the *ac*'s laws—such an approximation was increasingly inaccurate as you moved away from the *center* toward the *tails* of the distributions at issue. This predicament is especially conspicuous when p is near either to 0, or to 1. A Gaussian function is indeed perfectly symmetric around its center, while a binomial distribution shows the same feature only when $p = {}^1/_2$: if instead p departs from ${}^1/_2$ getting closer either to 0 or to 1 such a symmetry is lost. In these cases it is not reasonable to expect that a normal curve be a good approximation of a binomial distribution, except in the immediate vicinity of its maximum. These remarks suggest looking for a different asymptotic (for $n \to \infty$) approximation of the binomial distribution when p is close either to 0 or to 1. To discuss particular problems, on the other hand, we will be often obliged to produce probabilistic models rather different from that of Bernoulli: more precisely we could be required to suppose that the probability p has nor the same value for every n, and in particular that $p(n) \to 0$ for $n \to \infty$, as we will see in the subsequent example.

Example 4.29 Random instants: Suppose that a call center receives, at random times, phone calls with an average number proportional to the width of the time interval, and that in particular an average number $\lambda = 1.5$ of calls arrive every minute: namely an average of 90 calls per hour. If then S is the *rv* counting the random number of calls in an interval $T = 3$ minutes, we ask what is the distribution of S. To this end remark first that S takes unbounded integer values $k = 0, 1, \ldots$, namely the set of its possible values is $\boldsymbol{N} \cup \{0\}$. We then set up the following approximation procedure: since we have an average of $\lambda = 1.5$ calls per minute, we start by dividing T in a number n of equal sub-intervals small enough to find no more than 1 call in average. For example with $n = 9$ the average number of calls in every sub-interval is

$$\frac{\lambda T}{n} = 1.5 \times \frac{3}{9} = \frac{1}{2}$$

so that as a first approximation we can assume that there is no more than 1 random call per sub-interval. We then have a first model with $n = 9$ independent trials checking whether in every sub-interval a phone call is found or not: we can define 9 Bernoulli *rv*'s $X_j^{(9)}$ $(j = 1, \ldots, 9)$ taking value 1 if there is a call in the jth sub-interval, and 0 if there is none. Since apparently $X_j^{(9)} \sim \mathfrak{B}\,(1; p)$ and $\boldsymbol{E}\left[X_j^{(9)}\right] = {}^1/_2$, from (3.27) we also find that $p = p(9) = {}^1/_2$. As a consequence the *rv* $S_9 = X_1^{(9)} + \ldots + X_9^{(9)}$—our approximation for the number of calls in T—will be binomial $\mathfrak{B}\left(9; {}^1/_2\right)$, namely

$$\boldsymbol{P}\{S_9 = k\} = \binom{9}{k}\left(\frac{1}{2}\right)^k\left(\frac{1}{2}\right)^{9-k} = \binom{9}{k}\left(\frac{1}{2}\right)^9 \qquad k = 0, 1, \ldots, 9$$

The drawback of this first approximation is of course the hypothesis that in every sub-interval no more than 1 call can be found: we indeed approximated a *rv* S taking infinite values, with a *rv* $S_9 \sim \mathfrak{B}\left(9; {}^1/_2\right)$ taking only 10 values. This however also suggests how to improve the approximation: if the number n of the sub-intervals grows, we have at once *rv*'s S_n with a growing number of possible values, and—by making ever smaller the width of the sub-intervals—a growing probability of finding no more than 1 call per interval. By taking for instance $n = 18$ sub-intervals we get

$$p(18) = \frac{\lambda T}{n} = 1.5 \times \frac{3}{18} = \frac{1}{4}$$

so that $S_{18} \sim \mathfrak{B}\left(18; {}^1/_4\right)$, that is

$$\boldsymbol{P}\{S_{18} = k\} = \binom{18}{k}\left(\frac{1}{4}\right)^k\left(\frac{3}{4}\right)^{18-k} \qquad k = 0, 1, \ldots, 18$$

We can then continue to improve the approximation taking ever larger n, so that

$$p(n) = \frac{\lambda T}{n} = 1.5 \times \frac{3}{n} \xrightarrow{n} 0 \qquad np(n) = \lambda T = \alpha, \quad \forall n \in \boldsymbol{N}$$

and we must ask now to what limit distribution tends (for $n \to \infty$) the sequence of binomial laws $\mathfrak{B}(n; p(n))$

$$p_n(k) = \boldsymbol{P}\{S_n = k\} = \binom{n}{k} p(n)^k \big(1 - p(n)\big)^{n-k} \tag{4.32}$$

The answer is in the following theorem that we will give first in its classical, binomial form [5] before presenting it also in its more up-to-date versions.

Theorem 4.30 Poisson theorem for binomial *rv*'s: *Take a sequence of binomial rv's binomiali $S_n \sim \mathfrak{B}(n; p(n))$ as in in 4.32: if it exists a number $\alpha > 0$ such that*

$$p(n) \to 0 \qquad q(n) = 1 - p(n) \to 1 \qquad np(n) \to \alpha \qquad n \to \infty$$

then S_n converges in distribution to the Poisson law $\mathfrak{P}(\alpha)$, that is

$$S_n \xrightarrow{d} \mathfrak{P}(\alpha) \quad \textit{namely} \quad \lim_n p_n(k) = \frac{\alpha^k e^{-\alpha}}{k!}, \qquad k = 0, 1, \ldots$$

Proof Since for every $\alpha > 0$, from a certain n onward we have $\alpha/n < 1$, starting from there our hypotheses entitle us to write

$$p(n) = \frac{\alpha}{n} + o(n^{-1})$$

so that for $k = 0, 1, \ldots, n$ we will get

$$p_n(k) = \frac{n(n-1)\ldots(n-k+1)}{k!}\left[\frac{\alpha}{n} + o(n^{-1})\right]^k \left[1 - \frac{\alpha}{n} + o(n^{-1})\right]^{n-k}$$

From a well known limit result we then have

$$n(n-1)\dots(n-k+1)\left[\frac{\alpha}{n}+o(n^{-1})\right]^k = \frac{n(n-1)\dots(n-k+1)}{n^k}[\alpha+o(1)]^k$$
$$= \left(1-\frac{1}{n}\right)\dots\left(1-\frac{k-1}{n}\right)[\alpha+o(1)]^k \xrightarrow{n} \alpha^k$$
$$\left[1-\frac{\alpha}{n}+o(n^{-1})\right]^{n-k} = \left[1-\frac{\alpha}{n}+o(n^{-1})\right]^n\left[1-\frac{\alpha}{n}+o(n^{-1})\right]^{-k} \xrightarrow{n} e^{-\alpha}$$

and hence we easily find the result ■

Theorem 4.31 Poisson theorem for multinomial *r*-vec's: *Take a sequence of multinomial r-vec's* $\boldsymbol{S}_n = (X_1, \dots, X_r) \sim \mathfrak{B}(n; p_1, \dots, p_r)$ *with*

$$\boldsymbol{P}\{X_1 = k_1, \dots, X_r = k_r\} = \frac{n!}{k_0!k_1!\dots k_r!}\,p_0^{k_0}p_1^{k_1}\dots p_r^{k_r} \qquad \begin{cases} p_0+p_1+\dots+p_r = 1\\ k_0+k_1+\dots+k_r = n\end{cases}$$

If for $j = 1, \dots, r$ *and* $n \to \infty$ *there exist* $\alpha_j > 0$ *such that*

$$p_j = p_j(n) \to 0 \qquad p_0 = p_0(n) \to 1 \qquad np_j(n) \to \alpha_j$$

then we have

$$\boldsymbol{S}_n = (X_1, \dots, X_r) \xrightarrow{d} \mathfrak{P}(\alpha_1)\cdot\ldots\cdot\mathfrak{P}(\alpha_r).$$

Proof Omitted: the proof is lengthier, bus similar to that of the Theorem 4.30. See also [6] p. 58 ■

The Theorem 4.30 has been proved by making explicit use of the properties of the binomial laws resulting from the sum of *iid* Bernoulli *rv*'s. It can however be generalized to the sums of Bernoulli *rv*'s independent but *not identically distributed*: in this case the sums are no longer binomial and the previous proof cannot be adopted. To fix the ideas suppose to have a sequence of experiments, and for every n to have n independent Bernoulli *rv*'s $X_1^{(n)}, \dots, X_n^{(n)}$ with $X_k^{(n)} \sim \mathfrak{B}\left(1; p_k^{(n)}\right)$, i.e.

$$\boldsymbol{P}\{X_k^{(n)} = 1\} = p_k^{(n)} \qquad \boldsymbol{P}\{X_k^{(n)} = 0\} = q_k^{(n)} \qquad p_k^{(n)} + q_k^{(n)} = 1 \qquad k = 1, \dots, n$$

The sum $S_n = X_1^{(n)} + \dots + X_n^{(n)}$ will take then integer values from 0 to n, but in general it will not be binomial because the summands are not identically distributed. We will have indeed that

- for every fixed k: the $p_k^{(n)}$ depend on n and hence the *rv*'s $X_k^{(n)}$ change distribution according to n; in other words going from n to $n+1$ the *rv*'s at a place k are updated
- for every fixed n: the $X_k^{(n)}$ are not identically distributed, so that S_n is not binomial

In short we will have a triangular scheme of the type

$$\begin{array}{ll} X_1^{(1)} & p_1^{(1)} \\ X_1^{(2)}, X_2^{(2)} & p_1^{(2)}, p_2^{(2)} \\ \vdots & \vdots \\ X_1^{(n)}, \ldots, X_n^{(n)} & p_1^{(n)}, \ldots, p_n^{(n)} \\ \vdots & \vdots \end{array}$$

The $X_k^{(n)}$ in every row are independent but not identically distributed; along the columns instead the $p_k^{(n)}$ (to wit the laws) change in general with n. The next theorem fixes the conditions to allow the new S_n to converge again in distribution toward $\mathfrak{P}(\alpha)$

Theorem 4.32 *For every $n \in \boldsymbol{N}$ and $k = 1, \ldots, n$ take the independent rv's $X_k^{(n)}$ with*

$$\boldsymbol{P}\left\{X_k^{(n)} = 1\right\} = p_k^{(n)} \qquad \boldsymbol{P}\left\{X_k^{(n)} = 0\right\} = q_k^{(n)} \qquad p_k^{(n)} + q_k^{(n)} = 1$$

and posit $S_n = X_1^{(n)} + \cdots + X_n^{(n)}$: if

$$\max_{1 \le k \le n} p_k^{(n)} \xrightarrow{n} 0 \qquad \sum_{k=1}^{n} p_k^{(n)} \xrightarrow{n} \alpha > 0$$

then we have

$$S_n \xrightarrow{d} \mathfrak{P}(\alpha)$$

Proof From the independence of the $X_k^{(n)}$, and recalling (4.7), we have

$$\varphi_{S_n}(u) = \boldsymbol{E}\left[e^{iuS_n}\right] = \prod_{k=1}^{n}\left[p_k^{(n)} e^{iu} + q_k^{(n)}\right] = \prod_{k=1}^{n}\left[1 + p_k^{(n)}(e^{iu} - 1)\right]$$

Since by hypothesis $p_k^{(n)} \xrightarrow{n} 0$, from the series expansion of the logarithm we have

$$\ln \varphi_{S_n}(u) = \sum_{k=1}^{n} \ln\left[1 + p_k^{(n)}(e^{iu} - 1)\right] = \sum_{k=1}^{n}\left[p_k^{(n)}(e^{iu} - 1) + o(p_k^{(n)})\right] \xrightarrow{n} \alpha(e^{iu} - 1)$$

and given the continuity of the logarithm

$$\varphi_{S_n}(u) \xrightarrow{n} e^{\alpha(e^{iu}-1)}$$

Recalling then (4.9), from the Theorem 4.16 we find $S_n \xrightarrow{d} \mathfrak{P}(\alpha)$. ■

Theorem 4.33 *If $S \sim \mathfrak{P}(\alpha)$ is a Poisson rv, then*

$$S^* = \frac{S-\alpha}{\sqrt{\alpha}} \xrightarrow{d} \mathfrak{N}(0,1) \qquad \alpha \to +\infty$$

Proof If φ_α is the *chf* of S^*, from (4.9) and from the series expansion of an exponential we find for $\alpha \to +\infty$

$$\begin{aligned}\varphi_\alpha(u) &= \boldsymbol{E}\left[e^{iuS^*}\right] = e^{-iu\sqrt{\alpha}}\boldsymbol{E}\left[e^{iuS/\sqrt{\alpha}}\right] \\ &= \exp\left[-iu\sqrt{\alpha} + \alpha\left(e^{iu/\sqrt{\alpha}} - 1\right)\right] \\ &= \exp\left[-iu\sqrt{\alpha} - \alpha + \alpha\left(1 + \frac{iu}{\sqrt{\alpha}} - \frac{u^2}{2\alpha} + o\left(\frac{1}{\alpha}\right)\right)\right] \to e^{-u^2/2}\end{aligned}$$

The result follows then from the Theorem 4.16. ∎

4.6 Where the Classical Limit Theorems Fail

The results presented in this chapter are also known on the whole as the *classical* limit theorems and are rather general statements about the limit behavior of sums of independent *rv*'s, but we must not misread them by supposing that they can be applied in a totally indiscriminate manner. In particular we should be careful in checking their hypotheses, chiefly that about the required moments $\boldsymbol{E}[X_n]$ and $\boldsymbol{E}\left[X_n^2\right]$. To this end we will briefly discuss an example showing that problems can arise even in fairly ordinary contexts.

Example 4.34 Let us suppose, as in the Fig. 4.3, that a light beam from a source in A hit a mirror in C at a distance a free to wobble around a stud. The mirror position is taken at random in the sense that the reflection angle Θ is a uniform *rv* with distribution $\mathfrak{U}\left(-\frac{\pi}{2}, \frac{\pi}{2}\right)$. If now $X = a\tan\Theta$ is the distance from A of the point B where the reflected beam hit back the wall, it is easy to show that X follows a Cauchy law $\mathfrak{C}(a, 0)$: we have indeed

$$f_\Theta(\theta) = \begin{cases} 1/\pi & \text{if } |\theta| \le \pi/2 \\ 0 & \text{if } |\theta| > \pi/2 \end{cases}$$

while X results from Θ through the function $x = g(\theta) = a\tan\theta$ monotonic on $(-^{\pi}/_2, {}^{\pi}/_2)$. As a consequence, with

$$\theta_1(x) = g^{-1}(x) = \arctan\frac{x}{a} \qquad \theta_1'(x) = \frac{a}{a^2+x^2}$$

the transformation rule (3.60) entails that the law of X is the Cauchy $\mathfrak{C}(a, 0)$ with *pdf*

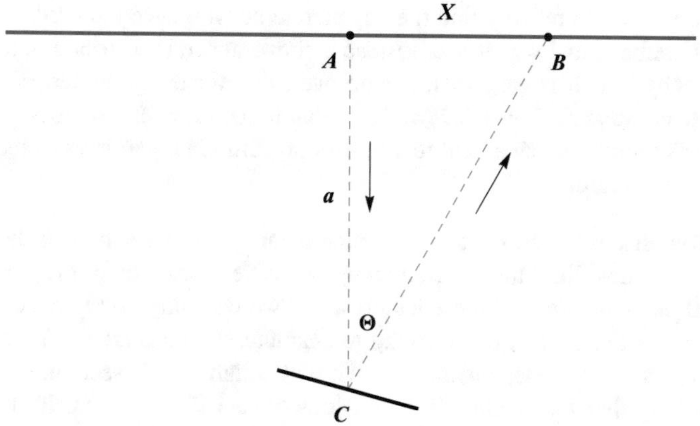

Fig. 4.3 How to produce a Cauchy *rv*

$$f_X(x) = f_\Theta\big(\theta_1(x)\big)|\theta_1'(x)| = \frac{1}{\pi}\frac{a}{a^2+x^2}$$

because f_Θ takes the constant value $^1/_\pi$ in $\left[-^\pi/_2, {}^\pi/_2\right]$, while apparently $\theta_1(x) \in \left(-^\pi/_2, {}^\pi/_2\right)$. This simple example shows that a Cauchy law can turn up in a natural way in realistic contexts, even if—as we are going to prove below—its behavior stands apart from that expected according to the limit theorems.

Let us suppose, indeed, to replicate a large number of independent measurements of X to get a sequence $(X_n)_{n\in N}$ of iid Cauchy *rv*'s with law $\mathfrak{C}(a, 0)$. From (4.16) we know that their *chf* is

$$\varphi(u) = \boldsymbol{E}\left[e^{iuX_n}\right] = e^{-a|u|}$$

so that, with $S_n = X_1 + \cdots + X_n$, the *chf* of the average S_n/n for every n is

$$\varphi_n(u) = \boldsymbol{E}\left[e^{iu\frac{S_n}{n}}\right] = \boldsymbol{E}\left[\prod_{k=0}^{n} e^{iu\frac{X_k}{n}}\right] = \left[\varphi\left(\frac{u}{n}\right)\right]^n = \left(e^{-\frac{a|u|}{n}}\right)^n = e^{-a|u|} = \varphi(u)$$

We then (trivially) have that

$$\varphi_n(u) \xrightarrow{n} \varphi(u) \qquad \forall u \in \boldsymbol{R}$$

and hence from the Lévy Theorem 4.16 it follows that

$$\frac{S_n}{n} \xrightarrow{d} X \sim \mathfrak{C}(a, 0)$$

In other words we in no way find the degenerate convergence required by the Law of Large Numbers, and we recover instead a convergence in distribution toward the initial Cauchy law. It is easy to see, moreover, that for every numerical sequence λ_n we find anyways $\lambda_n S_n \sim \mathfrak{C}(n\lambda_n a, 0)$, so that under no circumstances sums of *iid* Cauchy *rv*'s seem to show a bent to converge toward Gaussian laws as required by the Gaussian Theorems.

The previous discussion shows that our counterexample is indeed outside the jurisdiction of the classical limit theorems: a comprehensive discussion of this point would be beyond the boundaries of these lectures, and we will only briefly raise again this point later (see Sect. 7.1.3) while trying to characterize *the class of all the possible limit laws* of sums of independent *rv*'s. We will conclude this section, however, by just remarking that the seemingly anomalous behavior of the Cauchy *rv*'s in the Example 4.34 is essentially due to their ***non-compliance with the hypotheses*** of the classical limit theorems. As already remarked in the Example 3.25, the expectation of a Cauchy *rv* $X \sim \mathfrak{C}(a, 0)$ is not defined, while the existence of $\boldsymbol{E}[X]$ (in the sense of the Lebesgue integral) is a mandatory requirement that plays a crucial role in the proofs of both the Gaussian Theorems and the Laws of Large Numbers.

References

1. Shiryaev, A.N.: Probability. Springer, New York (1996)
2. Bouleau, N.: Probabilités de l'Ingénieur. Hermann, Paris (1986)
3. Feller, W.: An Introduction to Probability Theory and its Applications, vol. I-II. Wiley, New York (1968 and 1971)
4. Loève, M.: Probability Theory, vols. I-II. Springer, New York (1977 and 1978)
5. Poisson, S.D.: Recherches sur la ProbabilitÉ des Jugements en MatiÉre Criminelle et en MatiÉre Civile. Bachelier, Paris (1837)
6. Papoulis, A.: Probability, Random Variables and Stochastic Processes. McGraw Hill, Boston (2002)

Part II
Stochastic Processes

Chapter 5
General Notions

The notion of stochastic process (*sp*) $X(t) = X(\omega; t)$ on a probability space $(\Omega, \mathcal{F}, \boldsymbol{P})$ with $t > 0$ has already been introduced in the Sect. 3.2 where we pointed out that it can be considered from two complementary standpoints:

- as an application that to every given $t > 0$ associates a *rv* $X(\omega; t) = X(t)$ that represent the *state* of the system at the time t; in this sense the *sp* is a family of *rv*'s parameterized by t;
- as an application that to every given $\omega \in \Omega$ associates a whole *trajectory* (*sample*) $X(\omega; t) = x(t)$ of the process; in this second sense the *sp* consists of the set of all its possible trajectories.

These two perspectives are essentially equivalent and are adopted according to the needs. It will be expedient to remark immediately, however, that t is here considered as a *time* just to fix ideas: every family of *rv*'s $X(\alpha)$ classified by one or more parameters α is a *sp*: the instinctual view that the parameter t is a time originates only from the routine examples used to present this notion. As a rule our *sp*'s will be defined for $t \geq 0$, but the possibility of $t \in \boldsymbol{R}$ (*bilateral processes*, see also Appendix N) is not excluded. As we will see later, moreover, a *sp* can have more than one component: $\boldsymbol{X}(t) = \big(X_1(t), \ldots, X_M(t)\big)$, but for simplicity's sake we will initially confine ourselves just to the case $M = 1$. In the following we will also look at the ***increments*** $X(s) - X(t)$ of $X(t)$ on an interval $[t, s]$, taking often advantage of the notation $\Delta X(t) = X(t + \Delta t) - X(t)$ with $\Delta t = s - t > 0$, and at the ***increment process*** $\Delta X(t)$ with varying t and a fixed $\Delta t > 0$.

5.1 Identification and Distribution

As we know, to every *sp* it is possible to associate a hierarchy of finite dimensional laws that – as stated in the Kolmogorov Theorem 2.37—uniquely determine the ***global law of the sp***: given a finite number n of arbitrary instants $t_1, \ldots, t_n$, consider

N. Cufaro Petroni, *Probability and Stochastic Processes for Physicists*, UNITEXT for Physics, https://doi.org/10.1007/978-3-030-48408-8_5

all the joint laws of the *r-vec*'s $(X(t_1), \ldots, X(t_n))$ that can be represented through either their *cdf*'s or their *chf*'s

$$F(x_1, t_1; \ldots; x_n, t_n) \qquad \varphi(u_1, t_1; \ldots; u_n, t_n)$$

In the following, however, we will also use either their discrete distributions (usually with integer values $k, \ell \ldots$), or – when *ac*– their *pdf*'s respectively with the notations

$$p(k_1, t_1; \ldots; k_n, t_n) \qquad f(x_1, t_1; \ldots; x_n, t_n)$$

Taken then in the same way m other instants $s_1, \ldots, s_m$, we can also introduce the conditional *cdf*'s, probabilities and *pdf*'s according to the notations of Sect. 3.4.1.

$$\begin{aligned} F(x_1, t_1; \ldots; x_n, t_n | \, y_1, s_1; \ldots; y_m, s_m) &= \boldsymbol{P}\{X(t_1) \leq x_1 \ldots | \, X(s_1) = y_1 \ldots\} \\ p(k_1, t_1; \ldots; k_n, t_n | \, \ell_1, s_1; \ldots; \ell_m, s_m) &= \frac{p(k_1, t_1; \ldots; k_n, t_n; \ell_1, s_1; \ldots; \ell_m, s_m)}{p(\ell_1, s_1; \ldots; \ell_m, s_m)} \\ f(x_1, t_1; \ldots; x_n, t_n | \, y_1, s_1; \ldots; y_m, s_m) &= \frac{f(x_1, t_1; \ldots; x_n, t_n; y_1, s_1; \ldots; y_m, s_m)}{f(y_1, s_1; \ldots; y_m, s_m)} \end{aligned}$$

For the time being the ordering of the t_i, s_j is immaterial, but it is usually supposed that

$$t_n \geq \ldots \geq t_1 \geq s_m \geq \ldots \geq s_1 \geq 0$$

In particular we will call ***transition pdf's and probabilities*** the two-instant conditional *pdf*'s and probabilities $f(x, t|y, s)$ and $p(k, t|\ell, s)$

We can now go on to define in what sense we can speak of ***equality of two sp's*** $X(t)$ and $Y(t)$:

1. $X(t)$ and $Y(t)$ are said to be *indistinguishable* (or even *identical* $\boldsymbol{P}$-a.s.) if all the trajectories coincide for every t, with the possible exception of a negligible set of them, namely if
$$\boldsymbol{P}\{X(t) = Y(t), \ \forall t > 0\} = 1$$
2. $X(t)$ e $Y(t)$ are said to be *equivalent* (and we also say that $X(t)$ is a *modification* of $Y(t)$ and viceversa) if instead for every given t the states of the processes coincide $\boldsymbol{P}$-a.s., namely if
$$\boldsymbol{P}\{X(t) = Y(t)\} = 1, \qquad \forall t > 0$$
3. $X(t)$ and $Y(t)$ are said to be *wide sense equivalent*, or even *equal in distribution* if all their finite dimensional joint laws (and hence their global laws) coincide0.

These three definitions are rather different: it is easy to see for instance that two indistinguishable processes are also equivalent, but the reverse does not hold. Taken indeed

$$N_t = \overline{\{X(t) = Y(t)\}} \qquad t > 0$$
$$N = \overline{\{X(t) = Y(t), \ \forall t > 0\}} = \bigcup_{t>0} N_t$$

the indistinguishability requires $\boldsymbol{P}\{N\} = 0$, and hence entails $\boldsymbol{P}\{N_t\} = 0, \ \forall t > 0$, namely leads to the equivalence. The simple equivalence, however, only requires $\boldsymbol{P}\{N_t\} = 0, \ \forall t > 0$, and that is not enough to have also $\boldsymbol{P}\{N\} = \boldsymbol{P}\{\bigcup N_t\} = 0$ because the set of the instants $t > 0$ is uncountable. In the same way it is possible to show that equivalent *sp*'s also have the same finite dimensional joint distributions, but the reverse does not hold in general.

As already mentioned, the Kolmogorov Theorem 2.37 guarantees that the knowledge of the whole family of *consistent* finite dimensional *pdf*'s (let us suppose that our *sp* is *ac*) is enough to determine a probability measure on the whole space $(\boldsymbol{R}^T, \mathcal{B}(\boldsymbol{R}^T))$ with $T = [0, +\infty)$. We however also remarked in the point 4 of the Example 1.7 that $\mathcal{B}(\boldsymbol{R}^T)$ is not large enough for our needs: statements as *the process is continuous in a given instant*, for instance, require trajectories subsets that have no place in such a σ-algebra. We need then an extension of this probability space: without going into technical details, we will only recall here that this extension is always possible in an unambiguous way if our *sp* enjoys a property called *separability* (for further details see [1] p. 51), that we will refrain here from defining exactly. In this case in fact the finite dimensional *pdf*'s determine a probability measure extendable to all the trajectory subsets of practical interest, and what is more the required property is more a formal than a substantial limitation in the light of the following general result.

Theorem 5.1 *It is always possible to find a separable modification of a given sp: in other words, every sp is equivalent to some other separable sp.*

Proof Omitted: see [1] p. 57. ■

We will therefore always be able to behave as if all our *sp*'s are separable, and hence to suppose that the knowledge of all the finite dimensional *pdf*'s coherently determines the probability measure on a suitable trajectory space encompassing all the required events.

By generalizing the notion of a canonical *rv* presented in the Sect. 3.1.3, we will also say that, given – through a consistent family of joint finite dimensional laws – a probability $\boldsymbol{P}$ on a set $(\boldsymbol{R}^T, \mathcal{B}(\boldsymbol{R}^T))$ of trajectories, it will always be possible to find a *sp* having exactly $\boldsymbol{P}$ as its global law.

Definition 5.2 Given a probability $\boldsymbol{P}$ on the space $(\boldsymbol{R}^T, \mathcal{B}(\boldsymbol{R}^T))$ of the trajectories, we will call **canonical process** the sp $X(t)$ defined as the identical map from $(\boldsymbol{R}^T, \mathcal{B}(\boldsymbol{R}^T), \boldsymbol{P})$ to $(\boldsymbol{R}^T, \mathcal{B}(\boldsymbol{R}^T))$, whose distribution will coincide with the given $\boldsymbol{P}$.

These remarks explain why the *sp*'s (as for the *rv*'s) are essentially classified through their laws, even if to a given distribution can be associated many different *sp*'s sharing only their global law. Here you are an important class of *sp*'s:

Definition 5.3 We will say that $X(t)$ is a **Gaussian process** when, however taken $t_1, t_2, \dots, t_n$ with $n = 0, 1, \dots$, it is $(X(t_1), \dots, X(t_n)) \sim \mathfrak{N}(\boldsymbol{b}, \mathbb{A})$ where $\boldsymbol{b}$ and $\mathbb{A}$ are mean vectors and covariance matrices dependent on $t_1, t_2, \dots, t_n$.

5.2 Expectations and Correlations

A great deal of information about a *sp* $X(t)$ (with only one component for the time being) can be retrieved looking just to the expectations (when they exist) in one or more time instants, without providing all the details about the finite dimensional distributions. First it is expedient to introduce the ***expectation*** and the ***variance*** of a *sp* in every t:

$$m(t) = \boldsymbol{E}[X(t)] \qquad \sigma^2(t) = \boldsymbol{V}[X(t)] \tag{5.1}$$

Then the second order moments accounting for the correlations among the values of a *sp* in several time instants: to this end we define first the ***autocorrelation*** of the *sp* $X(t)$ as the function (symmetric in its two arguments)

$$R(s,t) = R(t,s) = \boldsymbol{E}[X(s)X(t)] \tag{5.2}$$

We must remark at once, however, that the word *correlation* is used here with a slightly different meaning w.r.t. what was done for two *rv*'s. We will in fact define also an ***autocovariance*** and a ***correlation coefficient*** of $X(t)$ respectively as

$$C(s,t) = \boldsymbol{cov}[X(s), X(t)] = R(s,t) - m(s)m(t) \tag{5.3}$$

$$\rho(s,t) = \rho[X(s), X(t)] = \frac{C(s,t)}{\sigma(s)\sigma(t)} \tag{5.4}$$

also noting that the autocovariance $C(s,t)$ includes the variance of the process because apparently

$$\boldsymbol{V}[X(t)] = \sigma^2(t) = C(t,t) \tag{5.5}$$

We will finally say that a *sp* is ***centered*** when

$$m(t) = 0 \qquad R(s,t) = C(s,t)$$

It is easy to see then that, given an arbitrary *sp* $X(t)$, the process

$$\widetilde{X}(t) = X(t) - m(t) = X(t) - \boldsymbol{E}[X(t)]$$

always turns out to be centered. The knowledge of the one- and two-times functions $m(t)$, $R(s,t)$, $C(s,t)$ and $\rho(s,t)$, albeit in general not exhaustive of the information pertaining to a *sp*, allows nonetheless to retrieve a fairly precise idea of its behavior, and in particular cases even a complete account of the process distribution.

5.3 Convergence and Continuity

The types of convergence for sequences of *rv*'s $(X_n)_{n\in N}$ presented in the Definition 4.1 can now be immediately extended to the *sp*'s in order to define the convergence of $X(t)$ toward some *rv* X_0 for $t \to t_0$. For instance the *mean square convergence* (*ms*)

$$X(t) \xrightarrow{ms} X_0 \qquad t \to t_0 \qquad \text{or else} \qquad \lim_{t\to t_0}\text{-}ms\; X(t) = X_0$$

will be defined as

$$\lim_{t\to t_0} \boldsymbol{E}\left[|X(t) - X_0|^2\right] = 0 \tag{5.6}$$

In the same way we can introduce the *convergences* $\boldsymbol{P}$-a.s.,*in probability, in* L^p *and in distribution* with the same mutual relations listed in the Theorem 4.4 for the sequences, and it is moreover possible to prove that the respective ***Cauchy convergence tests*** hold. It will be expedient finally to introduce a further notion: the convergence of a whole sequence of processes toward another process.

Definition 5.4 We will say that **a sequence of processes** $\{X_n(t)\}_{n\in N}$ **converges in distribution** toward the process $X(t)$ when, with $k = 1, 2, \ldots$, all the k-dimensional, joint distributions of the *r-vec's* $(X_n(t_1), \ldots, X_n(t_k))$ weakly converge toward the corresponding distributions of the *r-vec's* $(X(t_1), \ldots, X(t_k))$.

We are able now to introduce several notions of *process continuity* according to the adopted kind of convergence. It is important to make clear however that in this case we will be also obliged to discriminate between the continuity in a given, arbitrary instant t, and the global continuity of the process trajectories in every t.

Definition 5.5 Given a *sp* $X(t)$ with $t \geq 0$ on $(\Omega, \mathcal{F}, \boldsymbol{P})$ we will say that it is

- **continuous** $\boldsymbol{P}$-a.s., **in probability, in** L^p **or in distribution** when in every arbitrary, fixed $t \geq 0$, and for $s \to t$ it turns out that $X(s) \to X(t)$ $\boldsymbol{P}$-a.s., in probability, in L^p or in distribution respectively; a *sp* continuous in probability is also said **stochastically continuous**;
- **sample continuous** when almost every trajectory is continuous in every $t \geq 0$, that is if

$$\boldsymbol{P}\{\omega \in \Omega : x(t) = X(t;\omega) \text{ is continuous } \forall t \geq 0\} = 1$$

The sample continuity of the second point must not be misinterpreted as the $\boldsymbol{P}$-a.s. continuity of the previous point: a *sp* is $\boldsymbol{P}$-a.s. continuous if every instant t is almost surely a continuity point; it turns instead to be sample continuous if the set of the trajectories that are not continuous even in a single point t has zero probability. All these different varieties of continuity are not equivalent, but comply rather with the same kind of implications listed in the Theorem 4.4 for the sequences of *rv*'s. In particular the continuity in L^2 (in *ms*) is sufficient to entail the stochastic continuity. It is then useful to remark that the *ms* continuity can be scrutinized by looking at the continuity properties of the autocorrelation functions as stated in the following proposition.

Proposition 5.6 *A sp $X(t)$ is continuous in ms–and hence is stochastically continuous—iff the autocorrelation $R(s,t)$ is continuous for $s = t$.*

Proof Since it is

$$\begin{aligned} \boldsymbol{E}\left[|X(s) - X(t)|^2\right] &= \boldsymbol{E}\left[X^2(s)\right] + \boldsymbol{E}\left[X^2(t)\right] - 2\boldsymbol{E}\left[X(s)X(t)\right] \\ &= R(t,t) + R(s,s) - 2R(s,t) \end{aligned}$$

by definition and from (5.6) we find that the *ms* continuity in t is equivalent to the continuity of $R(s,t)$ in $s = t$. ∎

The conditions for the sample continuity of a Markov process will be subsequently presented in the Sect. 7.1.7; here it will be enough to add only that the sample continuity apparently entails all the other types of continuity listed above.

5.4 Differentiation and Integration in *ms*

Even the integration and differentiation of a *sp* require suitable limit procedures and must then be defined according to the adopted type of convergence. In the subsequent chapters we will go in further details about these topics, and here we will begin by confining ourselves to look just to the *ms* convergence (entailing anyhow also that in probability). First, according to our definitions, a *sp* $X(t)$ will be ***differentiable in ms*** if it exists another process $\dot{X}(t)$ such that for every $t \geq 0$

$$\frac{X(t+\Delta t) - X(t)}{\Delta t} \xrightarrow{ms} \dot{X}(t) \qquad \Delta t \to 0$$

that is if

$$\lim_{\Delta t \to 0} \boldsymbol{E}\left[\left|\frac{X(t+\Delta t) - X(t)}{\Delta t} - \dot{X}(t)\right|^2\right] = 0 \tag{5.7}$$

We can then show that, as for the *ms* continuity, also the *ms* differentiability can be verified by looking at the properties of the process autocorrelation:

Proposition 5.7 *A sp $X(t)$ is ms differentiable in t, iff the second mixed derivative $R_{1,1} = \partial_s \partial_t R$ of its autocorrelation $R(s,t)$ exists in $s = t$. In this case we also have $\boldsymbol{E}\left[\dot{X}(t)\right] = \dot{m}(t)$.*

Proof By applying the Cauchy convergence test to the limit (5.7), the *ms* differentiability in t requires

$$\lim_{\Delta s, \Delta t \to 0} \boldsymbol{E}\left[\left|\frac{X(t+\Delta s) - X(t)}{\Delta s} - \frac{X(t+\Delta t) - X(t)}{\Delta t}\right|^2\right] = 0$$

Since on the other hand, when the limits exist, we have

$$\begin{aligned}
&\lim_{\Delta s, \Delta t \to 0} \boldsymbol{E}\left[\frac{X(t+\Delta t) - X(t)}{\Delta t}\,\frac{X(t+\Delta s) - X(t)}{\Delta s}\right] \\
&\quad = \lim_{\Delta s, \Delta t \to 0} \frac{R(t+\Delta s, t+\Delta t) - R(t+\Delta s, t) - R(t, t+\Delta t) + R(t,t)}{\Delta t \Delta s} \\
&\quad = R_{1,1}(t,t) \\
&\lim_{\Delta t \to 0} \boldsymbol{E}\left[\left|\frac{X(t+\Delta t) - X(t)}{\Delta t}\right|^2\right] = \lim_{\Delta s \to 0} \boldsymbol{E}\left[\left|\frac{X(t+\Delta s) - X(t)}{\Delta s}\right|^2\right] \\
&\quad = \lim_{\Delta t \to 0} \frac{R(t+\Delta t, t+\Delta t) - R(t+\Delta t, t) - R(t, t+\Delta t) + R(t,t)}{\Delta t^2} \\
&\quad = R_{1,1}(t,t)
\end{aligned}$$

the Cauchy test is met, and $\dot{X}(t)$ exists, *iff* the derivative $R_{1,1}(t,t)$ exists because in this case

$$\begin{aligned}
&\lim_{\Delta s, \Delta t \to 0} \boldsymbol{E}\left[\left|\frac{X(t+\Delta s) - X(t)}{\Delta s} - \frac{X(t+\Delta t) - X(t)}{\Delta t}\right|^2\right] \\
&\qquad = R_{1,1}(t,t) - 2R_{1,1}(t,t) + R_{1,1}(t,t) = 0
\end{aligned}$$

That also $\boldsymbol{E}\left[\dot{X}(t)\right] = \dot{m}(t)$ holds is finally proved by checking the conditions to exchange limits and expectations. ■

As for the ***stochastic integrals***, that will be discussed in more detail in the Sect. 8.2, here we will just look at those of the type

$$\int_a^b X(t)\,dt \tag{5.8}$$

that at any rate—if they can be established in some suitable sense—apparently define a new *rv*. For the time being we will take them as defined through a Riemann procedure by looking at their convergence in *ms*, by postponing to a subsequent chapter a more detailed discussion of the stochastic integration in general. Taken to this end a partition of $[a, b]$ in intervals of width Δt_j, the arbitrary points τ_j belonging to every j^{th} interval, and $\delta = \max\{\Delta t_j\}$, we will say that the integral (5.8) exists in *ms* if it exists the limit.

$$\lim_{\delta \to 0}\text{-}ms \sum_j X(\tau_j)\,\Delta t_j \tag{5.9}$$

and its value is independent from the choice of the point τ_j inside the decomposition intervals.

Proposition 5.8 *The integral* (5.8) *exists in ms iff the autocorrelation* $R(s,t)$ *of* $X(t)$ *is integrable, that is*

$$\left|\int_a^b\int_a^b R(s,t)\,ds\,dt\right| < +\infty$$

In this case we also have

$$\int_a^b\int_a^b R(s,t)\,ds\,dt = \boldsymbol{E}\left[\left|\int_a^b X(t)\,dt\right|^2\right] \geq 0 \tag{5.10}$$

Proof According to the Cauchy convergence test the limit (5.9) exists if

$$\lim_{\gamma,\delta\to 0} \boldsymbol{E}\left[\left|\sum_j X(\rho_j)\Delta s_j - \sum_k X(\tau_k)\Delta t_k\right|^2\right] = 0$$

Since on the other hand, when the limits exist, we have

$$\begin{aligned}
\lim_{\gamma,\delta\to 0} \boldsymbol{E}\left[\sum_j X(\rho_j)\Delta s_j \cdot \sum_k X(\tau_k)\Delta t_k\right] &= \lim_{\gamma,\delta\to 0}\sum_{j.k} R(\rho_j,\tau_k)\,\Delta s_j\Delta t_k \\
&= \int_a^b\int_a^b R(s,t)\,ds\,dt \\
\lim_{\gamma\to 0} \boldsymbol{E}\left[\sum_j X(\rho_j)\Delta s_j \cdot \sum_i X(\rho_i)\Delta s_i\right] &= \lim_{\delta\to 0} \boldsymbol{E}\left[\sum_\ell X(\tau_\ell)\Delta t_\ell \cdot \sum_k X(\tau_k)\Delta t_k\right] \\
&= \lim_{\delta\to 0}\sum_{\ell.k} R(\tau_\ell,\tau_k)\,\Delta t_\ell\Delta t_k \\
&= \int_a^b\int_a^b R(s,t)\,ds\,dt
\end{aligned}$$

again the *ms* integrability of $X(t)$ coincides with the integrability of $R(s,t)$. The result (5.10) is then proved by checking the conditions to exchange limits and expectations. ■

5.5 Stationarity and Ergodicity

Definition 5.9 We will say that $X(t)$ is a **stationary process (strict-sense)** when however taken $s \in \boldsymbol{R}$ the two *sp*'s $X(t)$ and $X(t+s)$ are equal in distribution. We will instead say that the process has **stationary increments** when, for a given $\Delta t > 0$, it is

$$\Delta X(t) \stackrel{d}{=} \Delta X(s) \qquad \forall s, t \in \boldsymbol{R}$$

In other words, the global law of a stationary process must be invariant under arbitrary changes in the origin of the times, namely (if the process is *ac*) for every n, and for every choice of $t_1, \ldots, t_n$ and s we must find

$$f(x_1, t_1; \ldots; x_n, t_n) = f(x_1, t_1 + s; \ldots; x_n, t_n + s) \tag{5.11}$$

In particular from (5.11) it follows first that the one-time *pdf* of a stationary process must be constant in time, and second that its joint, two-times *pdf* must depend only on the time differences, that is:

$$f(x, t) = f(x) \tag{5.12}$$

$$f(x_1, t_1; x_2, t_2) = f(x_1, x_2; \tau) \qquad \tau = t_2 - t_1 \tag{5.13}$$

Remark on the other hand that the stationarity of the increments $\Delta X(t)$ only requires that their laws depend in fact on Δt, but not on t: this does not imply the stationarity of the *increment process* (that would require conditions also on the joint laws of the increments), even less that of the process $X(t)$ itself. Conversely the stationarity of a process $X(t)$ entails the stationarity of the increments as stated in the following proposition.

Proposition 5.10 *A stationary process $X(t)$ has stationary increments $\Delta X(t)$*

Proof We have indeed for the *cdf* of the increments $\Delta X(t)$ of width τ

$$\begin{aligned} F_{\Delta X}(x, t) &= \boldsymbol{P}\{\Delta X(t) \leq x\} = \boldsymbol{E}\left[\, \boldsymbol{P}\{\Delta X(t) \leq x \mid X(t)\}\,\right] \\ &= \int \boldsymbol{P}\{X(t+\tau) - X(t) \leq x \mid X(t) = y\}\, f(y)\, dy \\ &= \int \boldsymbol{P}\{X(t+\tau) \leq x + y \mid X(t) = y\}\, f(y)\, dy \\ &= \int F(x+y, t+\tau \mid y, t) f(y)\, dy \end{aligned}$$

so that differentiating and taking (5.13) in to account we get the *pdf*

$$f_{\Delta X}(x) = \int f(x+y, t+\tau \mid y, t) f(y)\, dy = \int f(x+y, y; \tau)\, dy \tag{5.14}$$

that depends only on τ while being independent from t: namely the increments are stationary. ∎

The expectation and the variance of a stationary process are patently constant, while the autocorrelation depends only on the time difference, that is

$$\boldsymbol{E}[X(t)] = m \qquad \boldsymbol{E}[X(t)X(t+\tau)] = R(\tau) \qquad \forall t \geq 0 \tag{5.15}$$

$$C(\tau) = R(\tau) - m^2 \qquad \sigma^2(t) = \sigma^2 = C(0) = R(0) - m^2 \qquad \rho(\tau) = \frac{R(\tau) - m^2}{R(0) - m^2}$$

and moreover, given the symmetry of $R(s, t)$ in its two arguments, the function $R(\tau)$ will turn out to be *even*. Remark however that, while the relations (5.15) are always true for a stationary process, the reverse does not hold: the conditions (5.15) alone are not enough to entail the (strict-sense) stationarity of the Definition 5.9. Given nevertheless their importance, when a process meets at least the conditions (5.15) it is usually said to be ***wide-sense stationary***.

Proposition 5.11 *A wide-sense stationary process $X(t)$ with autocorrelation $R(\tau)$ turns out to be (1) ms-continuous if $R(\tau)$ is continuous in $\tau = 0$; (2) ms-differentiable if the second derivative $R''(\tau)$ exists in $\tau = 0$; and (3) ms-integrable on $[-T, T]$ if*

$$\int_{-2T}^{2T} (2T - |\tau|)R(\tau)\, d\tau < +\infty$$

The abridged conditions for generic, non symmetric integration limits $[a, b]$ are more involved and will be left aside.

Proof These results are corollaries of the Propositions 5.6, 5.7 and 5.8. As for the integrability condition it follows from the Proposition 5.8 with a change of variables and an elementary integration according to a procedure that will be employed in the proof of the subsequent Theorem 5.12. ∎

For a stationary process it seems reasonable—as a sort of extension of the Law of Large Numbers—to surmise that its *expectations* could be replaced with some kind of limit on *time averages* along the trajectories. When this actually happens we say that the process is—in some suitable sense to be specified—***ergodic***. We will survey now the conditions sufficient to entail the ***ergodicity for the expectations and the autocorrelations*** of a wide-sense stationary process. To this end we preliminarily define, for an arbitrary $T > 0$, the *rv*'s

$$\overline{X}_T = \frac{1}{2T}\int_{-T}^{T} X(t)\, dt \qquad R_T(\tau) = \frac{1}{T}\int_{0}^{T} X(t)X(t+\tau)\, dt$$

namely the *time averages* of both the *sp* and its products at different times as first introduced by da G.I. Taylor in 1920, and for further convenience the function

$$r(\tau, \sigma) = \boldsymbol{E}\left[X(t+\sigma+\tau)X(t+\sigma)X(t+\tau)X(t)\right] - R^2(t)$$

Theorem 5.12 *For a wide-sense stationary sp $X(t)$ we find*

$$\lim_{T\to\infty}\text{-}ms\ \overline{X}_T = m \qquad\qquad \lim_{T\to\infty}\text{-}ms\ R_T(\tau) = R(\tau) \tag{5.16}$$

when the following conditions are respectively met

$$\lim_{T\to\infty} \frac{1}{T}\int_{-2T}^{2T}\left(1-\frac{|\tau|}{2T}\right)C(\tau)\,d\tau = 0 \tag{5.17}$$

$$\lim_{T\to\infty} \frac{1}{T}\int_{-2T}^{2T}\left(1-\frac{|\sigma|}{2T}\right)r(\tau,\sigma)\,d\sigma = 0 \tag{5.18}$$

We say then that the sp is ***expectation and autocorrelation ergodic****.*

Proof To prove the degenerate *ms*-limits (5.16) we can adopt the tests (4.1) from the Theorem 4.6 that for the expectation ergodicity read

$$\lim_{T\to\infty} \boldsymbol{E}\left[\overline{X}_T\right] = m \qquad\qquad \lim_{T\to\infty} \boldsymbol{V}\left[\overline{X}_T\right] = 0 \tag{5.19}$$

Neglecting once more to check that we can exchange expectations and integrations, the first condition in (5.19) is trivially fulfilled because for every $T > 0$ it is

$$\boldsymbol{E}\left[\overline{X}_T\right] = \frac{1}{2T}\int_{-T}^{T}\boldsymbol{E}\left[X(t)\right]dt = \frac{m}{2T}\int_{-T}^{T}dt = m$$

As for the second condition (5.19) we remark that

$$\begin{aligned}\boldsymbol{V}\left[\overline{X}_T\right] &= \boldsymbol{E}\left[\overline{X}_T^2\right] - \boldsymbol{E}\left[\overline{X}_T\right]^2 = \frac{1}{4T^2}\boldsymbol{E}\left[\int_{-T}^{T}\int_{-T}^{T}X(s)X(t)\,ds dt\right] - m^2 \\ &= \frac{1}{4T^2}\int_{-T}^{T}\int_{-T}^{T}\left[R(t-s)-m^2\right]ds dt = \frac{1}{4T^2}\iint_D C(t-s)\,ds dt\end{aligned}$$

and that with the following change of integration variables (see Fig. 5.1)

$$\tau = t - s \qquad\qquad \sigma = s \qquad\qquad |J| = 1 \tag{5.20}$$

we have

$$\begin{aligned}\boldsymbol{V}\left[\overline{X}_T\right] &= \frac{1}{4T^2}\iint_D C(t-s)\,ds dt = \frac{1}{4T^2}\iint_\Delta C(\tau)\,d\sigma d\tau \\ &= \frac{1}{4T^2}\left[\int_{-2T}^{0}d\tau\ C(\tau)\int_{-T-\tau}^{T}d\sigma + \int_{0}^{2T}d\tau\ C(\tau)\int_{-T}^{T-\tau}d\sigma\right]\end{aligned}$$

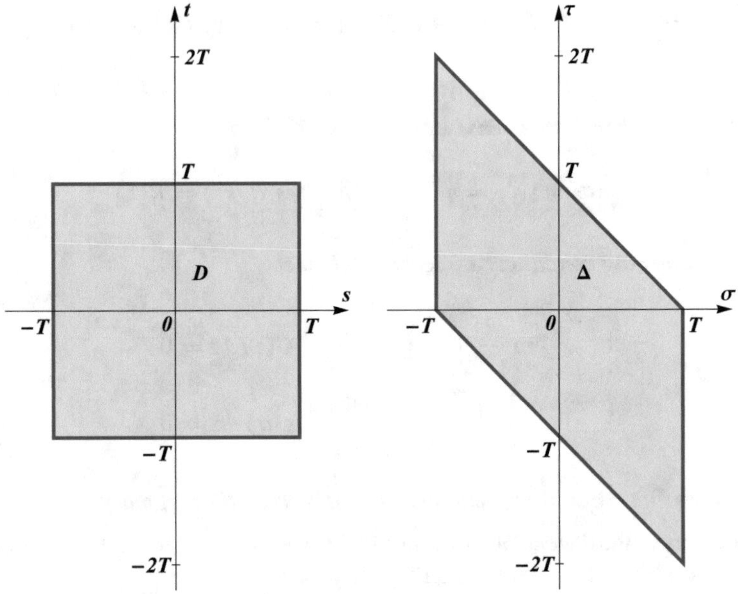

Fig. 5.1 Transformation of the integration domain produced by the change of variables (5.20)

$$= \frac{1}{4T^2}\left[\int_{-2T}^{0} C(\tau)(2T+\tau)\,d\tau + \int_{0}^{2T} C(\tau)(2T-\tau)\,d\tau\right]$$
$$= \frac{1}{2T}\int_{-2T}^{2T} C(\tau)\left(1-\frac{|\tau|}{2T}\right)d\tau$$

and hence the second relation in (5.19) holds if the hypothesis (5.17) is met. We leave aside instead the similar explicit proof of the autocorrelation ergodicity. ■

Corollary 5.13 *A wide-sense stationary sp $X(t)$ is expectation and autocorrelation ergodic if*

$$\int_0^{+\infty} |C(\tau)|\,d\tau < +\infty \tag{5.21}$$

Proof Since it is

$$\left|1-\frac{|\tau|}{2T}\right| \le 1 \qquad -2T \le \tau \le 2T$$

and $C(\tau)$ is an even function we easily find

$$\left|\frac{1}{T}\int_{-2T}^{2T}\left(1-\frac{|\tau|}{2T}\right)C(\tau)\,d\tau\right| \le \frac{1}{T}\int_{-2T}^{2T}|C(\tau)|\,d\tau = \frac{2}{T}\int_0^{2T}|C(\tau)|\,d\tau$$

so that (5.17) is met if (5.21) holds. We leave aside instead the more lengthy proof for the autocorrelation ergodicity. ■

5.6 Power Spectrum

We could hope of performing a frequency analysis of a *sp* $X(t)$ (typically when $X(t)$ is a *signal*) simply by calculating the Fourier transform of its trajectories, but it is easy to see that in general these sample functions are not square integrable on $[0, +\infty)$, so that the transform can not be calculated in a direct way. We resort then first to a *truncated* transform

$$\widehat{X}_T(\varpi) = \int_0^T X(t)e^{-i\varpi t}dt \tag{5.22}$$

that exists for every $T > 0$ and formally is a new *sp* with parameter ϖ. Taking then inspiration from the idea that the square modulus of a Fourier transform represents the energetic share allotted to every component of the signal, we initially define the *power spectrum* simply as

$$S(\varpi) = \lim_{T\to\infty}\text{-}ms\ \frac{1}{T}\left|\widehat{X}_T(\varpi)\right|^2 \tag{5.23}$$

The resulting $S(\varpi)$ is in principle again a *sp* with parameter ϖ, and it is then a remarkable result the fact that, for stationary and ergodic *sp*'s, $S(\varpi)$ turns out instead to be a deterministic (non random) function that can be calculated as the Fourier transform of the process autocorrelation $R(\tau)$.

Theorem 5.14 Wiener-Khinchin Theorem: *Take a stationary sp $X(t)$: if it is ergodic – in the sense that it fulfills the hypotheses of the Theorem 5.12 – then it turns out that*

$$S(\varpi) = \lim_{T\to\infty}\text{-}ms\ \frac{1}{T}\left|\widehat{X}_T(\varpi)\right|^2 = \int_{-\infty}^{+\infty} R(\tau)\,e^{-i\varpi\tau}d\tau \tag{5.24}$$

Proof The conditions imposed are sufficient to entail both the limit convergence and the legitimacy of all the subsequent formal steps: we will not bother however to explicitly check these points, and we will rather confine ourselves to show how the result (5.24) basically comes out from the Theorem 5.12. From the definitions (5.22) and (5.23), and with the cange (5.20) of the integration variables, we first of all have

$$
\begin{aligned}
S(\varpi) &= \lim_{T\to\infty} \text{-}ms\, \frac{1}{T}\int_0^T\int_0^T X(t)X(s)e^{-i\varpi(t-s)}dtds \\
&= \lim_{T\to\infty} \text{-}ms\, \frac{1}{T}\left[\int_{-T}^0 d\tau \int_{-\tau}^T d\sigma e^{-i\varpi\tau} X(\sigma)X(\sigma+\tau)\right. \\
&\qquad \left. + \int_0^T d\tau \int_0^{T-\tau} d\sigma e^{-i\varpi\tau} X(\sigma)X(\sigma+\tau)\right] \\
&= \lim_{T\to\infty} \text{-}ms\, \frac{1}{T}\int_0^T d\tau \left[e^{i\varpi\tau}\int_\tau^T X(\sigma)X(\sigma-\tau)\,d\sigma\right. \\
&\qquad \left. + e^{-i\varpi\tau}\int_0^{T-\tau} X(\sigma)X(\sigma+\tau)\,d\sigma\right]
\end{aligned}
$$

and since with a further change of variables ($\sigma' = \sigma - \tau$) it is

$$
\int_\tau^T X(\sigma)X(\sigma-\tau)\,d\sigma = \int_0^{T-\tau} X(\sigma')X(\sigma'+\tau)\,d\sigma'
$$

We finally get

$$
S(\varpi) = \lim_{T\to\infty} \text{-}ms \int_0^T d\tau \cos\varpi\tau\, \frac{2}{T}\int_0^{T-\tau} X(\sigma)X(\sigma+\tau)\,d\sigma
$$

To simplify the proof we will suppose now that the limit in point can be performed in two subsequent distinct steps

$$
S(\varpi) = \lim_{T\to\infty} \int_0^T d\tau \cos\varpi\tau \lim_{T'\to\infty} \text{-}ms\, \frac{2}{T'}\int_0^{T'-\tau} X(\sigma)X(\sigma+\tau)\,d\sigma
$$

Going first to the limit $T' \to \infty$, with an arbitrary τ fixed in $[0, T]$, we can take advantage of the hypothesized autocovariance ergodicity (see the second limit (5.16) in the Theorem 5.12) in order to find

$$
S(\varpi) = \lim_{T\to\infty} 2\int_0^T R(\tau)\cos\varpi\tau\, d\tau = \int_{-\infty}^{+\infty} R(\tau)\, e^{-i\varpi\tau} d\tau
$$

where we also took into account the fact that $R(\tau)$ is a real, even function. ■

The practical relevance of this result has meant that over the years the propensity to consider (5.24) as the very definition of power spectrum has prevailed, and its origin has been completely neglected. We too we will conform to this almost universal standpoint by adopting the following definition that can be applied to every wide-sense stationary process.

Definition 5.15 Given a wide-sense stationary process $X(t)$ we call **power spectrum** the Fourier transform (when it exists) of its autocorrelation with the reciprocity relations

$$S(\varpi) = \int_{-\infty}^{+\infty} R(\tau)\, e^{-i\varpi\tau} d\tau \qquad\qquad R(\tau) = \frac{1}{2\pi}\int_{-\infty}^{+\infty} S(\varpi)\, e^{i\varpi\tau} d\varpi \tag{5.25}$$

Remark that the ergodicity condition (5.21) requires that $C(\tau)$ vanishes for $\tau \to \pm\infty$, namely that $X(t)$ and $X(t+\tau)$ become uncorrelated for large separations τ. From the definitions we then find $R(\tau) \to m^2$ for $\tau \to \pm\infty$, and hence the Fourier transform (5.25) does not exist if $m \neq 0$. To elude this snag it is customary to give also a second definition that resorts to the autocovariance $C(\tau)$ instead of the autocorrelation $R(\tau)$.

Definition 5.16 Given a wide-sense stationary process $X(t)$ we call **covariance spectrum** the Fourier transform of its autocovariance with the reciprocity relations

$$S_c(\varpi) = \int_{-\infty}^{+\infty} C(\tau)\, e^{-i\varpi\tau} d\tau \qquad\qquad C(\tau) = \frac{1}{2\pi}\int_{-\infty}^{+\infty} S_c(\varpi)\, e^{i\varpi\tau} d\varpi \tag{5.26}$$

References

1. Doob, J.L.: Stochastic Processes. Wiley, New York (1953)
2. Papoulis, A.: Probability, Random Variables and Stochastic Processes. McGraw Hill, Boston (2002)

Chapter 6
Heuristic Definitions

6.1 Poisson Process

At this stage of the presentation we will introduce the Poisson process by first explicitly producing its trajectories, and then by analyzing its probabilistic properties. This illuminating and informative procedure, however, can not be easily replicated for other typical, non trivial processes whose trajectories, as we will see later, can only be defined either as limits of suitable approximations or by adopting a more general standpoint.

6.1.1 Point Processes and Renewals

Take *at random* some instants on a time axis, and look into the la *rv* enumerating the points falling in an interval $[s, t]$ of width $\Delta t = t - s > 0$. In this formulation, however, the question is rather hazy and careless: first it should be said what *at random* means; then, if a point is taken at random (whatever this means, but for very special situations) on an *infinite axis* the probability of falling into a finite interval $[s, t]$ will be zero; finally the *number of points* must be specified. To find suitable answers we will follow a successive approximation procedure.

Proposition 6.1 *If on an infinite axis we cast at random (in a sense to be specified later) an infinite number of independent points, and if their average number for unit interval (**intensity**) is λ, the rv*

$$N = \text{ number of points falling in an interval } [s, t] \text{ of width } \Delta t = t - s$$

N. Cufaro Petroni, *Probability and Stochastic Processes for Physicists*, UNITEXT for Physics, https://doi.org/10.1007/978-3-030-48408-8_6

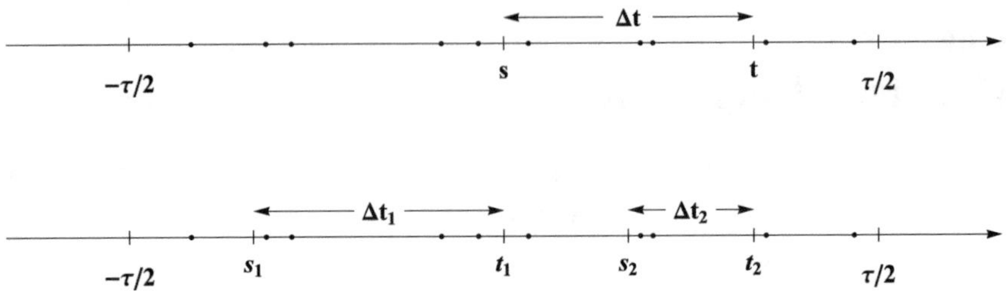

Fig. 6.1 Random instants taken on a finite interval $\left[-\frac{\tau}{2}, \frac{\tau}{2}\right]$

obeys to the Poisson distribution of parameter $\lambda\Delta t$*, namely* $N \sim \mathfrak{P}(\lambda\Delta t)$ *and*

$$\boldsymbol{P}\{N = k\} = e^{-\lambda\Delta t}\frac{(\lambda\Delta t)^k}{k!}$$

The analogous rv's $N_1 \sim \mathfrak{P}(\lambda\Delta t_1)$ *and* $N_2 \sim \mathfrak{P}(\lambda\Delta t_2)$ *for non superposed intervals are moreover independent.*

Proof In a step-by-step approach we start with a finite interval $[-\tau/2, \tau/2]$ with $\tau > 0$ including $[s, t]$ (see Fig. 6.1), and we cast n points *at random* in the sense that

- the position of a point in $[-\tau/2, \tau/2]$ is a *rv* independent from the other points
- the distribution of this random position is uniform in $[-\tau/2, \tau/2]$.

This precisely means that each of the n points will fall in $[s, t]$ with a probability

$$p = \frac{\Delta t}{\tau}$$

and since the n throws are independent, the *rv* $X =$ *number of points falling in* $[s, t]$ will be binomial $\mathfrak{B}\,(n;\, p)$. The law of X apparently depends on the arbitrary values of n and τ, and we will now drive both n and τ to the infinity, with constant Δt, requiring also that the ratio n/τ (number of points per unit interval) stay bounded and converges toward a positive number λ; namely we will suppose that

$$n \to \infty \qquad \tau \to +\infty \qquad \frac{n}{\tau} \to \lambda > 0 \tag{6.1}$$

$$p = \frac{\Delta t}{\tau} \to 0 \qquad np = \frac{n}{\tau}\,\Delta t \to \lambda\,\Delta t \tag{6.2}$$

With varying n and τ we then get a family of binomial *rv*'s $X \sim \mathfrak{B}\,(n;\, p)$ that in the limit (6.1) fulfil the conditions of the Poisson Theorem 4.30, and hence we will have (in distribution)

$$X \xrightarrow{d} N \sim \mathfrak{P}(\lambda\Delta t)$$

In conclusion: if we throw on an unbounded time axis an infinite number of independent points at random (in the sense specified above), and if the average number of points per unit interval is a constant $\lambda > 0$ (with the dimensions of a *frequency*), then the limit *rv*: $N =$ *number of points in an interval of width* Δt, follows a Poisson distribution $\mathfrak{P}(\lambda\Delta t)$, namely

$$\boldsymbol{P}\{N = k\} = e^{-\lambda\Delta t}\frac{(\lambda\Delta t)^k}{k!}$$

We retrace now again the same path, but taking, as in the Fig. 6.1, *two* disjoint intervals $[s_1, t_1]$ and $[s_2, t_2]$ in $[-\tau/2, \tau/2]$, with $\Delta t_1 = t_1 - s_1$ and $\Delta t_2 = t_2 - s_2$: if we cast n points with the same properties as before, and we take $X_1 =$ *number of points falling in* $[s_1, t_1]$, $X_2 =$ *number of points falling in* $[s_2, t_2]$, and $X_0 =$ *number of points falling elsewhere in* $[-\tau/2, \tau/2]$ we will find that the *r-vec* $\boldsymbol{X} = (X_1, X_2)$ follows a three-nomial (multinomial (3.1) with $r = 2$) distribution $\mathfrak{B}\,(n;\, p_1, p_2)$ where

$$p_1 = \frac{\Delta t_1}{\tau}, \qquad p_2 = \frac{\Delta t_2}{\tau}$$

If we now suppose as before that both n and τ grows to infinity complying with the conditions (6.1) and keeping constant Δt_1 and Δt_2, we will find

$$p_1 = \frac{\Delta t_1}{\tau} \to 0, \qquad np_1 = \frac{n}{\tau}\,\Delta t_1 \to \lambda\,\Delta t_1$$
$$p_2 = \frac{\Delta t_2}{\tau} \to 0, \qquad np_2 = \frac{n}{\tau}\,\Delta t_2 \to \lambda\,\Delta t_2$$

and hence according to the multinomial Poisson Theorem 4.31

$$\boldsymbol{X} = (X_1, X_2) \xrightarrow{d} (N_1, N_2) \sim \mathfrak{P}(\lambda\Delta t_1)\cdot\mathfrak{P}(\lambda\Delta t_2)$$

namely the two limit *rv*'s N_1 and N_2 will behave as two *independent* Poisson *rv*'s. Remark instead that, all along the limit procedure, for every finite n the *rv*'s X_1 and X_2 *are not* independent ■

In the Appendix G it will be shown how these results must be adapted when the point intensity λ is not constant. Here instead we will go on by introducing the sequence of *rv*'s T_n representing the time position of our random points. To this end we must establish an order among the points by choosing first an arbitrary non-random origin $T_0 = 0$, and setting then that T_1 is the instant of the first point *to the right* of the origin, T_2 that of the second and so on, while T_{-1} is that of the first *to the left* and so on. This produces a bilateral sequence T_n, with $n = 0, \pm1, \pm2, \ldots$, of *rv*'s that, however, *no longer are independent* because for one thing the point T_n can not come before T_{n-1}. We will see instead in the next proposition that the waiting times $\Delta T_n = T_{n+1} - T_n$ of the $(n+1)^{\text{th}}$ point are independent both from T_n and among themselves. The sequence of such T_n's that we will now briefly investigate is a typical example of ***point process*** while the waiting times ΔT_n' are called ***renewals***.

Proposition 6.2 *The T_n with $n \geq 1$, falling after $T_0 = 0$, are Erlang $\mathfrak{E}_n(\lambda)$ rv's; those falling before $T_0 = 0$ ($n \leq -1$) comply with the specular law: $T_{-n} \stackrel{d}{=} -T_n$. The waiting times $\Delta T_n = T_{n+1} - T_n$ are iid exponential $\mathfrak{E}(\lambda)$ rv's that are also independent from the respective T_n.*

Proof Take first the case $n \geq 1$ of the points falling *to the right* of (namely *after*) $T_0 = 0$: if N is the number of points in $[0, t]$ following the law $\mathfrak{P}(\lambda t)$, and $\vartheta(t)$ is the Heaviside function (2.13), the *cdf* of $T_n > 0$ will be

$$F_n(t) = \boldsymbol{P}\{T_n \leq t\} = \boldsymbol{P}\{N \geq n\} = 1 - \boldsymbol{P}\{N < n\} = \left[1 - e^{-\lambda t} \sum_{k=0}^{n-1} \frac{(\lambda t)^k}{k!}\right] \vartheta(t)$$

giving rise to the *pdf*

$$f_n(t) = F_n'(t) = \left[\sum_{k=0}^{n-1} \frac{(\lambda t)^k}{k!} - \sum_{k=1}^{n-1} \frac{(\lambda t)^{k-1}}{(k-1)!}\right] \lambda e^{-\lambda t} \vartheta(t) = \frac{(\lambda t)^{n-1}}{(n-1)!} \lambda e^{-\lambda t} \vartheta(t) \tag{6.3}$$

which coincides with the *pdf* (4.27) of an *Erlang distribution* $\mathfrak{E}_n(\lambda)$. In particular the law of T_1 is $\mathfrak{E}_1(\lambda)$, namely an exponential $\mathfrak{E}(\lambda)$ with *pdf* $f_1(t) = \lambda e^{-\lambda t} \vartheta(t)$. Similarly it is shown that $T_{-n} \stackrel{d}{=} -T_n$, that are by symmetry *reversed* Erlang distributions concentrated on the negative time axis: we will neglect to check that explicitly.

To study then the waiting times ΔT_n we will start by remarking that—because of the properties of the increments of the simple Poisson process $N(t)$—their conditional *cdf* is

$$\begin{aligned}
G_n(\tau \mid T_n = t) &= \boldsymbol{P}\{\Delta T_n \leq \tau \mid T_n = t\} = \boldsymbol{P}\{T_{n+1} \leq t + \tau \mid T_n = t\} \\
&= \boldsymbol{P}\{N(t+\tau) - N(t) \geq 1\} = 1 - \boldsymbol{P}\{N(t+\tau) - N(t) = 0\} \\
&= 1 - e^{-\lambda\tau}
\end{aligned}$$

namely it is an exponential $\mathfrak{E}(\lambda)$ dependent neither on n nor on t, and as a consequence it also coincides with the un-conditional *cdf* of ΔT_n

$$\begin{aligned}
G_n(\tau) &= \boldsymbol{P}\{\Delta T_n \leq \tau\} = \boldsymbol{E}\left[\boldsymbol{P}\{\Delta T_n \leq \tau \mid T_n\}\right] \\
&= \int_{-\infty}^{+\infty} \boldsymbol{P}\{\Delta T_n \leq \tau \mid T_n = t\} f_n(t)\, dt = (1 - e^{-\lambda\tau}) \int_{-\infty}^{+\infty} f_n(t)\, dt \\
&= 1 - e^{-\lambda\tau} = G_n(\tau \mid T_n = t)
\end{aligned}$$

showing that ΔT_n and T_n are independent. To prove finally that the ΔT_n also are mutually independent, remark first that apparently

$$T_n = T_1 + (T_2 - T_1) + \cdots (T_n - T_{n-1}) = \Delta T_0 + \Delta T_1 \cdots + \Delta T_{n-1}$$

with $T_n \sim \mathfrak{E}_n(\lambda)$ and $\Delta T_k \sim \mathfrak{E}(\lambda)$, and then that—according to the discussion in the Example 4.22—an Erlang $\mathfrak{E}_n(\lambda)$ *rv* always is decomposable into the sum of exponential $\mathfrak{E}(\lambda)$ *rv*'s when these are independent ■

The *rv*'s ΔT_n are a particular example of a sequence of *renewals*, namely of *iid rv*'s $Z_n > 0$, that can be used in their turn as the starting point to assemble a point process according to the reciprocal relations

$$T_n = \sum_{k=0}^{n-1} Z_k \qquad Z_n = \Delta T_n = T_{n+1} - T_n \tag{6.4}$$

As a rule every sequence of renewals Z_n produces a point process and vice versa, and it must also be said that in general the renewals can be distributed according to arbitrary laws different from $\mathfrak{E}(\lambda)$, provided that Z_n stay positive. It is important then to remark that a sequence of *exponential* renewals always produces both a point process T_n with Erlang laws (to this end see the Example 4.22), and numbers N of points falling into finite intervals distributed according to Poisson laws, as will be shown in the subsequent proposition.

Proposition 6.3 *Take a sequence $Z_n \sim \mathfrak{E}(\lambda)$ of exponential renewals $\mathfrak{E}(\lambda)$, and the corresponding point process T_n defined as in* (6.4) *and distributed according to the Erlang laws $\mathfrak{E}_n(\lambda)$: then the number N of the points T_n falling into* $[0, t]$ *is distributed according to the Poisson law $\mathfrak{P}(\lambda t)$.*

Proof With an exchange of the integration order on the integration domain D represented in the Fig. 6.2 we indeed find

$$\begin{aligned}
\boldsymbol{P}\{N = n\} &= \boldsymbol{P}\{T_n \le t, T_{n+1} > t\} = \boldsymbol{P}\{T_n \le t, T_n + Z_n > t\} \\
&= \boldsymbol{E}\left[\boldsymbol{P}\{T_n \le t, T_n + Z_n > t \mid Z_n\}\right] \\
&= \int_0^{+\infty} \boldsymbol{P}\{T_n \le t, T_n + Z_n > t \mid Z_n = z\}\lambda e^{-\lambda z}\,dz \\
&= \int_0^{+\infty} \boldsymbol{P}\{t - z < T_n \le t\}\lambda e^{-\lambda z}\,dz \\
&= \int_0^{+\infty} dz\,\lambda e^{-\lambda z} \int_{t-z}^{t} \frac{(\lambda s)^{n-1}}{(n-1)!}\lambda e^{-\lambda s}\vartheta(s)\,ds \\
&= \iint_D \lambda e^{-\lambda z} \frac{(\lambda s)^{n-1}}{(n-1)!}\lambda e^{-\lambda s}\vartheta(s)\,dz\,ds \\
&= \int_0^t ds\,\frac{(\lambda s)^{n-1}}{(n-1)!}\lambda e^{-\lambda s} \int_{t-s}^{+\infty} \lambda e^{-\lambda z}\,dz = \int_0^t \frac{(\lambda s)^{n-1}}{(n-1)!}\lambda e^{-\lambda s} e^{-\lambda(t-s)}\,ds \\
&= e^{-\lambda t}\frac{\lambda^n}{n!}\int_0^t n s^{n-1} ds = e^{-\lambda t}\frac{(\lambda t)^n}{n!}
\end{aligned} \tag{6.5}$$

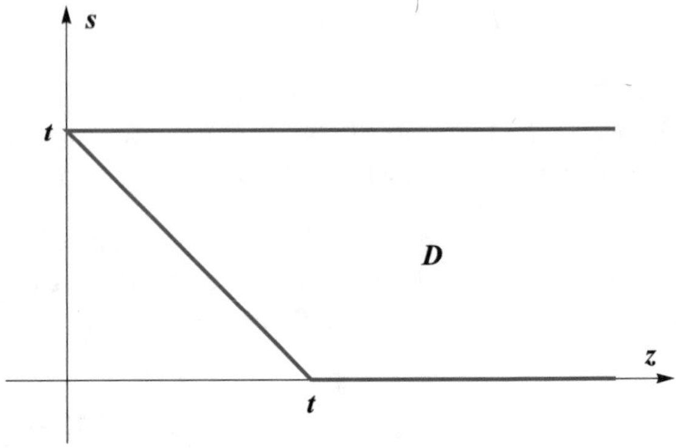

Fig. 6.2 Integration domain for the integral (6.5)

and hence N is distributed according to the Poisson law $\mathfrak{P}(\lambda t)$ as for the limit *rv*'s defined at the beginning of the present section ■

6.1.2 *Poisson Process*

Definition 6.4 Given a point process T_n of intensity λ, the **simple Poisson process of intensity** λ is the *sp* $N(t)$ with $t > 0$ **counting** the random number of points T_n falling in $[0, t]$, with the initial condition $N(0) = 0$, $\boldsymbol{P}$-a.s.. Taking advantage of the point process T_n the Poisson process $N(t)$ can also be represented as

$$N(t) = \sum_{k=1}^{\infty} \vartheta(t - T_k) \tag{6.6}$$

where ϑ is the Heaviside function (2.13). With a fixed $\Delta t > 0$ we can furthermore define the corresponding **process of the Poisson increments** $\Delta N(t) = N(t + \Delta t) - N(t)$ counting now the number of points T_n falling in an interval $[t, t + \Delta t]$

Proposition 6.5 *The Poisson process $N(t)$ has independent and stationary increments; the distributions and the chf's of $N(t)$ and $\Delta N(t)$ respectively are*

$$p_N(k, t) = \boldsymbol{P}\{N(t) = k\} = e^{-\lambda t}\frac{(\lambda t)^k}{k!} \tag{6.7}$$

$$p_{\Delta N}(k) = \boldsymbol{P}\{\Delta N(t) = k\} = e^{-\lambda \Delta t}\frac{(\lambda \Delta t)^k}{k!} \tag{6.8}$$

$$\varphi_N(u, t) = e^{\lambda t(e^{iu}-1)} \qquad \varphi_{\Delta N}(u, \Delta t) = e^{\lambda \Delta t(e^{iu}-1)} \tag{6.9}$$

while the ***transition probability*** *(namely the two-times conditional probability) of $N(t)$, with $\Delta t > 0$ and $k \geq \ell$, is*

$$p_N(k, t + \Delta t | \ell, t) = e^{-\lambda \Delta t} \frac{(\lambda \Delta t)^{k-\ell}}{(k-\ell)!} \tag{6.10}$$

Proof The increments on non-superposed intervals (with at most an extremal point in common) are independent *rv*'s by construction, and the increment $\Delta N(t)$ is also independent from $N(t)$: the Poisson process is then our first example of *independent increments process*, a class of *sp*'s that will be investigated in more detail in the Sect. 7.1.3. From the previous sections we know moreover that, for every $t > 0$, $N(t) \sim \mathfrak{P}(\lambda t)$, while $\Delta N(t) \sim \mathfrak{P}(\lambda \Delta t)$, and hence (6.7), (6.8) and (6.9) apparently hold. This in particular entails that the laws of the increments $\Delta N(t)$—at variance with the process $N(t)$ himself—do not change with t but depend only on Δt: as a consequence $N(t)$ has *stationary increments* (see also the Sect. 5.5). As for the transition probability, from the previous properties we finally have

$$\begin{aligned} p_N(k, t + \Delta t | \ell, t) &= \boldsymbol{P}\{N(t + \Delta t) = k \mid N(t) = \ell\} \\ &= \boldsymbol{P}\{N(t + \Delta t) - N(t) + N(t) = k \mid N(t) = \ell\} \\ &= \boldsymbol{P}\{\Delta N(t) = k - \ell\} = e^{-\lambda \Delta t} \frac{(\lambda \Delta t)^{k-\ell}}{(k-\ell)!} \end{aligned}$$

namely (6.10). Remark that this distribution depends only on Δt, but not on t, because of the increments stationarity (Sect. 5.5), and on $k - \ell$, but not separately on k and ℓ, because of the increments independence (see also the Sect. 7.1.3) ■

Proposition 6.6 *The main statistical properties of a simple Poisson process $N(t)$ are*

$$m_N(t) = \sigma_N^2(t) = \lambda t \tag{6.11}$$

$$R_N(s, t) = \lambda \min\{s, t\} + \lambda^2 st \tag{6.12}$$

$$C_N(s, t) = \lambda \min\{s, t\} \tag{6.13}$$

$$\rho_N(s, t) = \frac{\min\{s, t\}}{\sqrt{st}} = \begin{cases} \sqrt{s/t} & \text{if } s < t \\ \sqrt{t/s} & \text{if } t < s \end{cases} \tag{6.14}$$

Proof The results (6.11) immediately stem from (6.7). To prove (6.12) we start instead from the remark that from the previous results for $s = t$ it is

$$R_N(t, t) = \boldsymbol{E}\left[N^2(t)\right] = \boldsymbol{V}\left[N(t)\right] + \boldsymbol{E}\left[N(t)\right]^2 = \lambda t + \lambda^2 t^2$$

so that from the increments independence we will have for $s < t$

$$\begin{aligned} R_N(s, t) &= \boldsymbol{E}\left[N(s)N(t)\right] = \boldsymbol{E}\left[N(s)(N(t) - N(s) + N(s))\right] \\ &= \boldsymbol{E}\left[N(s)\right] \boldsymbol{E}\left[N(t) - N(s)\right] + R_N(s, s) \\ &= \lambda s \cdot \lambda(t - s) + \lambda s + \lambda^2 s^2 = \lambda s + \lambda^2 st \end{aligned}$$

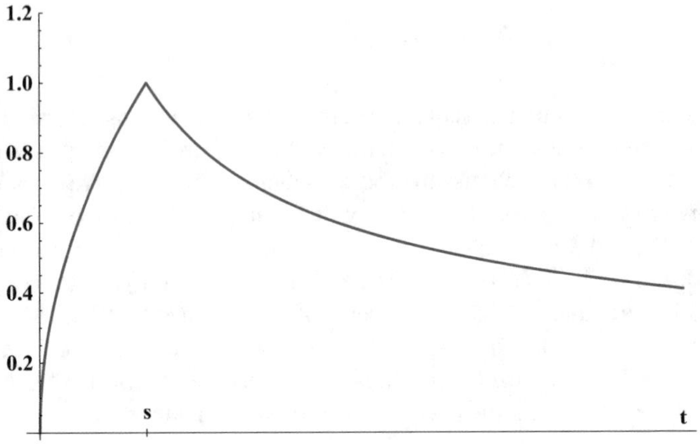

Fig. 6.3 The correlation coefficient $\rho_N(s,t)$ (6.14) of a simple Poisson process $N(t)$

and eventually the relation (6.12) for arbitrary s and t; (6.13) and (6.14) result then from the definition (5.3) and from (6.11). Notice that from (6.13) we also recover the variance (6.11) $\sigma_N^2(t) = C(t,t) = \lambda t$ ■

Remark that the variance in (6.11) linearly grows with the time: a property—sometimes also called *diffusion*—that is shared with other important processes. In the Fig. 6.3 (where, to fix the ideas, we kept s constant and t variable) the behavior of the correlation coefficient (6.14) is then displayed: the correlation apparently (slowly) decreases when s and t move away from each other, so that the process in t progressively forgets its state in s as time lapses away

The process $N(t)$ also enjoys a few other properties that we will find again later on: it is easy to check for instance that the distributions (6.7) are solutions of the equation

$$\partial_t p_N(n,t) = -\lambda\,[p_N(n,t) - p_N(n-1,t)] \qquad\qquad p_N(n,0) = \delta_{n0} \tag{6.15}$$

that is a first example of ***master equation***, an equation that we will study in more detail in the Sect. 7.2.3. Also the transition probabilities $p_N(k,t|\ell,s)$ in (6.10) turn out to be solutions of the same *master equation*, but for the different initial conditions $p_N(k,s^+) = \delta_{k\ell}$. From a power expansion of the exponential near $t = 0$ we then recover the following behaviors

$$p_N(n,t) = [1 - \lambda t + o(t)]\,\frac{(\lambda t)^n}{n!} = \begin{cases} 1 - \lambda t + o(t) & n = 0 \\ \lambda t + o(t) & n = 1 \\ o(t) & n \ge 2 \end{cases} \tag{6.16}$$

that constitute in fact a characteristic of the Poisson process: it would be possible to prove indeed that if the $p_N(n,t)$ of a *growth process* conforms to the (6.16), then its

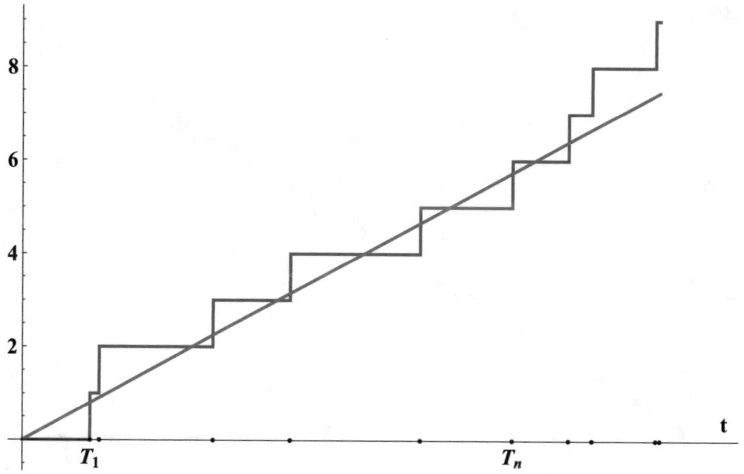

Fig. 6.4 An example of a 10 jumps trajectory of the Poisson process $N(t)$. In this plot also the point process T_n and the function $m_N(t) = \lambda t$ are displayed

laws are also solutions of the Eq. (6.15) and hence they must be of the Poisson type (6.7)

A typical ***trajectory*** of the Poisson process $N(t)$ for $t > 0$ is shown in the Fig. 6.4: it looks as infinite, climbing *stair* with a random steps length (ruled by the point process), and fixed unit height, representing the *enumeration* of the random points falling into the interval $[0, t]$. The relation between these trajectories and the linear function $m_N(t)$ is shown in the pictures: on a short time interval every trajectory in the Fig. 6.5 deviates little form the *average trend* and overall they are equally distributed

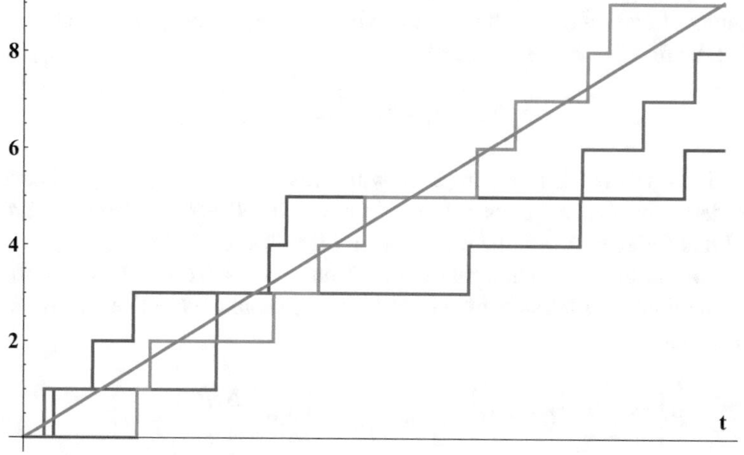

Fig. 6.5 A few trajectories of the Poisson process $N(t)$ scattered around its expectation $m_N(t) = \lambda t$

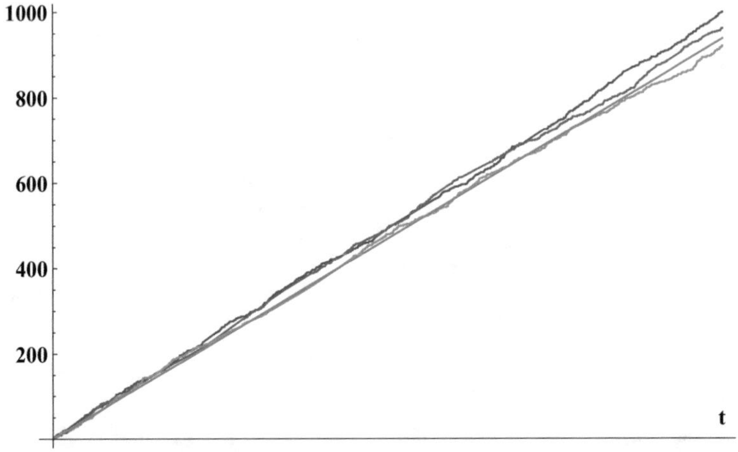

Fig. 6.6 Trajectories of a Poisson process $N(t)$ with 1 000 jumps. The standard deviation only grows as $\sqrt{\lambda t}$, and hence the paths appear to be little divergent from the average $m_N(t) = \lambda t$ because of a scale effect: $\sqrt{1\,000} \simeq 32$

around $m_N(t) = \lambda t$. For longer times, as in the Fig. 6.6, the trajectories continue to be equally distributed around λt, but they also progressively move away from it as an outcome of the variance growth. The fact that in the Fig. 6.6 the trajectories seem to be not too divergent from $m_N(t) = \lambda t$ depends mainly on a scale effect: the standard deviation only grows as $\sqrt{\lambda t}$, while of course the vertical axis scale goes up as λt.

Proposition 6.7 *The Poisson process $N(t)$ is ms-continuous, but not ms-differentiable in every $t > 0$; it is instead both $\boldsymbol{P}$-a.s.-continuous and $\boldsymbol{P}$-a.s.-differentiable for every $t > 0$, and in this sense we have $\dot{N}(t) = 0$. Finally $N(t)$ is not stationary.*

Proof According to the Proposition 5.6 the *ms*-continuity results from the continuity of the autocorrelation (6.12). It does not exist in $s = t$, instead, the mixed derivative $\partial_s \partial_t R_N$: the first derivative is indeed

$$\partial_t R_N(s, t) = \lambda \vartheta(s - t) + \lambda^2 s$$

where ϑ is the Heaviside function (2.13) with a discontinuity in $s = t$, and hence the second derivative $\partial_s \partial_t R_N$ does not exist in $s = t$. It follows then from the Proposition 5.7 that for every $t > 0$ the process is not *ms*-differentiable.

To prove on the other hand that the Poisson process $N(t)$ is $\boldsymbol{P}$-a.s. continuous and differentiable (with a vanishing derivative) we should respectively prove that for every $t > 0$ it is

$$\boldsymbol{P}\left\{\lim_{\Delta t \to 0} \Delta N(t) = 0\right\} = 1 \qquad\qquad \boldsymbol{P}\left\{\lim_{\Delta t \to 0} \frac{\Delta N(t)}{\Delta t} = 0\right\} = 1$$

This intuitively stems from the fact that the two limits could possibly not vanish *iff* $T_k = t$ for some k, that is *iff* t turns out to be one of the discontinuity instants of the point process T_n: this on the other hand happens with zero probability because the T_n are *ac* Erlang *rv*'s.

The Poisson process $N(t)$, finally, is not stationary (not even in the wide sense) because its expectation (6.11) is not constant and its autocorrelation (6.12) separately depends on s and t and not only on $t - s$ ■

Given the apparent *jumping* character of the Poisson trajectories, both the *ms* and $\boldsymbol{P}$-a.s. continuities stated in the Proposition 6.7 could be startling. We should however keep in mind that these continuities (as well as the stochastic continuity) just state that every t is a point of continuity in *ms* and $\boldsymbol{P}$-a.s. It is not at all asserted, instead, that the Poisson process is *sample continuous* in the sense of the Definition 5.5: as we will see later indeed this process does not meet the minimal requirements for this second—stronger—kind of continuity. In the same vein the $\boldsymbol{P}$-a.s. differentiability of $N(t)$ results from the remark that almost every trajectory of $N(t)$ is piecewise constant, but for the jumping points (a Lebesgue negligible set) where the discontinuities are located. As for the seeming incongruity between the *ms* non differentiability and the $\boldsymbol{P}$-a.s. differentiability we will only remark that this is a typical example of the different meaning of the two convergences: remember for instance that the possible existence of the *ms* derivative $\dot{N}(t) = 0$ would require that the expectation

$$\boldsymbol{E}\left[\left|\frac{\Delta N(t)}{\Delta t}\right|^2\right] = \frac{\boldsymbol{E}\left[(\Delta N(t))^2\right]}{\Delta t^2} = \frac{\lambda\Delta t + \lambda^2\Delta t^2}{\Delta t^2} = \frac{\lambda}{\Delta t} + \lambda^2$$

be infinitesimal for $\Delta t \to 0$, while apparently it is not. This discussion about the process differentiability will be resumed in the Sect. 6.3 devoted to the *white noise*.

Proposition 6.8 *The Poisson increments process $\Delta N(t)$ with a fixed $\Delta t > 0$ is wide sense stationary: we have indeed*

$$m_{\Delta N} = \sigma^2_{\Delta N} = \lambda\Delta t \tag{6.17}$$

$$R_{\Delta N}(\tau) = \begin{cases} \lambda^2\Delta t^2 & \text{if } |\tau| \geq \Delta t \\ \lambda^2\Delta t^2 + \lambda(\Delta t - |\tau|) & \text{if } |\tau| < \Delta t \end{cases} \tag{6.18}$$

$$C_{\Delta N}(\tau) = \begin{cases} 0 & \text{if } |\tau| \geq \Delta t \\ \lambda(\Delta t - |\tau|) & \text{if } |\tau| < \Delta t \end{cases} \tag{6.19}$$

$$\rho_{\Delta N}(\tau) = \begin{cases} 0 & \text{if } |\tau| \geq \Delta t \\ 1 - \frac{|\tau|}{\Delta t} & \text{if } |\tau| < \Delta t \end{cases} \tag{6.20}$$

$$S_{\Delta N}(\varpi) = 2\lambda(\Delta t)^2\,\frac{1 - \cos\varpi\Delta t}{(\varpi\Delta t)^2} \tag{6.21}$$

where $S_{\Delta N}$ is the covariance spectrum (5.26).

Proof We already knew that $N(t)$ has *independent and stationary increments*: we moreover show here that $\Delta N(t)$ also is *wide sense stationary* as a *sp*. To begin

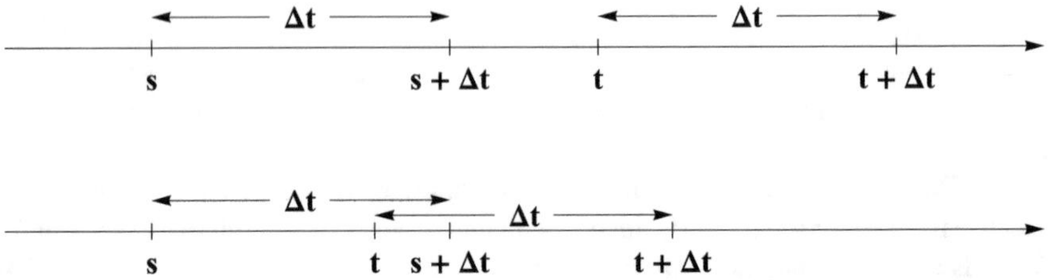

Fig. 6.7 Possible interval arrangements to calculate the autocorrelation (6.18) of the Poisson increments with a given $\Delta t > 0$

with the (6.17) immediately results form the remark that the increments $\Delta N(t)$ are distributed according to the Poisson law $\mathfrak{P}(\lambda\Delta t)$. As for the autocorrelation function

$$R_{\Delta N}(s,t) = \boldsymbol{E}\left[\Delta N(s)\Delta N(t)\right]$$

we must remember that the increments on non superposed time intervals (see Fig. 6.7) are independent, so that if $|t-s| \geq \Delta t$

$$R_{\Delta N}(s,t) = \boldsymbol{E}\left[\Delta N(s)\right]\boldsymbol{E}\left[\Delta N(t)\right] = \lambda^2\Delta t^2 \qquad |t-s| \geq \Delta t$$

When instead $|t-s| < \Delta t$, we first take $t > s$ and remark that (see Fig. 6.7)

$$\begin{aligned}\Delta N(s)\Delta N(t) &= [N(s+\Delta t) - N(s)][N(t+\Delta t) - N(t)] \\ &= [N(s+\Delta t) - N(t) + N(t) - N(s)][N(t+\Delta t) - N(t)] \\ &= [N(t) - N(s)][N(t+\Delta t) - N(t)] + [N(s+\Delta t) - N(t)]^2 \\ &\qquad + [N(s+\Delta t) - N(t)][N(t+\Delta t) - N(s+\Delta t)]\end{aligned}$$

From the intervals arrangement we then find

$$\begin{aligned}R_{\Delta N}(s,t) &= \lambda(t-s)\cdot\lambda\Delta t + \lambda(\Delta t - t + s) + \lambda^2(\Delta t - t + s)^2 \\ &\qquad + \lambda(\Delta t - t + s)\cdot\lambda(t-s) \\ &= \lambda^2\Delta t^2 + \lambda[\Delta t - (t-s)]\end{aligned}$$

If instead $t < s$ we just swap s and t: summing up all the cases we then find (6.18) with $\tau = t - s$. From these results also immediately stem the autocovariance (6.19) and the correlation coefficient (6.20), while the covariance spectrum (6.21) result from an elementary Fourier transform ■

6.1.3 Compensated Poisson Process

The Poisson process and its corresponding point process also constitute the first step in the definition of other important processes: to begin with the *compensated* Poisson process is defined as

$$\widetilde{N}(t) = N(t) - \lambda t \tag{6.22}$$

Since λt is the expectation of $N(t)$, it is apparent that $\widetilde{N}(t)$ is just a *centered* Poisson process whose sample trajectories are displayed in the Fig. 6.8. Keeping (6.12) into account we then have

$$m_{\widetilde{N}}(t) = 0 \qquad \sigma^2_{\widetilde{N}}(t) = \lambda t$$

$$R_{\widetilde{N}}(s,t) = C_{\widetilde{N}}(s,t) = \lambda \min\{s,t\} \qquad \rho_{\widetilde{N}}(s,t) = \frac{\min\{s,t\}}{\sqrt{st}}$$

so that again the variance grows linearly in time while the trajectories will be evenly arranged around the horizontal axis steadily drifting away as in the Fig. 6.9: in other words the process $\widetilde{N}(t)$ too diffuses around its zero expectation. Remark however that now, at variance with the Fig. 6.6, the *diffusion* is more appreciable, the scale effect having been eliminated by the centering. Since moreover the autocorrelation of $\widetilde{N}(t)$ essentially coincides with that of $N(t)$, the Proposition 6.7 still holds for the compensated Poisson process. From (6.9) we finally find its *chf*

$$\varphi_{\widetilde{N}}(u,t) = \varphi_N(u,t)e^{-iu\lambda t} = e^{\lambda t(e^{iu}-iu-1)}$$

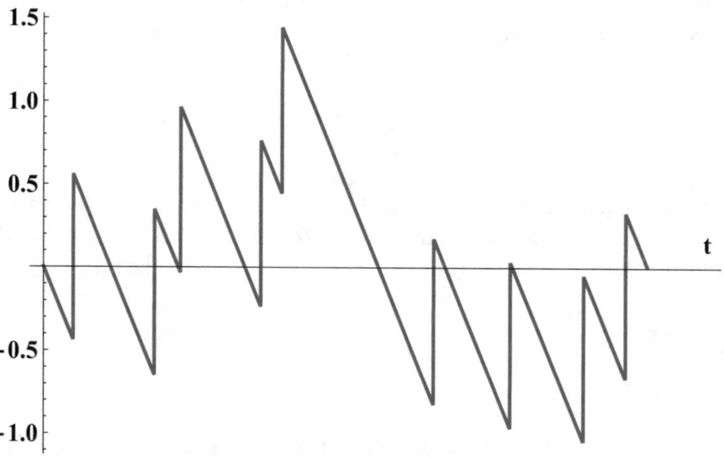

Fig. 6.8 Sample trajectory of the compensated Poisson process (6.22)

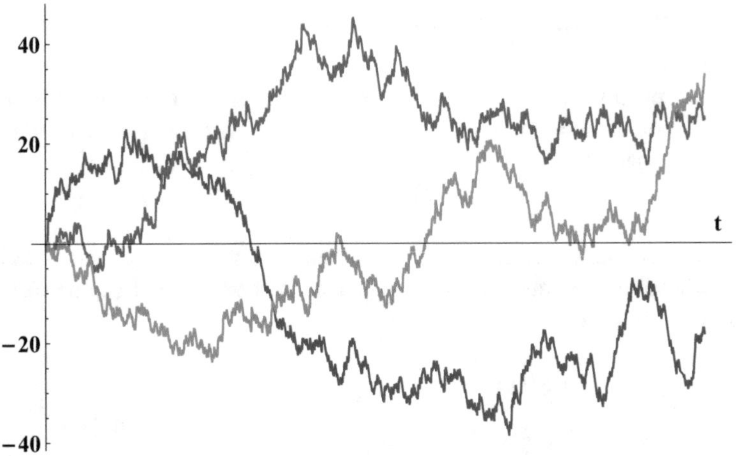

Fig. 6.9 1 000 steps trajectories of the compensated Poisson process (6.8)

The compensated Poisson process plays an important role in the theory of the stochastic differential equations of the jump process, a topic that however will exceed the scope of these lectures.

6.1.4 Compound Poisson Process

Definition 6.9 By extending the definition (6.6), given a point process T_n, a **compound Poisson process** is the *sp*

$$X(t) = \sum_{k=1}^{\infty} X_k \vartheta(t - T_k) \tag{6.23}$$

where X_k is a sequence of *rv*'s independent from T_n. If moreover $N(t)$ is the simple Poisson process associated to T_n, the process (6.23) can also be represented as the sum of the random number $N(t)$ of *rv*'s X_k turned up within the time t

$$X(t) = \sum_{k=1}^{N(t)} X_k \tag{6.24}$$

All in all the compound Poisson process follows ***trajectories*** that are akin to that of the simple Poisson process, but for the fact that in every instant T_k, instead of jumping deterministically ahead of a unit length, it now takes a leap of random

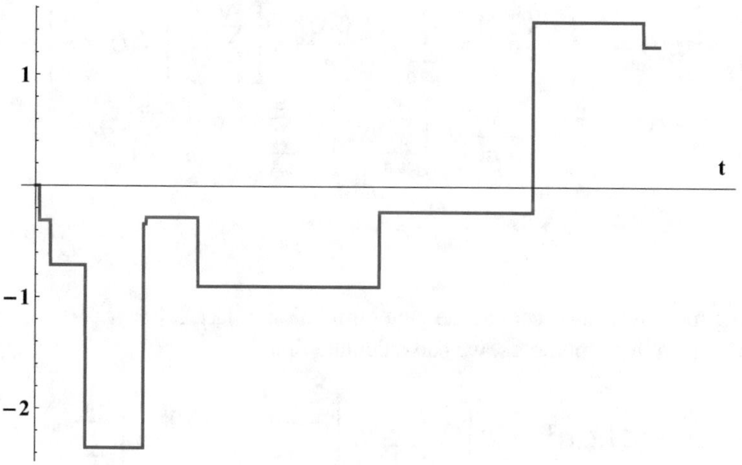

Fig. 6.10 Sample trajectory of a compound Poisson process (6.23) with $\mathfrak{N}(0, 1)$ distributed *iid* components X_k

length X_k. In the Fig. 6.10 an example is displayed where the X_k are independent standard Gaussians.

The simple Poisson process itself is also apparently a particular case of the compound process: it would be enough to take $X_k = 1$, $\boldsymbol{P}$-a.s.. In the Fig. 6.11 a few examples of longer spanning trajectories are presented, and to a purely qualitative observation they look not very different from those of a compensated Poisson process. In the next proposition, moreover, we will find that the representation (6.24) of the *sp* $X(t)$ turns out to be especially advantageous to calculate its main statistical characteristics.

Proposition 6.10 *If $X(t)$ is the compound Poisson process* (6.23), *and if the rv's X_k are iid with $\boldsymbol{E}[X_k] = \mu$ and $\boldsymbol{V}[X_k] = \sigma^2$, then it is*

$$m_X(t) = \lambda\mu t \tag{6.25}$$

$$\sigma_X^2(t) = \lambda(\mu^2 + \sigma^2)t \tag{6.26}$$

$$R_X(s, t) = \lambda(\mu^2 + \sigma^2)\min\{s, t\} + \lambda^2\mu^2 st \tag{6.27}$$

$$C_X(s, t) = \lambda(\mu^2 + \sigma^2)\min\{s, t\} \tag{6.28}$$

$$\rho_X(s, t) = \frac{\min\{s, t\}}{\sqrt{st}} \tag{6.29}$$

Proof The (6.25) results from the representation (6.24) and the usual properties of the conditional expectations:

$$
\begin{aligned}
m_X(t) &= \boldsymbol{E}\left[\sum_{k=1}^{N(t)} X_k\right] = \sum_{n=0}^{\infty} e^{-\lambda t}\frac{(\lambda t)^n}{n!}\,\boldsymbol{E}\left[\sum_{k=1}^{N(t)} X_k \,\middle|\, N(t)=n\right] \\
&= \sum_{n=0}^{\infty} e^{-\lambda t}\frac{(\lambda t)^n}{n!}\boldsymbol{E}\left[\sum_{k=1}^{n} X_k\right] = \sum_{n=1}^{\infty} e^{-\lambda t}\frac{(\lambda t)^n}{(n-1)!}\,\mu \\
&= \lambda\mu t\sum_{n=0}^{\infty} e^{-\lambda t}\frac{(\lambda t)^n}{n!} = \lambda\mu t
\end{aligned}
$$

Keeping moreover into account the general relation $\boldsymbol{E}\left[X_k X_\ell\right] = \mu^2 + \sigma^2\delta_{k\ell}$ easily deduced from the hypotheses, we can calculate first

$$
\begin{aligned}
R_X(t,t) &= \boldsymbol{E}\left[X(t)^2\right] = \boldsymbol{E}\left[\sum_{k,\ell=1}^{N(t)} X_k X_\ell\right] = \sum_{n=0}^{\infty} e^{-\lambda t}\frac{(\lambda t)^n}{n!}\boldsymbol{E}\left[\sum_{k,\ell=1}^{n} X_k X_\ell\right] \\
&= \sum_{n=0}^{\infty} e^{-\lambda t}\frac{(\lambda t)^n}{n!}\sum_{k,\ell=1}^{n}(\mu^2+\sigma^2\delta_{k\ell}) = \sum_{n=0}^{\infty} e^{-\lambda t}\frac{(\lambda t)^n}{n!}(n^2\mu^2+n\sigma^2) \\
&= \mu^2\sum_{n=1}^{\infty} n e^{-\lambda t}\frac{(\lambda t)^n}{(n-1)!} + \sigma^2\sum_{n=1}^{\infty} e^{-\lambda t}\frac{(\lambda t)^n}{(n-1)!} \\
&= \mu^2\lambda t\sum_{n=0}^{\infty}(n+1)e^{-\lambda t}\frac{(\lambda t)^n}{n!} + \sigma^2\lambda t\sum_{n=0}^{\infty} e^{-\lambda t}\frac{(\lambda t)^n}{n!} \\
&= (\lambda t)^2\mu^2 + \lambda t(\mu^2+\sigma^2)
\end{aligned}
$$

and then—from the increments independence—the autocorrelation (6.27): taking indeed $s < t$ we get

$$
\begin{aligned}
R_X(s,t) &= \boldsymbol{E}\left[X(s)X(t)\right] = \boldsymbol{E}\left[X(s)(X(t)-X(s)+X(s))\right] \\
&= \boldsymbol{E}\left[X(s)\right]\boldsymbol{E}\left[X(t)-X(s)\right] + R(s,s) \\
&= \lambda^2\mu^2 s(t-s) + (\lambda s)^2\mu^2 + \lambda s(\mu^2+\sigma^2) = \lambda^2\mu^2 st + \lambda s(\mu^2+\sigma^2)
\end{aligned}
$$

and hence (6.27) in the general case with arbitrary s, t. The autocovariance (6.28), the variance (6.26) and the correlation coefficient (6.29) trivially result then from the definitions (5.3) and (5.5) ■

The variance is then again diffusive in the sense that it linearly grows in time as can be guessed also from the Fig. 6.11. Remark that the correlation coefficient (6.29) exactly concurs with those of both the simple and compensated Poisson processes (6.14), while the correspondent autocorrelation and autocovariance (6.12) are recovered for $\mu = 1, \sigma = 0$. As a consequence also the stationarity, continuity and differentiability properties of a compound Poisson process coincide with that of the simple Poisson process summarized in the Proposition 6.7. Remark finally that, if $\varphi(u)$ is the common *chf* of the X_k, the *chf* of the compound Poisson process is

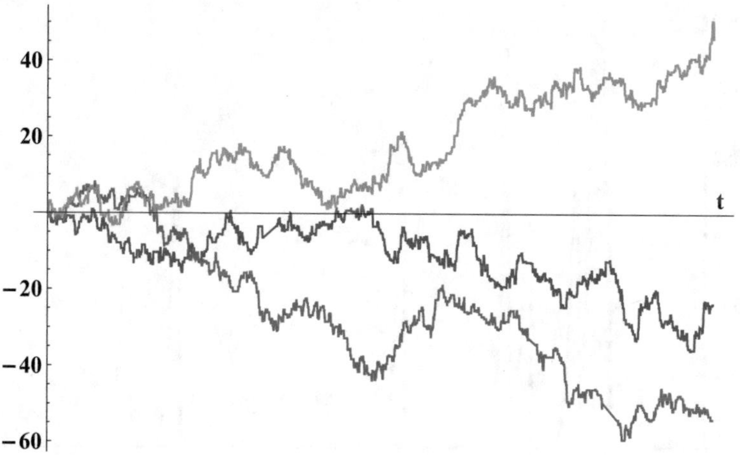

Fig. 6.11 1 000 steps sample trajectories of the compound Poisson process (6.23) with $\mathfrak{N}(0, 1)$ *iid* X_k's

$$\varphi_X(u,t) = \boldsymbol{E}\left[e^{iuX(t)}\right] = \sum_{n=0}^{\infty} e^{-\lambda t}\frac{(\lambda t)^n}{n!}\boldsymbol{E}\left[e^{iu\sum_{k=1}^{n}X_k}\right]$$
$$= \sum_{n=0}^{\infty} e^{-\lambda t}\frac{(\lambda t)^n}{n!}\varphi(u)^n = e^{\lambda t[\varphi(u)-1]} \qquad (6.30)$$

Also this *chf* is reduced to that of the simple Poisson process (6.9) when $X_k = 1$, $\boldsymbol{P}$-a.s. for every k because now $\varphi(u) = e^{iu}$

6.1.5 Shot Noise

Definition 6.11 Take a point process with intensity λ: we call **shot noise** the *sp*

$$X(t) = \sum_{k=1}^{\infty} h(t - T_k) \qquad (6.31)$$

where $h(t)$ is an arbitrary integrable function that as a rule (but not necessarily) is non zero only for $t > 0$.

A typical example of $h(t)$ is

$$h(t) = q\,at\,e^{-at}\vartheta(t) \qquad a > 0,\ q > 0 \qquad (6.32)$$

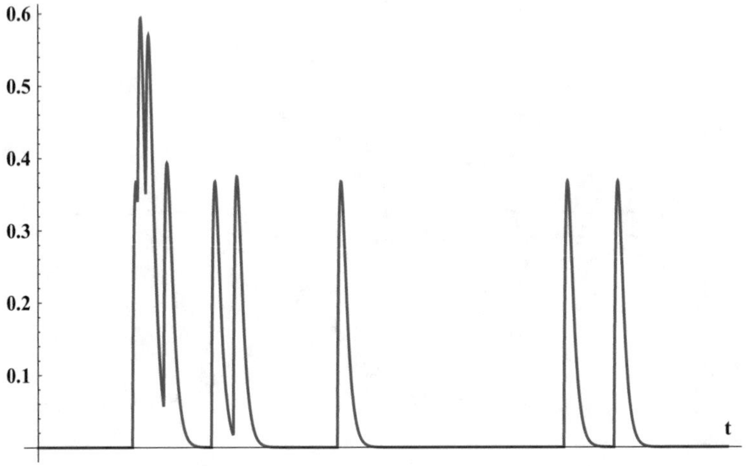

Fig. 6.12 Sample trajectory of the shot noise process (6.31) with $h(t)$ chosen as in (6.32)

that yields sample trajectories like that in the Fig. 6.12. To fix the ideas we could imagine that this *sp* describes the current impulses produced by the random arrival of isolated thermal electrons on the cathode of a vacuum tube: in this case the arrival times apparently constitute a point process, and every electron elicit in the circuit a current impulse of the form $h(t)$ that as a rule exponentially vanishes. Of course close arrivals produce superpositions with the effects shown in the Fig. 6.12.

Proposition 6.12 *If the function $h(x)$ is integrable and square integrable, the shot noise $X(t)$ (6.31) is wide sense stationary, and with $\tau = t - s$ we find*

$$m_X(t) = \lambda H \qquad \sigma_X^2(t) = \lambda g(0) \tag{6.33}$$

$$R_X(\tau) = \lambda g(|\tau|) + \lambda^2 H^2 \qquad C_X(\tau) = \lambda g(|\tau|) \qquad \rho_X(\tau) = \frac{g(|\tau|)}{g(0)} \tag{6.34}$$

$$H = \int_{-\infty}^{+\infty} h(t)\,dt \qquad g(t) = \int_{-\infty}^{+\infty} h(t+s)h(s)\,ds \tag{6.35}$$

Proof Omitted: see [1] p. 360. Remark that the stationarity is now consistent with the traits of the new trajectories consisting of a sequence of impulses of the form h with intensity λ, and no longer showing a bent to diffuse drifting away from the horizontal axis ■

Exemple 6.13 The general results of the Proposition (6.12) for a shot noise $X(t)$ are explicitly implemented according to the choice of the function h: with an $h(t)$ of the form (6.32) we get in particular (see also Fig. 6.13)

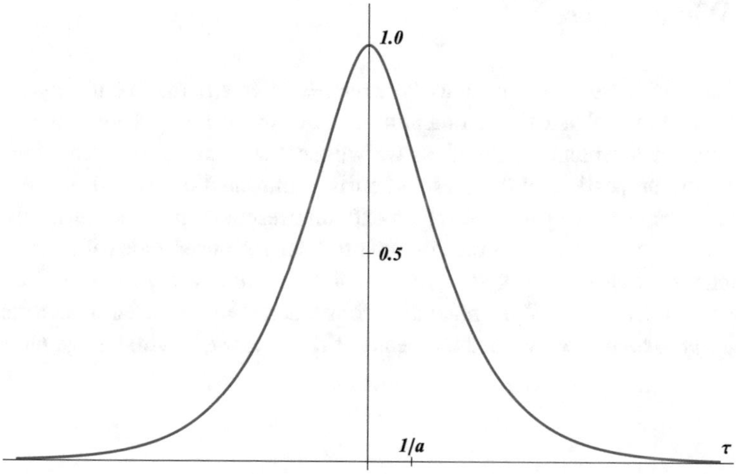

Fig. 6.13 Correlation coefficient $\rho_X(\tau)$ (6.36) of a shot noise with $h(t)$ chosen as in (6.32)

$$H = \frac{q}{a} \qquad g(t) = \frac{q^2}{4a}(1 + a|t|)\, e^{-a|t|}$$
$$m_X(t) = \frac{\lambda q}{a} \qquad \sigma_X^2(t) = \frac{\lambda q^2}{4a} \qquad R_X(\tau) = \frac{\lambda q^2}{4a}(1 + a|\tau|)\, e^{-a|\tau|} + \frac{\lambda^2 q^2}{a^2}$$
$$C_X(\tau) = \frac{\lambda q^2}{4a}(1 + a|\tau|)\, e^{-a|\tau|} \qquad \rho_X(\tau) = (1 + a|\tau|)\, e^{-a|\tau|} \tag{6.36}$$

and hence, taking also into account the results of the Sect. 5, we can say that

- $X(t)$ is *ms*-continuous because its autocorrelation $R_X(\tau)$ is continuous in $\tau = 0$ (Proposition 5.6)
- $X(t)$ *ms*-differentiable because it is possible to show explicitly from (6.36) that now its second derivative $R''_X(\tau)$ exists in $\tau = 0$ (Proposition 5.11)
- $X(t)$ is ergodic for expectation and autocorrelation because the condition (5.21) is apparently met by our $C_X(\tau)$ (Corollary 5.13)
- The covariance spectrum of the shot noise in our example can be calculated from (6.36) and (5.26) and is

$$S_X(\varpi) = \frac{\lambda a^2 q^2}{2\pi \left(a^2 + \varpi^2\right)^2} \tag{6.37}$$

6.2 Wiener Process

The Wiener process (also known as *Brownian motion*, a name that nevertheless we will reserve for the physical phenomenon discussed in the Sect. 6.4 and in the Chap. 9) can be more conveniently defined, as we will see later on, starting from its formal probabilistic properties. In the present heuristic introduction we will however first follow a more intuitive path through an explicit presentation of its trajectories, in a way similar to that adopted for the Poisson process. Yet, at variance with this last one, the Wiener process stems only as a limit in distribution of a *sequence of elementary processes* known as *random walks*: as a consequence its sketched trajectories will only be *approximations* attained by means of *random walks* with a large number of steps.

6.2.1 Random Walk

Definition 6.14 Take $s > 0,\ \tau > 0$, and the sequence $(X_j)_{j\geq 0}$ of *iid rv*'s

$$X_0 = 0,\ \ \boldsymbol{P}\text{-a.s.} \qquad X_j = \begin{cases} +s, & \text{with probability } p \\ -s, & \text{with probability } q = 1 - p \end{cases} \qquad j = 1, 2, \dots$$

then we will call **random walk** the *sp*

$$X(t) = \sum_{j=0}^{\infty} X_j \vartheta(t - j\tau) \tag{6.38}$$

that is, in a different layout,

$$X(t) = \begin{cases} X_0 = 0 & 0 \leq t < \tau \\ X_1 + \cdots + X_n & n\tau \leq t < (n+1)\tau, \quad n = 1, 2, \dots \end{cases}$$

The possible sample trajectories of a *random walk* are then (ascending and descending) stairs like to that in the Fig. 6.14 that at first sight resemble those of a compound Poisson process (6.23). At variance with them however the steps length in not random and is instead always τ, while their height takes only two possible values $\pm s$. Since moreover

$$\boldsymbol{E}\left[X_j\right] = (p - q)s \qquad \boldsymbol{E}\left[X_j^2\right] = s^2 \qquad \boldsymbol{V}\left[X_j\right] = 4pqs^2$$

it is easy to see that for every $n = 0, 1, 2, \dots$ we have

$$\boldsymbol{E}\left[X(t)\right] = (p - q)ns \qquad \boldsymbol{V}\left[X(t)\right] = 4pqns^2 \qquad n\tau \leq t < (n+1)\tau \tag{6.39}$$

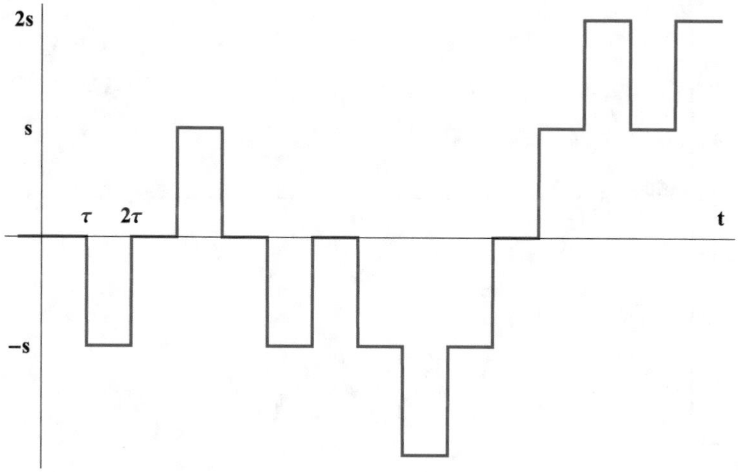

Fig. 6.14 Typical trajectory of a symmetric *random walk* with 15 steps

so that, if for instance $p \neq q$, the absolute value of the expectation grows with n, namely with t. When instead $p = q = {}^1/_2$ we have a ***symmetric random walk***, and in this case $\boldsymbol{E}[X(t)] = 0$ for every t. The variance on the other hand in any event grows with n, and hence with t. Furthermore also this process has by construction independent increments $\Delta X(t)$ for non overlapping intervals.

6.2.2 Wiener Process

Definition 6.15 Take the family of all the symmetric random walks $X(t)$ with $p = q = {}^1/_2$, $s > 0$ and $\tau > 0$, then the **Wiener process** $W(t)$, with $W(0) = 0$, $\boldsymbol{P}$-a.s., and **diffusion coefficient** $D > 0$ is the limit (in distribution according to the Definition 5.4) of the random walks $X(t)$ when

$$\tau \to 0 \qquad s \to 0 \qquad \frac{s^2}{\tau} \to D > 0 \tag{6.40}$$

When $D = 1$ we also call it **standard Wiener process**. We finally define the **Wiener increments process** $\Delta W(t) = W(t + \Delta t) - W(t)$ for every given $\Delta t > 0$.

We will take for granted without proof that the limits in distribution of the previous definition actually exist, namely that—under the conditions (6.40)—all the finite joint distributions of $X(t)$ converge toward a consistent family of finite joint distributions defining the global law of $W(t)$. It is furthermore apparent from our definition that the trajectories of $W(t)$ (at variance with those of a Poisson process, or of a *random walk*) can not be graphically represented in an exact way because we are dealing

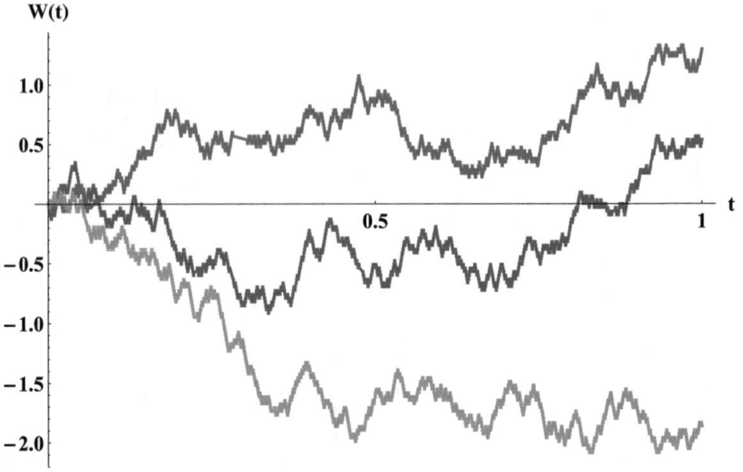

Fig. 6.15 Typical 1 000 steps trajectory of a symmetric *random walk*: it can be considered as an approximation of a Wiener process $W(t)$ (here $D = 1$)

with a *limit process*. We can expect however that the trajectories of a *random walk* with a large number of steps will constitute a good approximation for the trajectories of $W(t)$, as happens in the Fig. 6.15 for the samples of a *random walk* with 1 000 steps. Remark that for long enough times these trajectories qualitatively resemble those of both a compensated (Fig. 6.9) and a compound Poisson process (Fig. 6.11). The main difference between the real Wiener trajectories and its approximations is that if we would look more closely (at a shorter time scale) at the approximate trajectories of Fig. 6.15 we would immediately find the underlying *random walk* of the Fig. 6.14, while if we zoom in on a trajectory of $W(t)$ at any level we always find the same kind of irregular behavior. In other words the samples of a Wiener process are ***self-similar*** in the sense that they always show the same irregular look regardless of the space-time scale of our investigation.

In the following we will also adopt the shorthand notation

$$\phi_{a^2}(x) = \frac{e^{-x^2/2a^2}}{\sqrt{2\pi a^2}} \qquad \Phi_{a^2}(x) = \int_{-\infty}^{x} \phi_{a^2}(y)\, dy \tag{6.41}$$

respectively for the *pdf* and the *cdf* of a *centered* $\mathfrak{N}(0, a^2)$ law.

Proposition 6.16 *A Wiener process $W(t)$ has stationary and independent increments; we have moreover with $\Delta t > 0$*

$$W(t) \sim \mathfrak{N}(0, Dt) \qquad \Delta W(t) \sim \mathfrak{N}(0, D\Delta t) \tag{6.42}$$

and hence the pdf's and the chf's respectively are

$$f_W(x,t) = \frac{e^{-x^2/2Dt}}{\sqrt{2\pi Dt}} \qquad f_{\Delta W}(x) = \frac{e^{-x^2/2D\Delta t}}{\sqrt{2\pi D\Delta t}} \tag{6.43}$$

$$\varphi_W(u,t) = e^{-Dtu^2/2} \qquad \varphi_{\Delta W}(u,\Delta t) = e^{-D\Delta t\, u^2/2} \tag{6.44}$$

*The **transition** pdf (conditional pdf) with $\Delta t > 0$ finally is $\mathfrak{N}(y, D\Delta t)$, that is*

$$f_W(x, t+\Delta t|y,t) = f_{\Delta W}(x-y) = \frac{e^{-(x-y)^2/2D\Delta t}}{\sqrt{2\pi D\Delta t}} \tag{6.45}$$

Proof Since $W(t)$ has been defined as the limit process of a family of *random walks* $X(t)$ that have independent increments by definition, it is apparent that also the increments $\Delta W(t)$ on non overlapping intervals are independent: this is the second example of a *process with independent increments* after that of Poisson. The result (6.42) ensues on the other hand from the Central Limit Theorem 4.27: for every arbitrary but fixed t take indeed the sequence of symmetric *random walks* $X(t)$ with

$$\tau = \frac{t}{n} \qquad s^2 = \frac{Dt}{n} \qquad n = 1, 2, \dots$$

so that the requirements (6.40) are met for $n \to \infty$. As a consequence

$$X(t) = X(n\tau) = X_0 + X_1 + \cdots + X_n = S_n \qquad n = 0, 1, 2, \dots$$

with $\boldsymbol{E}[S_n] = 0$ and $\boldsymbol{V}[S_n] = Dt$, as follows from (6.39) for $p = q = {}^1/_2$. From the Central Limit Theorem 4.27 we then have for $n \to \infty$

$$S_n^* = \frac{X(t)}{\sqrt{Dt}} \xrightarrow{d} \mathfrak{N}(0,1)$$

If now $W(t)$ is the limit in distribution of $X(t)$ we can say that

$$\frac{W(t)}{\sqrt{Dt}} \sim \mathfrak{N}(0,1)$$

namely $W(t) \sim \mathfrak{N}(0, Dt)$, that is (6.42) so that the *pdf* of the Wiener process will be (6.43). The *chf* (6.44) is then easily calculated from (4.13) keeping into account (6.42). The law (6.42) of the increments $\Delta W(t)$, its *pdf* (6.43) and *chf* (6.44) are derived in a similar way. As for the transition *pdf*, from (6.42) and the increments independence, we finally have

$$\begin{aligned} F_W(x, t+\Delta t|y, t) &= \boldsymbol{P}\{W(t+\Delta t) \le x \mid W(t) = y\} \\ &= \boldsymbol{P}\{W(t+\Delta t) - W(t) + W(t) \le x \mid W(t) = y\} \\ &= \boldsymbol{P}\{\Delta W(t) \le x - y \mid W(t) = y\} \\ &= \boldsymbol{P}\{\Delta W(t) \le x - y\} \;=\; \Phi_{D\Delta t}(x - y) \end{aligned}$$

where $\Phi_{D\Delta t}(x)$ is the *cdf* (6.41) of the law $\mathfrak{N}(0, D\Delta t)$ so that (6.45) immediately follows from an x-differentiation ■

Proposition 6.17 *The main statistical properties of a Wiener process $W(t)$ are*

$$m_W(t) = 0 \qquad \sigma_W^2(t) \;=\; Dt \tag{6.46}$$

$$R_W(s, t) = C_W(s, t) \;=\; D\min\{s, t\} \tag{6.47}$$

$$\rho_W(s, t) = \frac{\min\{s, t\}}{\sqrt{st}} \;=\; \begin{cases} \sqrt{s/t} & \text{if } s < t \\ \sqrt{t/s} & \text{if } t < s \end{cases} \tag{6.48}$$

Proof The formulas (6.46) immediately result from (6.42). As for the autocorrelation (6.47) (coincident with the autocovariance because the expectation is zero) consider first $s = t$ so that from (6.42) we get

$$R_W(t, t) = \boldsymbol{E}\left[W^2(t)\right] = \boldsymbol{V}\left[W(t)\right] = Dt \tag{6.49}$$

Then, with $s < t$, remark that the increments $W(s) = W(s) - W(0)$ and $W(t) - W(s)$ are independent and respectively distributed according to $\mathfrak{N}(0, Ds)$ and $\mathfrak{N}(0, D(t-s))$, so that

$$\begin{aligned} R_W(s, t) &= \boldsymbol{E}\left[W(s)W(t)\right] = \boldsymbol{E}\left[W(s)\big(W(t) - W(s) + W(s)\big)\right] \\ &= \boldsymbol{E}\left[W^2(s)\right] = R_W(s, s) = Ds \end{aligned}$$

In conclusion, however chosen s and t, we retrieve (6.47) and hence also (6.48). ■

Despite their obvious differences, the statistical properties of the Wiener and of the simple Poisson processes—as listed in the previous proposition and in the Proposition 6.6—are rather comparable, with the diffusion coefficient D playing a role analogous to that of the Poisson intensity λ: the coefficient D, whose existence we surmised in the definition of $W(t)$, by construction has dimensions $\mathrm{m}^2/\mathrm{sec}$ and is the main characteristic parameter of the Wiener process. Remark finally that—as for the Poisson process—here too the process variance linearly grows with t, so that also the Wiener process is considered a *diffusion*.

Proposition 6.18 *The Wiener process $W(t)$ is sample continuous, but almost every trajectory is nowhere differentiable. Moreover $W(t)$ is non stationary (not even in wide sense), but it is Gaussian and, however taken $t_1 < t_2 < \cdots < t_n$ (the ordering has been fixed here only for convenience), we have $(W(t_1), \ldots, W(t_n)) \sim \mathfrak{N}(0, \mathbb{A})$ with covariance matrix*

$$\mathbb{A} = D \begin{pmatrix} t_1 & t_1 & t_1 & \cdots & t_1 \\ t_1 & t_2 & t_2 & & t_2 \\ t_1 & t_2 & t_3 & & t_3 \\ \vdots & & & \ddots & \vdots \\ t_1 & t_2 & t_3 & \cdots & t_n \end{pmatrix} \tag{6.50}$$

Proof A more detailed discussion of the sample continuity of $W(t)$ according to the Definition 5.5 will be postponed until the Sect. 7.1.7 (see Proposition 7.23): here we will confine ourselves only to remark that this result also apparently entails all the other, weaker continuities listed in the Sect. 5.3. It would be easy to show, however, that the *ms*-continuity for every t could be independently proved in the same way as that of the simple Poisson process in the Proposition 6.7 because the autocorrelation functions of the two processes essentially coincide. Also the proof of the non stationarity is clearly the same.

Neglecting then a proof of the global non differentiability stated in the theorem, we will only show the weaker result that for every $t > 0$ the Wiener process is not differentiable in probability, and hence—according to the negative of the point 1 of the Theorem 4.4—it is also non differentiable in *ms* and $\boldsymbol{P}$-a.s. We are reduced then to prove that for every $t > 0$

$$\left|\frac{\Delta W(t)}{\Delta t}\right| \xrightarrow{P} +\infty \qquad \text{for } \Delta t \to 0$$

namely

$$\lim_{\Delta t \to 0} \boldsymbol{P}\left\{\left|\frac{\Delta W(t)}{\Delta t}\right| > M\right\} = 1 \qquad \forall M > 0 \tag{6.51}$$

Take indeed $M > 0$: from (6.43) it is

$$\boldsymbol{P}\left\{\left|\frac{\Delta W(t)}{\Delta t}\right| > M\right\} = 1 - \boldsymbol{P}\left\{\left|\frac{\Delta W(t)}{\Delta t}\right| \leq M\right\} = 1 - \int_{-M|\Delta t|}^{M|\Delta t|} \frac{e^{-x^2/2D|\Delta t|}}{\sqrt{2\pi D|\Delta t|}}\, dx$$

and (6.51) follows from the remark that with $y = x/\sqrt{|\Delta t|}$ we get

$$\lim_{\Delta t \to 0} \int_{-M|\Delta t|}^{M|\Delta t|} \frac{e^{-x^2/2D|\Delta t|}}{\sqrt{2\pi D|\Delta t|}}\, dx = \lim_{\Delta t \to 0} \int_{-M\sqrt{|\Delta t|}}^{M\sqrt{|\Delta t|}} \frac{e^{-y^2/2D}}{\sqrt{2\pi D}}\, dy = 0$$

From the Proposition 6.16 we already know that the *pdf* and the transition *pdf* of $W(t)$ are Gaussian: we will show now that also the higher order joint *pdf*'s are Gaussian. Take first the *r-vec* $(W(s), W(t))$ with $s < t$ for convenience: its joint *pdf* then is[1]

[1]Remark that to retrieve the marginals (6.43) from the joint *pdf* (6.52) the integral

$$f_W(x,t;y,s) = f_W(x,t|y,s)f_W(y,s) = \phi_{D(t-s)}(x-y)\phi_{Ds}(y) \tag{6.52}$$

If we now write down this *pdf* and compare it with the general *pdf*(2.24) of a bivariate normal law—we will skip the explicit calculation—we find that (6.52) exactly conforms to the Gaussian $\mathfrak{N}(0, \mathbb{A})$ with the covariance matrix

$$\mathbb{A} = D\begin{pmatrix} s & s \\ s & t \end{pmatrix}$$

To go on to the joint *pdf*'s with $n = 3, 4, \dots$ instants we iterate the procedure: take $t_1 < t_2 < t_3$ and—retracing the path leading to (6.45), but we will neglect the details – calculate first the conditional *pdf*

$$f_W(x_3, t_3 \,|\, x_2, t_2; x_1, t_1) = \phi_{D(t_3-t_2)}(x_3 - x_2)$$

and then, keping (6.52) into account, the joint *pdf*

$$\begin{aligned} f_W(x_3, t_3; x_2, t_2; x_1, t_1) &= f_W(x_3, t_3 \,|\, x_2, t_2; x_1, t_1) f_W(x_2, t_2; x_1, t_1) \\ &= \phi_{D(t_3-t_2)}(x_3 - x_2)\phi_{D(t_2-t_1)}(x_2 - x_1)\phi_{Dt_1}(x_1) \end{aligned}$$

that again by inspection turns out to be Gaussian with a covariance matrix of the form (6.50). Iterating the procedure for arbitrary $t_1 < \cdots < t_n$ we find that all the joint *pdf*'s of the *r-vec*'s $(W(t_1), \dots, W(t_n))$ are Gaussian $\mathfrak{N}(0, \mathbb{A})$ with covariance matrices (6.50), and hence that $W(t)$ is a *Gaussian process* ■

It is easy to check by direct calculation that the transition *pdf*'s $f_W(x, t|\, y, s)$ (6.45) of a Wiener process are solutions of the equation

$$\partial_t f(x,t) = \frac{D}{2}\,\partial_x^2 f(x,t), \qquad f(x, s^+) = \delta(x - y) \tag{6.53}$$

that represents a first example of a ***Fokker-Planck equation*** to be discussed in further details in the Sect. 7.2.3.

Proposition 6.19 *The Wiener increments process $\Delta W(t)$ with $\Delta t > 0$ is wide sense stationary with*

$$f_W(x,t) = \int_{-\infty}^{+\infty} f_W(x,t;y,s)\,dy = [\phi_{D(t-s)} * \phi_{Ds}](x) = \phi_{Dt}(x)$$

is handily performed by taking advantage of the reproductive properties (3.67) of the Gaussian distributions, namely

$$\mathfrak{N}(0, D(t-s)) * \mathfrak{N}(0, Ds) = \mathfrak{N}(0, Dt).$$

$$m_{\Delta W} = 0 \qquad \sigma^2_{\Delta W} = D\Delta t \tag{6.54}$$

$$R_{\Delta W}(\tau) = C_{\Delta W}(\tau) = \begin{cases} 0 & \textit{if } |\tau| \geq \Delta t \\ D(\Delta t - |\tau|) & \textit{if } |\tau| < \Delta t \end{cases} \tag{6.55}$$

$$\rho_{\Delta W}(\tau) = \begin{cases} 0 & \textit{if } |\tau| \geq \Delta t \\ 1 - \frac{|\tau|}{\Delta t} & \textit{if } |\tau| < \Delta t \end{cases} \tag{6.56}$$

$$S_{\Delta W}(\varpi) = 2D(\Delta t)^2 \frac{1 - \cos \varpi \Delta t}{(\varpi \Delta t)^2} \tag{6.57}$$

Proof We have already seen that the increments $\Delta W(t)$ are independent and stationary: here it will be shown moreover that the *increment process* is wide sense stationary. The results (6.54) directly follow from (6.42), namely from the remark that $\Delta W(t) \sim \mathfrak{N}(0, D\Delta t)$. As for the autocorrelation and autocovariance we will retrace the procedure adopted for the Poisson increments: by recalling that the increments on non overlapping intervals are independent, when $|t - s| \geq \Delta t$ we have

$$R_{\Delta W}(s, t) = \boldsymbol{E}[\Delta W(s)\Delta W(t)] = \boldsymbol{E}[\Delta W(s)]\, \boldsymbol{E}[\Delta W(t)] = 0$$

If instead $|t - s| < \Delta t$, take first $t > s$ to have (see Fig. 6.7)

$$\begin{aligned} \Delta W(s)\Delta W(t) &= [W(s + \Delta t) - W(s)]\,[W(t + \Delta t) - W(t)] \\ &= [W(s + \Delta t) - W(t) + W(t) - W(s)]\,[W(t + \Delta t) - W(t)] \\ &= [W(t) - W(s)][W(t + \Delta t) - W(t)] + [W(s + \Delta t) - W(t)]^2 \\ &\quad + [W(s + \Delta t) - W(t)][W(t + \Delta t) - W(s + \Delta t)] \end{aligned}$$

and hence—because of the increments independence and the vanishing of their expectations—we will have with $\tau = t - s > 0$

$$R_{\Delta W}(\tau) = \boldsymbol{E}\left[[W(s + \Delta t) - W(t)]^2\right] = D(\Delta t - \tau)$$

For $t < s$ it would be enough to swap s and t, and by gathering all the results we immediately get (6.55). The results (6.56) and (6.57) will finally follow from their respective definitions ∎

6.2.3 *Geometric Wiener Process*

As for the Poisson process, a number of different *sp*'s can be derived from the Wiener process, but here we will only briefly linger on the so called ***geometric Wiener process (geometric Brownian motion)*** defined as

$$X(t) = e^{W(t)} \qquad X(0) = 1 \tag{6.58}$$

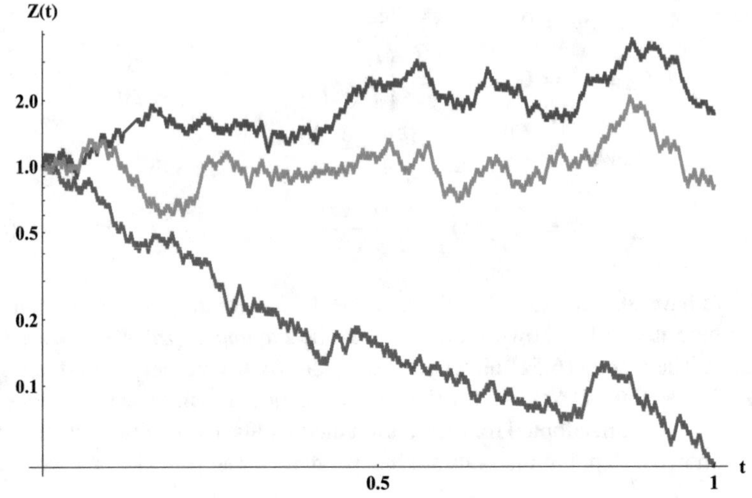

Fig. 6.16 Sample trajectories (logarithmic scale and $D = 1$) of a geometric Wiener process (6.60) approximated as the exponential of 1 000 steps random walks

where $W(t) \sim \mathfrak{N}(0, Dt)$ is a Wiener process. Such a process is especially relevant in the field of the mathematical finance, and we already said in the Sect. 3.47 that its law is *log-normal* with

$$f_X(x,t) = \frac{e^{-\ln^2 x/2Dt}}{x\sqrt{2\pi Dt}} \qquad x > 0 \tag{6.59}$$

while its expectation and variance are

$$m_X(t) = e^{Dt/2} \qquad \sigma_X^2(t) = \left(e^{Dt} - 1\right) e^{Dt}$$

At variance with those of the Wiener process $W(t)$, the sample trajectories of $X(t)$ apparently never go negative (see Fig. 6.16 with a logarithmic scale), and this is one of the main reasons why the geometric Brownian motion, unlike the simple Wiener process, is considered a good model to describe the price trend on the market. In the applications, however, it is customary to adopt a de-trended variant of $X(t)$ centered around 1 and defined as

$$Z(t) = \frac{X(t)}{m_X(t)} = e^{W(t)-Dt/2} \tag{6.60}$$

so that we immediately find that

$$m_Z(t) = 1 \qquad \sigma_Z^2(t) = e^{Dt} - 1$$

6.3 White Noise

From a strictly formal standpoint the *white noise* does not exist as a *sp*, in the same sense in which the *Dirac delta* $\delta(x)$ does not exist as a function. Neglecting however for short the rigorous definitions giving a precise meaning to this idea, we will limit ourselves here just to a few heuristic remarks to put in evidence its opportunities and pitfalls.

Definition 6.20 We will call **white noise** every process $B(t)$ with autocovariance

$$C_B(s,t) = q(t)\delta(t-s) \tag{6.61}$$

where $q(s) > 0$ is its **intensity**. A white noise is **stationary** when its intensity q is constant, and in this case with $\tau = t - s$ we will have

$$C_B(\tau) = q\delta(\tau) \qquad\qquad S_B(\varpi) = q \tag{6.62}$$

namely its covariance spectrum is flat and hence motivates the name of the process.

A white noise is then a *singular process*, as it is apparently disclosed by the occurrence of a Dirac delta in its definition. Their distinctively irregular character is substantiated by the remark that (6.61) entails in particular the non correlation of the process $B(t)$ in arbitrary separate instants, so that it takes values that are totally not predictable from previous observations. In the following we will survey a few examples of white noises in order to link their singular behavior to the use of processes allegedly derived as derivatives of other non-differentiable processes, in the same way that the Dirac δ can be considered as the derivative of the famously non-differentiable Heaviside function.

Exemple 6.21 Poisson white noise: Take first the shot noise (6.31) obtained by choosing $h(t) = \delta(t)$: in this case the process consists in a sequence of δ-like impulses at the random times T_k, and could be formally considered as the derivative, trajectory by trajectory, of a Poisson process in the form (6.6), namely

$$\dot{N}(t) = \sum_{k=1}^{\infty} \delta(t - T_k) \tag{6.63}$$

also called **process of the Poisson impulses**. To see that (6.63) is indeed a stationary white noise with intensity λ it is enough to remark that from (6.33) with $h(t) = \delta(t)$ we find $H = 1$ and $g(t) = \delta(t)$ and hence

$$m_{\dot{N}} = \lambda \qquad R_{\dot{N}}(\tau) = \lambda^2 + \lambda\delta(\tau) \qquad C_{\dot{N}}(\tau) = \lambda\delta(\tau) \tag{6.64}$$

A slightly different stationary white noise with zero expectation can be obtained as the derivative of a compensated Poisson process (6.22)

$$\tilde{\dot{N}}(t) = \dot{N}(t) - \lambda \tag{6.65}$$

It is apparent then that we are dealing here with a centered process of impulses because from (6.64) it is easy to see that

$$m_{\tilde{\dot{N}}} = m_{\dot{N}} - \lambda = 0$$

That this too is a stationary white noise follows from (6.64) and (6.65) since

$$R_{\tilde{\dot{N}}}(\tau) = C_{\tilde{\dot{N}}}(\tau) = R_{\dot{N}}(\tau) - \lambda^2 = \lambda\delta(\tau)$$

It is illuminating finally to remark that all the other shot noises (6.31) with arbitrary $h(t)$ different from $\delta(t)$ can be obtained from a convolution of the process of the impulses with the function $h(t)$ according to the relation

$$X(t) = [\dot{N} * h](t) \tag{6.66}$$

Exemple 6.22 Wienerian white noise: Another kind of white noise is associated to the Wiener process $W(t)$: we know that $W(t)$ is not differentiable and we can then surmise that its formal derivative $\dot{W}(t)$ could display the singular properties of a white nose. If we in fact consider, with a fixed Δt, the process of the difference quotients

$$Z(t) = \frac{\Delta W(t)}{\Delta t} \tag{6.67}$$

it is easy to see from the Proposition 6.19 about the increments process $\Delta W(t)$ that

$$m_Z = 0 \qquad R_Z(\tau) = C_Z(\tau) = \begin{cases} 0 & \text{if } |\tau| \geq |\Delta t| \\ \frac{D}{|\Delta t|}\left(1 - \frac{|\tau|}{|\Delta t|}\right) & \text{if } |\tau| < |\Delta t| \end{cases} \tag{6.68}$$

A plot of $R_Z(\tau) = C_Z(\tau)$ is displayed in the Fig. 6.17 and shows that

$$C_Z(\tau) \to D\,\delta(\tau) \qquad \Delta t \to 0$$

As a consequence, if in a sense whatsoever we accept that the derivative $\dot{W}(t)$ exists as the limit

$$Z(t) = \frac{\Delta W(t)}{\Delta t} \to \dot{W}(t) \qquad \Delta t \to 0$$

we also expect that

$$m_{\dot{W}} = 0 \qquad R_{\dot{W}}(\tau) = C_{\dot{W}}(\tau) = D\,\delta(\tau) \tag{6.69}$$

namely that $\dot{W}(t)$ is a stationary white noise of intensity D. The present discussion about the Wienerian white noise will be resumed in the later Sect. 8.1.

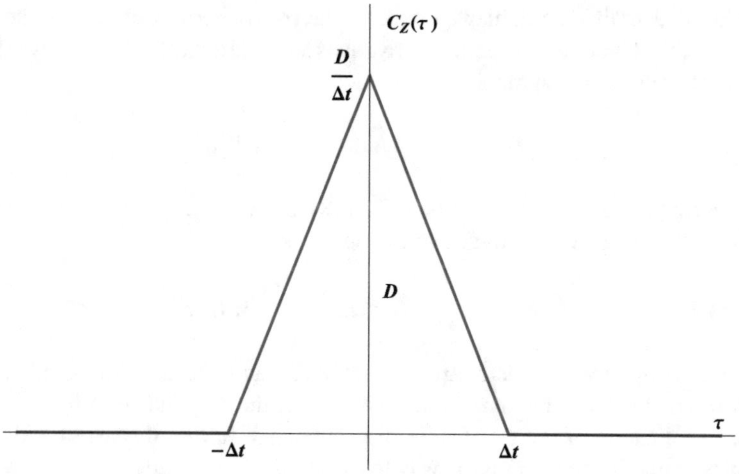

Fig. 6.17 Autocovariance $C_Z(\tau)$ (6.68) of the ratio (6.67) for a Wiener process: the triangle area always is D for every value of Δt, but its shape is higher and narrower for $\Delta t \to 0$

The previous examples hint to a few increment features that we will also resume later in further detail. Consider the increment processes $\Delta N(t)$, $\Delta\widetilde{N}(t)$ and $\Delta W(t)$, with $m_{\Delta\widetilde{N}}(t) = m_{\Delta W}(t) = 0$ from (6.17) e (6.54): we know then that

$$\boldsymbol{E}\left[\Delta\widetilde{N}^2(t)\right] = \sigma^2_{\Delta\widetilde{N}} = \lambda|\Delta t| \qquad \boldsymbol{E}\left[\Delta W^2(t)\right] = \sigma^2_{\Delta W} = D|\Delta t| \tag{6.70}$$

$$\boldsymbol{E}\left[\Delta N^2(t)\right] = \sigma^2_{\Delta N} + m^2_{\Delta N} = \lambda|\Delta t| + \lambda^2\Delta t^2 \tag{6.71}$$

that is—but only in a symbolic sense for the time being—in the limit $\Delta t \to 0$

$$\boldsymbol{E}\left[d\widetilde{N}^2(t)\right] = \boldsymbol{E}\left[dN^2(t)\right] = \lambda|dt| \qquad \boldsymbol{E}\left[dW^2(t)\right] = D|dt| \tag{6.72}$$

This prompts the idea that the infinitesimal increments of our processes are not in fact of the order dt, but rather of the order $\sqrt{dt}$, namely, symbolically again,

$$dN(t) = o\left(\sqrt{dt}\right) \qquad d\widetilde{N}(t) = o\left(\sqrt{dt}\right) \qquad dW(t) = o\left(\sqrt{dt}\right)$$

Albeit stated in a rather inaccurate form, this remark intuitively accounts for the non existence of the limits of the difference quotients (6.67), and then explains the non-differentiability of our processes. Remark moreover that, while strictly speaking the white noises $\dot{N}(t)$, $\dot{\widetilde{N}}(t)$ and $\dot{W}(t)$ do not exist as *sp*'s, the finite increments $\Delta N(t)$, $\Delta\widetilde{N}(t)$ and $\Delta W(t)$ always are well defined, and in a suitable sense (as we will see later in the framework of the stochastic calculus) do exist—and play a decisive role—also their *stochastic differentials* $dN(t)$, $d\widetilde{N}(t)$ and $dW(t)$ that for the time being we intuitively understand just as infinitesimals of the order $\sqrt{dt}$. What in any case is not possible to generalize straightaway without the risk of serious errors are the usual formulas of the calculus connecting derivatives and differentials. In other words, for a stochastic process the symbols

$$dN(t) \qquad d\widetilde{N}(t) \qquad dW(t)$$

can be correctly defined end used (see Chap. 8), but we can not right away identify them with the respective familiar expressions

$$\dot{N}(t)\,dt \qquad \dot{\widetilde{N}}(t)\,dt \qquad \dot{W}(t)\,dt$$

both because the involved derivatives do not exist, and because the increments are infinitesimals of the order $\sqrt{dt}$ rather than dt. In the Appendice H the risks of a careless application of the usual rules of calculus are further discussed with a few examples, while in the Chap. 8 the way to overcome these hurdles will be presented in a more systematic way.

6.4 Brownian Motion

After Robert Brown [2] observed in 1827 the flutter of pollen particles in a fluid (known since then as *Brownian motion*) a long debate started about the nature of this phenomenon, and even that these corpuscles were *living beings* was conjectured. Subsequent experiments showed that this is not the case, but the origin of the movement remained puzzling. We had to wait for the papers of Einstein [3] and Smoluchowski [4] to get a theory giving a satisfactory account. Einstein was able in particular to manufacture a physical model based on the interactions of the pollen grains with the surrounding fluid molecules: he showed that the movement is characterized by a diffusion coefficient D depending on the temperature, and also anticipated that the mean square displacement in a time t in every direction is proportional to $\sqrt{Dt}$. These statements—that today would ring trivial—were rather contentious at that time, and it is important to remember that the success of the Einstein model was a major factor in establishing the idea that matter is composed of atoms and molecules (as was proved for good by Jean Perrin [5]): an idea far from being generally shared at that time.

We must remember moreover that even a rigorous and coherent theory of the stochastic processes is quite new. The first pioneering ideas about models that can be traced back to the Wiener process are contained in a book by T.N. Thiele [6] on the method of least squares, and chiefly in the doctoral thesis of L. Bachelier [7]. The latter work however has long been ignored because his argument was a description of the prices behavior in the financial markets, a problem that has only recently become popular in the physical and mathematical context. For this reason the first works that actually opened the way to the modern study of the *sp*'s are those of Einstein in 1905,

Smoluchowski in 1906 and Langevin [8]. In particular, this group of articles identifies from the beginning the two paths that can be followed to examine the evolution of a random phenomenon:

1. We can first study the evolution of the *laws*, namely—in our notation—of the *pdf*'s $f_X(x, t)$ and of the transition *pdf*'s $f_X(x, t|y, s)$ by means of suitable partial differential equations to be satisfied by these functions; in this case the focus apparently is on the process distributions, and not on the process itself with its trajectories
2. Alternatively we can investigate the trajectories $x(t)$ of the process considering them as generalizations of the traditional functions, and in this case we will have to deal with differential equations on the process $X(t)$ more or less as we do in the traditional Newtonian mechanics. We will have to exercise, however, particular care to correctly define what these equations can mean: while in fact the process *pdf*'s are ordinary functions, the processes $X(t)$ we are dealing with are not in general differentiable.

Historically the articles by Einstein and Smoluchowski follow the first line of thought, while that of Langevin opened the second path. We will now briefly examine the problems posed by these articles to introduce the topic in an intuitive way, referring then to the Chap. 9 for a more in-depth discussion of the Brownian motion.

6.4.1 Einstein (1905)

Let's try to retrace—within a notation and a language adapted to ours—the arguments of Einstein's paper: let us first consider a time interval τ that is simultaneously *quite small* with respect to the macroscopic times of observation, and *quite large* with respect to the microscopic times of the movement of the molecules in the fluid. This choice, as we will see later, is instrumental: on the one hand it reflects the observation that pollen particles are *small* on macroscopic scales, but they are also *large* on molecular scales; on the other hand it allows us to realistically conjecture that two displacements in subsequent intervals τ actually are independent. If indeed τ is large with respect to the characteristic times of the molecular thermal agitation, we can think that the corresponding displacements resulting from the sum of many individual impacts are altogether independent of each other. The smallness of τ on macroscopic scales, on the other hand, allows us to use some convenient series expansions. The scale of a parameter such as τ, that is both small on macroscopic scales and large on microscopic scales, is also called *mesoscopic*. It must be said, however, that a description of the Brownian motion at smaller scales (we will present it in the Chap. 9) was subsequently proposed in 1930 by Ornstein and Uhlenbeck in another celebrated paper [9] in which a new process was defined that takes its name from its two proponents and that we will discuss at length in the next chapters.

Take then a *rv* Z, representing the pollen grain displacement in τ, and its *pdf* $g(z)$ that we will suppose symmetric and clustered near to $z = 0$ with

$$\int_{-\infty}^{+\infty} g(z)\,dz = 1, \qquad g(-z) = g(z)$$
$$g(z) \neq 0 \qquad \text{only for small values of} z$$

Einstein starts by proving that, if $X(t)$ is the position of the pollen grain at the time t, then its *pdf* $f(x,t)$ must comply with the equation

$$f(x, t+\tau) = \int_{-\infty}^{+\infty} f(x+z, t)g(z)\,dz \tag{6.73}$$

We will see later that this amounts to a particular form of the Markov property, but for the time being we will prove it just in this form. His line of reasoning is typically physical, but we prefer to give a justification in a language more adherent to our notations. We know indeed from the rules of conditioning that for the two *rv*'s $X(t)$ and $X(t+\tau)$ we can always write

$$\begin{aligned} f(x, t+\tau)\,dx &= \boldsymbol{P}\{x \leq X(t+\tau) < x+dx\} \\ &= \int_{-\infty}^{+\infty} \boldsymbol{P}\{x \leq X(t+\tau) < x+dx \mid X(t) = y\}\, f(y,t)\,dy \end{aligned}$$

On the other hand from our hypotheses we can say that $Z = X(t+\tau) - X(t)$ is independent from $X(t)$ and hence

$$\begin{aligned} \boldsymbol{P}\{x \leq X(t+\tau) < x+dx \mid X(t) = y\} &= \boldsymbol{P}\{x \leq X(t) + Z < x+dx \mid X(t) = y\} \\ &= \boldsymbol{P}\{x \leq y + Z < x+dx \mid X(t) = y\} \\ &= \boldsymbol{P}\{x - y \leq Z < x - y + dx\} \\ &= g(x-y)\,dx \end{aligned}$$

Taking then $z = y - x$, from the symmetry properties of $g(z)$ it follows that

$$\begin{aligned} f(x, t+\tau)\,dx &= dx \int_{-\infty}^{+\infty} g(x-y) f(y,t)\,dy \\ &= dx \int_{-\infty}^{+\infty} f(x+z, t)g(z)\,dz \end{aligned}$$

that is (6.73). Since now τ is small and $g(z)$ is non vanishing only for small z values, we are entitled to adopt in (6.73) the following Taylor expansions

$$\begin{aligned} f(x, t+\tau) &= f(x,t) + \tau\partial_t f(x,t) + o(\tau) \\ f(x+z, t) &= f(x,t) + z\partial_x f(x,t) + \frac{z^2}{2}\partial_x^2 f(x,t) + o(z^2) \end{aligned}$$

so that (in a slightly condensed notation)

$$f + \tau\partial_t f = f\int g(z)\,dz + \partial_x f\int zg(z)\,dz + \partial_x^2 f\int \frac{z^2}{2}g(z)\,dz$$

But from our hypotheses it is

$$\int_{-\infty}^{+\infty} g(z)\,dz = 1 \qquad \int_{-\infty}^{+\infty} zg(z)\,dz = 0$$

and hence we finally have

$$\partial_t f(x,t) = \partial_x^2 f(x,t)\,\frac{1}{2\tau}\int_{-\infty}^{+\infty} z^2 g(z)\,dz$$

It is apparent moreover that $g(z)$ (the *pdf* of the displacement Z in τ) is contingent on τ, and that by symmetry $\boldsymbol{E}[Z] = 0$, so that

$$\int_{-\infty}^{+\infty} z^2 g(z)\,dz$$

is the variance of Z that we can reasonably suppose to be infinitesimal for $\tau \to 0$ (in the sense that the increment Z tends to be invariably zero for $\tau \to 0$). If furthermore we suppose that

$$\lim_{\tau\to 0}\frac{1}{\tau}\int_{-\infty}^{+\infty} z^2 g(z)\,dz = D$$

our equation will be

$$\partial_t f(x,t) = \frac{D}{2}\partial_x^2 f(x,t) \tag{6.74}$$

namely will coincide with the Eq. (6.53) satisfied by the *pdf* of a Wiener process: its solution, with the initial condition $f(x, 0^+) = \delta(x)$ (that is $X(0) = 0$, $\boldsymbol{P}$-a.s.), will then be the normal *pdf* $\mathfrak{N}(0, Dt)$ with

$$\boldsymbol{E}[X(t)] = 0 \qquad \boldsymbol{V}[X(t)] = Dt \tag{6.75}$$

The first important outcome of this Einstein model is then that the Brownian particle **position** is well described by a **Wiener process** with diffusion coefficient D and a variance linearly growing with t. From further thermodynamical argumentations Einstein was also able to calculate the diffusion constant from the fluid temperature according to the formula

$$D = \frac{kT}{3\pi\eta a} \tag{6.76}$$

where k is the Boltzmann constant, T the temperature, η the viscosity and a the diameter of the supposedly spherical particle. It will be shown in the next section that the formula (6.76) and the other results listed above can also be derived, and in a simpler way, from in the Langevin model.

6.4.2 Langevin (1908)

In 1908 Langevin obtained basically the same results as Einstein directly handling the particle trajectories with a shrewd (though not very rigorous) generalization of the Newton equations of motion. In his model $X(t)$ is the particle position while $V(t) = \dot{X}(t)$ is its velocity that is supposed to be subjected to two kinds of forces due to the surrounding fluid:

- the deterministic force due to the viscous drag that, within our notations, is proportional to the velocity according to the formula $-6\pi\eta a V(t)$
- a random force $B(t)$ due to the collisions with the molecules, having zero expectation $\boldsymbol{E}[B(t)] = 0$ and uncorrelated to $X(t)$.

The Newton equation of motion then is

$$m\ddot{X}(t) = -6\pi\eta a\dot{X}(t) + B(t) \tag{6.77}$$

that, with $V(t) = \dot{X}(t)$, can also be reformulated as a first order equation for the velocity

$$m\dot{V}(t) = -6\pi\eta a V(t) + B(t) \tag{6.78}$$

known today as the ***Langevin equation***.

It will be argued in the Sect. 8.1 that the force $B(t)$ effectively behaves as the Wiener white noise of the Example 6.22: this will also be the starting point to introduce the Itō stochastic calculus. For the time being however we will just derive the behavior of $\boldsymbol{E}\left[X^2(t)\right]$ amounting to the position variance because $X(t)$ will be supposed always centered around the origin. To this end multiply then (6.77) by $X(t)$

$$mX(t)\ddot{X}(t) = -6\pi\eta a X(t)\dot{X}(t) + X(t)B(t)$$

and remarking that from the usual rules of calculus

$$\frac{d}{dt}\left[X^2(t)\right] = 2X(t)\dot{X}(t) \tag{6.79}$$

$$\frac{d^2}{dt^2}\left[X^2(t)\right] = 2\dot{X}^2(t) + 2X(t)\ddot{X}(t) = 2V^2(t) + 2X(t)\ddot{X}(t) \tag{6.80}$$

we could also write

$$\frac{m}{2}\frac{d^2}{dt^2}\left[X^2(t)\right] - mV^2(t) = -3\pi\eta a\frac{d}{dt}\left[X^2(t)\right] + X(t)B(t)$$

Take now the expectations of both the sides, and remembering that by hypothesis it is $\boldsymbol{E}[X(t)B(t)] = \boldsymbol{E}[X(t)]\,\boldsymbol{E}[B(t)] = 0$, we find

$$\frac{m}{2}\frac{d^2}{dt^2}\boldsymbol{E}\left[X^2(t)\right] - \boldsymbol{E}\left[mV^2(t)\right] = -3\pi\eta a\frac{d}{dt}\boldsymbol{E}\left[X^2(t)\right]$$

From the equipartition law of the statistical mechanics we moreover know that, at the thermal equilibrium, the average kinetic energy of a (one-dimensional) particle is

$$\boldsymbol{E}\left[\frac{m}{2}V^2(t)\right] = \frac{kT}{2} \tag{6.81}$$

where k is the Boltzmann constant and T the temperature. Our equation then reads as

$$\frac{m}{2}\frac{d^2}{dt^2}\boldsymbol{E}\left[X^2(t)\right] + 3\pi\eta a\frac{d}{dt}\boldsymbol{E}\left[X^2(t)\right] = kT$$

From a first integration we then have

$$\frac{d}{dt}\boldsymbol{E}\left[X^2(t)\right] = \frac{kT}{3\pi\eta a} + Ce^{-6\pi\eta at/m}$$

where C represents an integration constant, so that, after a short transition (the exponential vanishes with a characteristic time of the order of 10^{-8} sec), and taking into account (6.76), we find

$$\frac{d}{dt}\boldsymbol{E}\left[X^2(t)\right] = \frac{kT}{3\pi\eta a} = D$$

and then with a second integration we get again the Einstein result

$$\boldsymbol{V}[X(t)] = \boldsymbol{E}\left[X^2(t)\right] = Dt \tag{6.82}$$

where we have supposed $X(0) = 0$ in order to have $\boldsymbol{E}\left[X^2(0)\right] = 0$. The convenience of this Langevin treatment is that it is rather intuitive being based on a physical model with a simple equation of motion; also the calculations are rather elementary. However, it also presents some risks due of its shaky mathematical basis as it is elucidated in the Appendix H. The more rigorous foundations for a convincing reformulation of this Langevin model are instead postponed to the Chap. 8.

References

1. Papoulis, A.: Probability, Random Variables and Stochastic Processes. McGraw Hill, Boston (2002)
2. Brown, R.: A brief account of microscopical observations made in the months of June, July and August, 1827, on the particles contained in the pollen of plants; and on the general existence of

active molecules in organic and inorganic bodies. Phil. Mag. **4**, 161 (1828)
3. Einstein, A.: Über die von der molekularkinetischen Theorie der Wärme geforderte Bewegung von in ruhenden Flüssigkeiten suspendierten Teilchen. Ann. Phys. **17**, 549 (1905)
4. von Smoluchowski, M.: Zur kinetischen Theorie der Brownschen Molekularbewegung und der suspensionen. Ann. Phys. **21**, 757 (1906)
5. Perrin, J.: Mouvement brownien et réalité moléculaire. Ann. Chim. Phys. 8-ième série **18**, 5 (1909)
6. Thiele, T.N.: Om Anvendelse af mindste Kvadraters Methode i nogle Tilfaelde, hvor en Komplikation af visse Slags uensartede tilfaeldige Fejlkilder giver Fejlene en "systematisk" Karakter, B. Lunos Kgl. Hof.-Bogtrykkeri, (1880)
7. Bachelier, L.: Théorie de la Spéculation. Ann. Sci. de l'É.N.S. 3e série, tome **17**, 21 (1900)
8. Langevin, P.: On the theory of Brownian motion. C. R. Acad. Sci. (Paris) **146**, 530 (1908)
9. Ornstein, L.S., Uhlenbeck, G.E.: On the theory of Brownian motion. Phys. Rev. **36**, 823 (1930)

Chapter 7
Markovianity

7.1 Markov Processes

7.1.1 *Markov Property*

Although Markov property is generally given with a preferential time orientation—that from the *past* to the *future*—its statement is actually symmetric in both directions, and it could be intuitively expressed by saying that *the events of the future and those of the past result mutually independent conditionally to the knowledge of the information available at present.* To emphasize this symmetry we will start by giving the following definition of the ***Markov property***, briefly postponing a proof of its equivalence with the other, more familiar formulations.

Definition 7.1 We will say that an M components process $\boldsymbol{X}(t) = \big(X_1(t), \dots, X_M(t)\big)$ is a **Markov process** if, for every choice of n, of the instants $s_1 \le \dots \le s_m \le s \le t_1 \le \dots \le t_n$ and of the vectors $\boldsymbol{y}_1, \dots, \boldsymbol{y}_m, \boldsymbol{y}, \boldsymbol{x}_1, \dots, \boldsymbol{x}_n$, it is

$$\begin{aligned} &F(\boldsymbol{x}_n, t_n; \dots; \boldsymbol{x}_1, t_1; \boldsymbol{y}_m, s_m; \dots; \boldsymbol{y}_1, s_1 \mid \boldsymbol{y}, s) \\ &\qquad = F(\boldsymbol{x}_n, t_n; \dots; \boldsymbol{x}_1, t_1 \mid \boldsymbol{y}, s)\, F(\boldsymbol{y}_m, s_m; \dots; \boldsymbol{y}_1, s_1 \mid \boldsymbol{y}, s) \end{aligned} \tag{7.1}$$

where the F are the conditional *cdf*'s of the process.

It is important to remark that in this definition, more than the actual *ordering* of the time instants, it is important their *separation* (produced by a *present* s) into two groups (a *past* $s_1, \dots, s_m$, and a *future* $t_1, \dots, t_n$) that however play symmetrical roles: the ordering of the instants within the two groups is instead inconsequential. Of course, if $\boldsymbol{X}(t)$ is *ac* with its *pdf*'s then (7.1) will become

$$\begin{aligned} &f(\boldsymbol{x}_n, t_n; \dots; \boldsymbol{x}_1, t_1; \boldsymbol{y}_m, s_m; \dots; \boldsymbol{y}_1, s_1 \mid \boldsymbol{y}, s) \\ &\qquad = f(\boldsymbol{x}_n, t_n; \dots; \boldsymbol{x}_1, t_1 \mid \boldsymbol{y}, s)\, f(\boldsymbol{y}_m, s_m; \dots; \boldsymbol{y}_1, s_1 \mid \boldsymbol{y}, s) \end{aligned} \tag{7.2}$$

N. Cufaro Petroni, *Probability and Stochastic Processes for Physicists*, UNITEXT for Physics, https://doi.org/10.1007/978-3-030-48408-8_7

if instead it is discrete with integer values $\boldsymbol{k}, \boldsymbol{\ell}$ and distribution p we will have

$$\begin{aligned} & p(\boldsymbol{k}_n, t_n; \ldots; \boldsymbol{k}_1, t_1; \boldsymbol{\ell}_m, s_m; \ldots; \boldsymbol{\ell}_1, s_1 \mid \boldsymbol{\ell}, s) \\ & \quad = p(\boldsymbol{k}_n, t_n; \ldots; \boldsymbol{k}_1, t_1 \mid \boldsymbol{\ell}, s)\, p(\boldsymbol{\ell}_m, s_m; \ldots; \boldsymbol{\ell}_1, s_1 \mid \boldsymbol{\ell}, s) \end{aligned} \tag{7.3}$$

Proposition 7.2 *$\boldsymbol{X}(t)$ is a Markov process iff*

$$F(\boldsymbol{x}_n, t_n; \ldots; \boldsymbol{x}_1, t_1 \mid \boldsymbol{y}, s; \boldsymbol{y}_m, s_m; \ldots; \boldsymbol{y}_1, s_1) = F(\boldsymbol{x}_n, t_n; \ldots; \boldsymbol{x}_1, t_1 \mid \boldsymbol{y}, s) \tag{7.4}$$

for every choice of $s_1 \le \cdots \le s_m \le s \le t_1 \le \cdots \le t_n$ and $\boldsymbol{y}_1, \ldots, \boldsymbol{y}_m, \boldsymbol{y}, \boldsymbol{x}_1, \ldots, \boldsymbol{x}_n$; the roles of the past $s_1, \ldots, s_m$ and the future $t_1, \ldots, t_n$ in (7.4) can moreover be interchanged.

Proof Remark first that when $\boldsymbol{X}(t)$ is either *ac*, or discrete with integer values $\boldsymbol{k}, \boldsymbol{\ell}$, the condition (7.4) respectively become

$$f(\boldsymbol{x}_n, t_n; \ldots; \boldsymbol{x}_1, t_1 \mid \boldsymbol{y}, s; \boldsymbol{y}_m, s_m; \ldots; \boldsymbol{y}_1, s_1) = f(\boldsymbol{x}_n, t_n; \ldots; \boldsymbol{x}_1, t_1 \mid \boldsymbol{y}, s) \tag{7.5}$$

$$p(\boldsymbol{k}_n, t_n; \ldots; \boldsymbol{k}_1, t_1 \mid \boldsymbol{\ell}, s; \boldsymbol{\ell}_m, s_m; \ldots; \boldsymbol{\ell}_1, s_1) = p(\boldsymbol{k}_n, t_n; \ldots; \boldsymbol{k}_1, t_1 \mid \boldsymbol{\ell}, s) \tag{7.6}$$

For convenience we will however prove the proposition only in the form (7.5): first, if (7.5) holds we have

$$\begin{aligned} & f(\boldsymbol{x}_n, t_n; \ldots; \boldsymbol{x}_1, t_1; \boldsymbol{y}_m, s_m; \ldots; \boldsymbol{y}_1, s_1 \mid \boldsymbol{y}, s) \\ & \quad = \frac{f(\boldsymbol{x}_n, t_n; \ldots; \boldsymbol{x}_1, t_1; \boldsymbol{y}, s; \boldsymbol{y}_m, s_m; \ldots; \boldsymbol{y}_1, s_1)}{f(\boldsymbol{y}, s)} \\ & \quad = \frac{f(\boldsymbol{x}_n, t_n; \ldots; \boldsymbol{x}_1, t_1; \boldsymbol{y}, s; \boldsymbol{y}_m, s_m; \ldots; \boldsymbol{y}_1, s_1)}{f(\boldsymbol{y}, s; \boldsymbol{y}_m, s_m; \ldots; \boldsymbol{y}_1, s_1)} \, \frac{f(\boldsymbol{y}, s; \boldsymbol{y}_m, s_m; \ldots; \boldsymbol{y}_1, s_1)}{f(\boldsymbol{y}, s)} \\ & \quad = f(\boldsymbol{x}_n, t_n; \ldots; \boldsymbol{x}_1, t_1 \mid \boldsymbol{y}, s; \boldsymbol{y}_m, s_m; \ldots; \boldsymbol{y}_1, s_1)\, f(\boldsymbol{y}_m, s_m; \ldots; \boldsymbol{y}_1, s_1 \mid \boldsymbol{y}, s) \\ & \quad = f(\boldsymbol{x}_n, t_n; \ldots; \boldsymbol{x}_1, t_1 \mid \boldsymbol{y}, s)\, f(\boldsymbol{y}_m, s_m; \ldots; \boldsymbol{y}_1, s_1 \mid \boldsymbol{y}, s) \end{aligned}$$

that is (7.2) and the process is Markovian. Conversely, if the process is Markovian, namely if (7.2) holds, we have

$$\begin{aligned} & f(\boldsymbol{x}_n, t_n; \ldots; \boldsymbol{x}_1, t_1 \mid \boldsymbol{y}, s; \boldsymbol{y}_m, s_m; \ldots; \boldsymbol{y}_1, s_1) \\ & \quad = \frac{f(\boldsymbol{x}_n, t_n; \ldots; \boldsymbol{x}_1, t_1; \boldsymbol{y}, s; \boldsymbol{y}_m, s_m; \ldots; \boldsymbol{y}_1, s_1)}{f(\boldsymbol{y}, s; \boldsymbol{y}_m, s_m; \ldots; \boldsymbol{y}_1, s_1)} \\ & \quad = \frac{f(\boldsymbol{x}_n, t_n; \ldots; \boldsymbol{x}_1, t_1; \boldsymbol{y}, s; \boldsymbol{y}_m, s_m; \ldots; \boldsymbol{y}_1, s_1)}{f(\boldsymbol{y}, s)} \, \frac{f(\boldsymbol{y}, s)}{f(\boldsymbol{y}, s; \boldsymbol{y}_m, s_m; \ldots; \boldsymbol{y}_1, s_1)} \\ & \quad = \frac{f(\boldsymbol{x}_n, t_n; \ldots; \boldsymbol{x}_1, t_1; \boldsymbol{y}_m, s_m; \ldots; \boldsymbol{y}_1, s_1 \mid \boldsymbol{y}, s)}{f(\boldsymbol{y}_m, s_m; \ldots; \boldsymbol{y}_1, s_1 \mid \boldsymbol{y}, s)} = f(\boldsymbol{x}_n, t_n; \ldots; \boldsymbol{x}_1, t_1 \mid \boldsymbol{y}, s) \end{aligned}$$

and we recover (7.4). Given moreover the past-future symmetry of (7.1), the ordering of the instants is not relevant for the proof and can be modified by interchanging past and future, but always in compliance with their separation. ■

In this second formulation—in the version from the *past* to the *future*—the Markov property states that the information afforded by the last available observation (the *present*, here the time s) summarizes all that is worthwhile in the *past* (the instants $s_1, \ldots, s_m$) in order to forecast the *future*: to this end it is relevant indeed to know *where* the process is at the present s, but there is no need to know *how* (namely, along which path) it arrived there. Of course the future is not altogether independent from the past, but the latter is superfluous because the previous history is summarized in the present s. This is in fact the simplest way to introduce a non trivial dependence among the values of the process in different instants. Remark that if the *future* is reduced to a single instant t then (7.5) and (7.6) take a simplified form highlighting the role of the *transition probabilities and pdf's*

$$f(\boldsymbol{x}, t \mid \boldsymbol{y}, s; \boldsymbol{y}_m, s_m; \ldots; \boldsymbol{y}_1, s_1) = f(\boldsymbol{x}, t \mid \boldsymbol{y}, s) \tag{7.7}$$

$$p(\boldsymbol{k}, t \mid \boldsymbol{\ell}, s; \boldsymbol{\ell}_m, s_m; \ldots; \boldsymbol{\ell}_1, s_1) = p(\boldsymbol{k}, t \mid \boldsymbol{\ell}, s) \tag{7.8}$$

Corollary 7.3 *$\boldsymbol{X}(t)$ is a Markov process iff, for every choice of the instants $s_1 \le \cdots \le s_m \le s \le t_1 \le \cdots \le t_n$ and of an arbitrary bounded Borel function $g(\boldsymbol{x}_1, \ldots, \boldsymbol{x}_n)$, it is*

$$\begin{aligned} &\boldsymbol{E}\left[g(\boldsymbol{X}(t_1), \ldots, \boldsymbol{X}(t_n)) \mid \boldsymbol{X}(s), \boldsymbol{X}(s_m), \ldots, \boldsymbol{X}(s_1)\right] \\ &\qquad = \boldsymbol{E}\left[g(\boldsymbol{X}(t_1), \ldots, \boldsymbol{X}(t_n)) \mid \boldsymbol{X}(s)\right] \qquad \boldsymbol{P}\text{-}a.s. \end{aligned} \tag{7.9}$$

that is iff, with arbitrary $\boldsymbol{y}_1, \ldots, \boldsymbol{y}_m, \boldsymbol{y}$, it turns out that

$$\begin{aligned} &\boldsymbol{E}\left[g(\boldsymbol{X}(t_1), \ldots, \boldsymbol{X}(t_n)) \mid \boldsymbol{X}(s) = \boldsymbol{y}, \boldsymbol{X}(s_m) = \boldsymbol{y}_m, \ldots, \boldsymbol{X}(s_1) = \boldsymbol{y}_1\right] \\ &\qquad = \boldsymbol{E}\left[g(\boldsymbol{X}(t_1), \ldots, \boldsymbol{X}(t_n)) \mid \boldsymbol{X}(s) = \boldsymbol{y}\right] \qquad \boldsymbol{P}_X\text{-}a.s. \end{aligned} \tag{7.10}$$

where $\boldsymbol{P}_X$ is here a shorthand for the joint distribution of $\boldsymbol{X}(s), \boldsymbol{X}(s_m), \ldots, \boldsymbol{X}(s_1)$. From the Definition 3.41 it follows that the condition (7.9) is also equivalent to require that for every $g(\boldsymbol{x}_1, \ldots, \boldsymbol{x}_n)$ is is possible to find another Borel function $h(\boldsymbol{x})$ such that

$$\boldsymbol{E}\left[g(\boldsymbol{X}(t_1), \ldots, \boldsymbol{X}(t_n)) \mid \boldsymbol{X}(s), \boldsymbol{X}(s_m), \ldots, \boldsymbol{X}(s_1)\right] = h(\boldsymbol{X}(s)) \qquad \boldsymbol{P}\text{-}a.s. \tag{7.11}$$

Of course even here past and future can be exchanged.

Proof If (7.10) holds, taking $g = \chi_A$ the indicator of the event $A = (-\infty, \boldsymbol{x}_1] \times \cdots \times (-\infty, \boldsymbol{x}_n]$ we find

$$\begin{aligned}
&F(\boldsymbol{x}_n, t_n, \dots, \boldsymbol{x}_1, t_1 \mid \boldsymbol{y}, s; \boldsymbol{y}_m, s_m \dots; \boldsymbol{y}_1, s_1) \\
&\quad = \boldsymbol{P}\big\{(\boldsymbol{X}(t_1), \dots, \boldsymbol{X}(t_n)) \in A \mid \boldsymbol{X}(s) = \boldsymbol{y}; \dots; \boldsymbol{X}(s_1) = \boldsymbol{y}_1\big\} \\
&\quad = \boldsymbol{E}\left[\chi_A(\boldsymbol{X}(t_1), \dots, \boldsymbol{X}(t_n)) \mid \boldsymbol{X}(s) = \boldsymbol{y}; \dots; \boldsymbol{X}(s_1) = \boldsymbol{y}_1\right] \\
&\quad = \boldsymbol{E}\left[\chi_A(\boldsymbol{X}(t_1), \dots, \boldsymbol{X}(t_n)) \mid \boldsymbol{X}(s) = \boldsymbol{y}\right] \\
&\quad = \boldsymbol{P}\{(\boldsymbol{X}(t_1), \dots, \boldsymbol{X}(t_n)) \in A \mid \boldsymbol{X}(s) = \boldsymbol{y}\} \;=\; F(\boldsymbol{x}_n, t_n, \dots, \boldsymbol{x}_1, t_1 \mid \boldsymbol{y}, s)
\end{aligned}$$

namely (7.4). if conversely (7.4) holds, the (7.10) easily follows because the conditional expectations are calculated from the conditional cdf's. ■

It is important to point out now that the Markovianity of a vector process $\boldsymbol{X}(t)$ in no way entails the Markovianity of its individual components $X_j(t)$ (or of a subset of them): these components in fact provide less information that the whole vector, and hence are often non Markovian. We can look at this remark, however, also from a reverse standpoint: if for instance a process with just one component $X_1(t)$ is not Markovian, it is in general possible to add further components to assemble a Markovian vector. This apparently considerably widens the scope of the Markov property, because in many practical cases we will be entitled to consider our possibly non Markovian processes just as components of some suitable Markovian vector process (see in particular the analysis of the Brownian motion in the Sect. 9.3).

Proposition 7.4 Markov chain rule: *The law of a Markov process $\boldsymbol{X}(t)$ is completely specified by its one-time distribution plus its transition distribution, that is—respectively in the ac and discrete cases—by*

$$f(\boldsymbol{x}, t) \quad \textit{and} \quad f(\boldsymbol{x}, t \mid \boldsymbol{y}, s) \tag{7.12}$$

$$p(\boldsymbol{k}, t) \quad \textit{and} \quad p(\boldsymbol{k}, t \mid \boldsymbol{\ell}, s) \tag{7.13}$$

according to the chain rules

$$f(\boldsymbol{x}_n, t_n; \dots; \boldsymbol{x}_1, t_1) = f(\boldsymbol{x}_n, t_n \mid \boldsymbol{x}_{n-1}, t_{n-1}) \dots f(\boldsymbol{x}_2, t_2 \mid \boldsymbol{x}_1, t_1) f(\boldsymbol{x}_1, t_1) \tag{7.14}$$

$$p(\boldsymbol{k}_n, t_n; \dots; \boldsymbol{k}_1, t_1) = p(\boldsymbol{k}_n, t_n \mid \boldsymbol{k}_{n-1}, t_{n-1}) \dots p(\boldsymbol{k}_2, t_2 \mid \boldsymbol{k}_1, t_1) p(\boldsymbol{k}_1, t_1) \tag{7.15}$$

where the time ordering must be either ascending ($t_1 \leq \dots \leq t_n$) or descending.

Proof As recalled in the Sect. 5.1, the global law of a process is specified when we know all its joint laws in an arbitrary (finite) number of arbitrary instants; but it is easy to see—for instance in the *ac* case—that if $\boldsymbol{X}(t)$ is Markovian such joint laws can be retrieved from the (7.12). Take indeed the arbitrary instants $t_1 \leq \dots \leq t_n$ (possibly also in the reverse order): from the definition (3.53) of conditional *pdf* and from (7.7) we have in fact

$$
\begin{aligned}
f(\boldsymbol{x}_n, t_n; \ldots; \boldsymbol{x}_1, t_1) = & f(\boldsymbol{x}_n, t_n \,|\, \boldsymbol{x}_{n-1}, t_{n-1}; \ldots; \boldsymbol{x}_1, t_1) \cdot \\
& f(\boldsymbol{x}_{n-1}, t_{n-1} \,|\, \boldsymbol{x}_{n-2}, t_{n-2}; \ldots; \boldsymbol{x}_1, t_1) \cdot \ldots \\
& \cdot f(\boldsymbol{x}_2, t_2 \,|\, \boldsymbol{x}_1, t_1)\, f(\boldsymbol{x}_1, t_1) \\
= & f(\boldsymbol{x}_n, t_n \,|\, \boldsymbol{x}_{n-1}, t_{n-1})\, f(\boldsymbol{x}_{n-1}, t_{n-1} \,|\, \boldsymbol{x}_{n-2}, t_{n-2}) \cdot \ldots \\
& \cdot f(\boldsymbol{x}_2, t_2 \,|\, \boldsymbol{x}_1, t_1)\, f(\boldsymbol{x}_1, t_1)
\end{aligned}
$$

that is the chain rule (7.14), so that all the joint *pdf*'s are recovered from (7.12). In the discrete case the proof is the same. Remark that to find (7.14) from the Markov property (7.7) the time instants must be put either in an ascending or in a descending order: any other possibility is excluded. ■

7.1.2 Chapman-Kolmogorov Equations

The importance of Markovianity is immediately appreciated if one reflects on the fact that, because of the Proposition 7.4, this property allows to completely reconstruct the whole hierarchy of the joint *pdf*'s of $\boldsymbol{X}(t)$—that is its global law—starting just from the knowledge of its one-time laws and of its transition laws (namely either (7.12) or (7.13)): this is a considerable simplification certainly not available for all other kinds of process. In this regard, however, it should be noted immediately that the functions (7.12) and (7.13) are always well defined for any type of process, even for the non-Markovian ones: but in this latter eventuality they are no longer sufficient to fully determine the law of the process. Furthermore it must be remembered that in general there can be different processes (not all Markovian, of course) sharing both the same one time laws, and the same transition laws displayed in (7.12) or (7.13). In other words, given (for example in the *ac* case) the pair of functions (7.12), in general there will be several different processes that admit them as one time *pdf* and transition *pdf*: if among them there is a Markovian one—but this is in no way assured—this is unique and retrievable through the chain rule (7.14) from the (7.12) only. These considerations suggest then that the simple a priori assignment of a pair of functions (7.12) does not guarantee at all that it is possible to assemble a Markov process from them: such functions could in fact be associated to a plurality of processes of which no one is Markovian. It is therefore important to be able first to find whether a given pair (7.12) may or may not generate a Markov process, and we will see that, while $f(\boldsymbol{x}, t)$ is totally arbitrary, not every possible transition *pdf* $f(\boldsymbol{x}, t, |\boldsymbol{y}, s)$ can do the job. For short in the following discussion the integrals and the sums without further indications are understood to be performed on the whole available domain, for example $\boldsymbol{R}^M$, $\boldsymbol{N}$, while $d\boldsymbol{x}$ is a simplification of $d^M\boldsymbol{x}$.

Proposition 7.5 Chapman-Kolmogorov equations: *If* $X(t)$ *is a Markov process the following equations hold in the ac case*

$$f(\boldsymbol{x},t) = \int f(\boldsymbol{x},t\,|\,\boldsymbol{y},s)f(\boldsymbol{y},s)\,d\boldsymbol{y} \tag{7.16}$$

$$f(\boldsymbol{x},t\,|\,\boldsymbol{y},s) = \int f(\boldsymbol{x},t\,|\,\boldsymbol{z},r)f(\boldsymbol{z},r\,|\,\boldsymbol{y},s)\,d\boldsymbol{z} \qquad s \le r \le t \tag{7.17}$$

while in the discrete case we will have

$$p(\boldsymbol{k},t) = \sum_{\boldsymbol{\ell}} p(\boldsymbol{k},t\,|\,\boldsymbol{\ell},s)p(\boldsymbol{\ell},s) \tag{7.18}$$

$$p(\boldsymbol{k},t\,|\,\boldsymbol{\ell},s) = \sum_{\boldsymbol{j}} p(\boldsymbol{k},t\,|\,\boldsymbol{j},r)p(\boldsymbol{j},r\,|\,\boldsymbol{\ell},s) \qquad s \le r \le t \tag{7.19}$$

The ordering of s, r, t can be reversed, but r must always fall between s and t.

Proof We remark first that the Eqs. (7.16) and (7.18) are satisfied by every process (even non Markovian): if indeed for instance $\boldsymbol{X}(t)$ has a *pdf*, (7.16) immediately derives from the definitions:

$$f(\boldsymbol{x},t) = \int f(\boldsymbol{x},t;\,\boldsymbol{y},s)\,d\boldsymbol{y} = \int f(\boldsymbol{x},t\,|\,\boldsymbol{y},s)f(\boldsymbol{y},s)\,d\boldsymbol{y}$$

In the discrete case (7.18) follows in the same way. On the other hand only a Markov process can satisfy the Eqs. (7.17) and (7.19): to deduce (7.17) it is enough to remark for example that

$$\begin{aligned} f(\boldsymbol{x},t\,|\,\boldsymbol{y},s) &= \int f(\boldsymbol{x},t;\,\boldsymbol{z},r\,|\,\boldsymbol{y},s)\,d\boldsymbol{z} = \int f(\boldsymbol{x},t\,|\,\boldsymbol{z},r;\,\boldsymbol{y},s)f(\boldsymbol{z},r\,|\,\boldsymbol{y},s)\,d\boldsymbol{z} \\ &= \int f(\boldsymbol{x},t\,|\,\boldsymbol{z},r)f(\boldsymbol{z},r\,|\,\boldsymbol{y},s)\,d\boldsymbol{z} \end{aligned}$$

where we used again the definitions, but also the Markov property (7.7) which holds both for $s \le r \le t$, and for $t \le r \le s$. ■

As a consequence, while on the one hand the Eqs. (7.16) and (7.18) allow to calculate the one time law at the instant t from those in previous times (the transition laws playing the role of *propagators*), on the other hand the Eqs. (7.17) and (7.19) are authentic ***Markovian compatibility conditions*** for the transition laws: if (7.17) and (7.19) are not met there is no hope to use either $f(\boldsymbol{x},t\,|\,\boldsymbol{y},s)$ or $p(\boldsymbol{k},t\,|\,\boldsymbol{\ell},s)$ to assemble a Markov process. When instead these equations are satisfied a Markov process can always be manufactured taking advantage of the chain rule of the Proposition 7.4.

Definition 7.6 We will call **Markovian transition laws** those whose $f(\boldsymbol{x},t\,|\,\boldsymbol{y},s)$ (or $p(\boldsymbol{k},t\,|\,\boldsymbol{\ell},s)$) satisfy (7.17) (respectively (7.19)); taken then an arbitrary but fixed conditioning instant $s \ge 0$, we will furthermore tell apart the **advanced** ($0 \le t < s$) from the **retarded region** ($s \le t$).

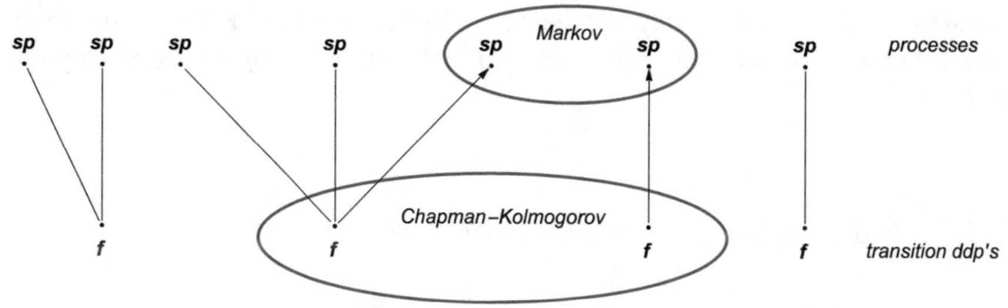

Fig. 7.1 Possible relations between the process Markovianity and the Markovianity of the transition *pdf*'s. The two arrow heads convey the notion that only the laws of Markov processes can be deduced from the initial and transition *pdf*'s only

Proposition 7.7 *To every Markovian transition law, known at least in the retarded region,*

$$f(\boldsymbol{x}, t \mid \boldsymbol{y}, s) \quad \text{or} \quad p(\boldsymbol{k}, t \mid \boldsymbol{\ell}, s) \qquad 0 \leq s \leq t$$

is associated a family of Markov processes, one for every initial law $f_0(\boldsymbol{x})$ *(or* $p_0(\boldsymbol{\ell})$*).*

Proof Confining ourselves for short to the *ac* case, take an arbitrary initial *pdf* $f(\boldsymbol{x}, 0) = f_0(\boldsymbol{x})$: then the Markovian transition *pdf* $f(\boldsymbol{x}, t \mid \boldsymbol{y}, s)$ in the retarded region enables us to calculate the one time *pdf* at every instant

$$f(\boldsymbol{x}, t) = \int f(\boldsymbol{x}, t \mid \boldsymbol{y}, 0) f_0(\boldsymbol{y})\, d\boldsymbol{y} \tag{7.20}$$

and hence to supplement the pair (7.12) needed to calculate the law of the Markov process from the chain rule (7.14). Remark that to perform this last step it is enough to know the transition *pdf* only in the *retarded region* when the chain rule is chosen with ascending times. ∎

Everything said so far has been settled in the perspective of *assembling* a Markov process from given $f_0(\boldsymbol{x})$ and $f(\boldsymbol{x}, t, \mid \boldsymbol{y}, s)$: we have shown that this is always possible if $f(\boldsymbol{x}, t, \mid \boldsymbol{y}, s)$ is Markovian, and known at least in the retarded region. There is however a reverse standpoint that is just as relevant: given a *sp*, how can we *check* whether it is Markovian or not? Can we confine ourselves to inspect its transition distributions? In this second perspective the Chapman-Kolmogorov equations are only a *necessary* condition of Markovianity: they effectively afford us just a way to say with certainty whether a process *is not* Markovian. There exist indeed examples of *sp*'s—see the Appendix I for details—that are not Markovian despite having Markovian transition distribution. This seeming inconsistency (see also Fig. 7.1) is suitably investigated only in the framework of the *non uniqueness* of the processes having the same transition laws. If a *sp* is given and, for example, its transition *pdf*'s satisfy the Chapman-Kolmogorov equations, this is not sufficient to affirm that our process is Markovian: starting from the given Markovian transition *pdf* a Markov process can always be built in a unique way, but there can also be other non Markovian processes featuring the same transition *pdf*, and in particular our initial *sp* can be exactly one of them.

7.1.3 Independent Increments Processes

Definition 7.8 We will say that $\boldsymbol{X}(t)$ is an **independent increments process** if, for every choice of $0 \le t_0 < t_1 < \cdots < t_{n-1} < t_n$, the *rv*'s $\boldsymbol{X}(t_0),\ \boldsymbol{X}(t_1) - \boldsymbol{X}(t_0), \ldots,$ $\boldsymbol{X}(t_n) - \boldsymbol{X}(t_{n-1})$ are independent. In particular an increment $\Delta\boldsymbol{X}(t) = \boldsymbol{X}(t+\Delta t) - \boldsymbol{X}(t)$ with $\Delta t > 0$ will be independent[1] from every $\boldsymbol{X}(s)$ with $s \le t$.

We have already met several examples of *sp*'s with independent increments by construction (for instance the Poisson and the Wiener processes, and several of their byproducts): we will show now that the processes enjoying this property represent in fact an especially important class of Markov processes.

Proposition 7.9 *Every independent increments process* $\boldsymbol{X}(t)$ *is Markovian and its distribution is completely specified (but for an initial condition) by the law of their increments* $\Delta\boldsymbol{X}(t) = \boldsymbol{X}(t+\Delta t) - \boldsymbol{X}(t)$ *with* $\Delta t > 0$. *Moreover it is*

$$\boldsymbol{E}\left[g(\boldsymbol{X}(t+\Delta t)) \,|\, \boldsymbol{X}(t) = \boldsymbol{x}\right] = \boldsymbol{E}\left[g(\Delta\boldsymbol{X}(t) + \boldsymbol{x})\right] \tag{7.21}$$

Proof To keep the things short we will prove the Markovianity in the form (7.4) with just one future time $t + \Delta t$, and a slightly different notation: take the instants $t_1 \le \cdots \le t_m \le t \le t + \Delta t$, then—from the increments independence and a straightforward application of the point 3 of the Proposition 3.42—it is

$$\begin{aligned}
&F_{\boldsymbol{X}}(\boldsymbol{x}, t+\Delta t \mid \boldsymbol{y}, t; \boldsymbol{y}_m, t_m; \ldots; \boldsymbol{y}_1, t_1) \\
&\quad = \boldsymbol{P}\big\{\boldsymbol{X}(t+\Delta t) \le \boldsymbol{x} \mid \boldsymbol{X}(t) = \boldsymbol{y}; \boldsymbol{X}(t_m) = \boldsymbol{y}_m; \ldots; \boldsymbol{X}(t_1) = \boldsymbol{y}_1\big\} \\
&\quad = \boldsymbol{P}\big\{\Delta\boldsymbol{X}(t) + \boldsymbol{X}(t) \le \boldsymbol{x} \mid \boldsymbol{X}(t) = \boldsymbol{y}; \boldsymbol{X}(t_m) = \boldsymbol{y}_m; \ldots; \boldsymbol{X}(t_1) = \boldsymbol{y}_1\big\} \\
&\quad = \boldsymbol{P}\big\{\Delta\boldsymbol{X}(t) + \boldsymbol{y} \le \boldsymbol{x} \mid \boldsymbol{X}(t) = \boldsymbol{y}; \boldsymbol{X}(t_m) = \boldsymbol{y}_m; \ldots; \boldsymbol{X}(t_1) = \boldsymbol{y}_1\big\} \\
&\quad = \boldsymbol{P}\{\Delta\boldsymbol{X}(t) + \boldsymbol{y} \le \boldsymbol{x} \mid \boldsymbol{X}(t) = \boldsymbol{y}\} = \boldsymbol{P}\{\Delta\boldsymbol{X}(t) + \boldsymbol{X}(t) \le \boldsymbol{x} \mid \boldsymbol{X}(t) = \boldsymbol{y}\} \\
&\quad = \boldsymbol{P}\{\boldsymbol{X}(t+\Delta t) \le \boldsymbol{x} \mid \boldsymbol{X}(t) = \boldsymbol{y}\} = F_{\boldsymbol{X}}(\boldsymbol{x}, t+\Delta t \mid \boldsymbol{y}, t)
\end{aligned}$$

In the same way we have moreover that

---

[1] With $0 \le s \le t \le t + \Delta t$, the *r-vec*'s

$$\boldsymbol{X}(0) \qquad \boldsymbol{X}(s) - \boldsymbol{X}(0) \qquad \boldsymbol{X}(t) - \boldsymbol{X}(s) \qquad \Delta\boldsymbol{X}(t) = \boldsymbol{X}(t+\Delta t) - \boldsymbol{X}(t)$$

are all independent by definition, and hence $\Delta\boldsymbol{X}(t)$ and $\boldsymbol{X}(s) = [\boldsymbol{X}(s) - \boldsymbol{X}(0)] + \boldsymbol{X}(0)$ are independent too.

$$
\begin{aligned}
F_X(\boldsymbol{x}, t+\Delta t \mid \boldsymbol{y}, t) &= \boldsymbol{P}\{\boldsymbol{X}(t+\Delta t) \leq \boldsymbol{x} \mid \boldsymbol{X}(t) = \boldsymbol{y}\} \\
&= \boldsymbol{P}\{\Delta\boldsymbol{X}(t) + \boldsymbol{X}(t) \leq \boldsymbol{x} \mid \boldsymbol{X}(t) = \boldsymbol{y}\} \\
&= \boldsymbol{P}\{\Delta\boldsymbol{X}(t) \leq \boldsymbol{x} - \boldsymbol{y} \mid \boldsymbol{X}(t) = \boldsymbol{y}\} \\
&= \boldsymbol{P}\{\Delta\boldsymbol{X}(t) \leq \boldsymbol{x} - \boldsymbol{y}\} = F_{\Delta X}(\boldsymbol{x} - \boldsymbol{y}, \Delta t; t)
\end{aligned}
$$

where $F_{\Delta X}$ is the *cdf* of the increment $\Delta\boldsymbol{X}(t)$ that apparently is now all we need to find the global law of the process: we know indeed from the Proposition 7.7 that—but for an initial distribution—this global law is completely determined from the retarded transition distribution that here coincides with the increment distribution. Of course, when the process is *ac*, we also can calculate the *pdf*'s by differentiation and in particular we find

$$
f_X(\boldsymbol{x}, t+\Delta t \mid \boldsymbol{y}, t) = f_{\Delta X}(\boldsymbol{x} - \boldsymbol{y}, \Delta t; t) \qquad \Delta t > 0 \tag{7.22}
$$

As for (7.21) we finally have

$$
\boldsymbol{E}\left[g(\boldsymbol{X}(t+\Delta t)) \mid \boldsymbol{X}(t) = \boldsymbol{x}\right] = \boldsymbol{E}\left[g(\Delta\boldsymbol{X}(t) + \boldsymbol{x}) \mid \boldsymbol{X}(t) = \boldsymbol{x}\right] = \boldsymbol{E}\left[g(\Delta\boldsymbol{X}(t) + \boldsymbol{x})\right]
$$

just by retracing the previous lines of reasoning. ■

This result is very important to understand the deep connection existing between the theory of independent increments processes and the *limit theorems* of the Chap. 4. If indeed $\boldsymbol{X}(t)$ is an independent increment process, we have just seen that the knowledge of the increment distribution is paramount: to study these increments $\Delta\boldsymbol{X}(t) = \boldsymbol{X}(t) - \boldsymbol{X}(s)$ we can now decompose the interval $[s, t]$ in n sub-intervals by means of the points

$$
s = t_0 < t_1 < \cdots < t_{n-1} < t_n = t
$$

and remark that the n increments $\boldsymbol{X}(t_k) - \boldsymbol{X}(t_{k-1})$ with $k = 1, \ldots, n$ are all independent. As a consequence the increment $\Delta\boldsymbol{X}(t)$ is the sum of n independent *rv*'s, and since n and the separation points are arbitrary it is easy to understand that in general the law of the increment $\Delta\boldsymbol{X}(t)$ will be the limit law of some suitable sequence of sums of independent *rv*'s. It is then cardinal for a complete understanding of the independent increment processes (a large class of Markov processes) to be able to identify all the possible limit laws of sums of independent *rv*'s. We already know some of them: the Gaussian laws (Central limit Theorem), the degenerate laws (Law of large numbers) and the Poisson laws (Poisson theorem). It would be possible to show however that these are only the most widespread examples of a much larger class of possible limit laws known as *infinitely divisible laws*, first suggested in 1929 by B. de Finetti and then completely classified in the 30's with the results of P. Lévy, A. Khinchin and others. We will not have here the leisure for a detailed discussion of these distributions that are a cornerstone of the *Lévy processes* (see later Sect. 7.1.6, and [1] for further details) and we will confine ourselves to give below just their definition.

Definition 7.10 We will say that a law with *chf* $\varphi(\boldsymbol{u})$ is **infinitely divisible** when for every $n = 1, 2, \ldots$ we can find another *chf* $\varphi_n(\boldsymbol{u})$ such that $\varphi(\boldsymbol{u}) = [\varphi_n(\boldsymbol{u})]^n$.

It is easy to see then from (4.5) that a *rv* with an infinitely divisible law φ can always be decomposed in the sum of an arbitrary number n of other *iid*rv's with law φ_n, and this apparently accounts for the chosen name. Beyond the three mentioned families of laws (Gauss, degenerate and Poisson), are infinitely divisible also the laws of Cauchy and Student, the exponentials and many other families of discrete and continuous laws. There are conversely several important families of laws that are not infinitely divisible: in particular it can be shown that no distribution with the probability concentrated in a bounded interval (as the uniform, the beta or the Bernoulli) can be infinitely divisible. Of course the distributions that are not infinitely divisible can not be the laws of the independent increments of a Markov processes.

7.1.4 Stationarity and Homogeneity

We already know that generally speaking the increments stationarity of a process does not entail its global stationarity in the strict sense (see Definition 5.9); but in fact even its wide sense stationarity (5.15) is not assured. There are indeed examples (the Poisson and Wiener processes are cases in point) with stationary increments, but non constant expectation and/or variance. On the other hand even the requirements (5.12) and (5.13) on the one- and two-times laws, albeit entailing at least the wide sense stationarity, are not sufficient for the strict sense stationarity. The case of Markov processes, however, is rather peculiar: for short in the following we will confine the exposition to *ac* processes with a *pdf*.

Proposition 7.11 *A Markov process* $\boldsymbol{X}(t)$ *is strict sense stationary iff*

$$f(\boldsymbol{x}, t) = f(\boldsymbol{x}) \tag{7.23}$$

$$f(\boldsymbol{x}, t; \boldsymbol{y}, s) = f(\boldsymbol{x}, \boldsymbol{y}; \tau) \qquad \tau = t - s \tag{7.24}$$

In this case its transition pdf's will depend only on τ *according to the notation*

$$f(\boldsymbol{x}, t \mid \boldsymbol{y}, s) = f(\boldsymbol{x}, \tau \mid \boldsymbol{y}) \tag{7.25}$$

and if moreover $\boldsymbol{X}(t)$ *also has independent increments these transition pdf's will coincide with the stationary increment laws*

$$f(\boldsymbol{x}, \tau \mid \boldsymbol{y}) = f_{\Delta X}(\boldsymbol{x} - \boldsymbol{y}, \tau) \tag{7.26}$$

Proof The strict sense stationarity (5.11)

$$f(\boldsymbol{x}_1, t_1; \ldots; \boldsymbol{x}_n, t_n) = f(\boldsymbol{x}_1, t_1 + s; \ldots; \boldsymbol{x}_n, t_n + s) \tag{7.27}$$

for every $t_1, \ldots, t_n$ and s, follows from (7.23) and (7.25) when the joint *pdf* in the r.h.s. of (7.27) is calculated with the Markov chain rule of Proposition 7.4 because only the time differences are taken into account. The reverse statement is trivial. Since $\boldsymbol{X}(t)$ is strict sense stationary also its increments $\Delta\boldsymbol{X}(t)$ will be stationary (see Sect. 5.5) and their *pdf* $f_{\Delta X}(\boldsymbol{x}, \tau)$ will depend only on τ: if they are also independent, from (7.22), (7.23) and (7.24) we have

$$f(\boldsymbol{x}, \boldsymbol{y}; \tau) = f(\boldsymbol{x}, t+\tau|\boldsymbol{y}, t)f(\boldsymbol{y}, t) = f_{\Delta X}(\boldsymbol{x}-\boldsymbol{y}, \tau)f(\boldsymbol{y}) \tag{7.28}$$

namely (7.26) because of the relation $f(\boldsymbol{x}, \boldsymbol{y}; \tau) = f(\boldsymbol{x}, \tau \mid \boldsymbol{y})f(\boldsymbol{y})$ between the joint and the conditional densities. ■

The ***Chapman-Kolmogorov equations for stationary Markov processes*** (here $s > 0, t > 0$ are the interval widths) are then reduced to

$$f(\boldsymbol{x}) = \int f(\boldsymbol{x}, t \mid \boldsymbol{y})f(\boldsymbol{y})\,d\boldsymbol{y} \tag{7.29}$$

$$f(\boldsymbol{x}, t+s \mid \boldsymbol{y}) = \int f(\boldsymbol{x}, t \mid \boldsymbol{z})f(\boldsymbol{z}, s \mid \boldsymbol{y})\,d\boldsymbol{z} \tag{7.30}$$

where the first equation (7.29) just states that $f(\boldsymbol{x})$ is an invariant *pdf* that is determined by the initial condition; the second equation (7.30) instead is the Markovianity condition for a stationary $f(\boldsymbol{x}, t \mid \boldsymbol{y})$.

Corollary 7.12 *The* ***Markovianity of stationary and independent increments*** *can also be expressed in terms either of the convolutions of their pdf's*

$$f_{\Delta X}(\boldsymbol{x}, t+s) = [f_{\Delta X}(t) * f_{\Delta X}(s)](\boldsymbol{x}) \qquad s > 0,\ t > 0 \tag{7.31}$$

or of the products of the corresponding chf's

$$\varphi_{\Delta X}(\boldsymbol{u}, t+s) = \varphi_{\Delta X}(\boldsymbol{u}, t)\,\varphi_{\Delta X}(\boldsymbol{u}, s) \qquad s > 0,\ t > 0 \tag{7.32}$$

and it amounts to require that an increment on an interval of width $s+t$ be the sum of two independent increments on intervals of widths s and t. In this case we also speak of ***Markovian increments****.*

Proof When the increments are independent too, we see from (7.26) that the stationary Chapman-Kolmogorov equations (7.29) and (7.30) become

$$f(\boldsymbol{x}) = \int f_{\Delta X}(\boldsymbol{x}-\boldsymbol{y}, t)f(\boldsymbol{y})\,d\boldsymbol{y} \tag{7.33}$$

$$f_{\Delta X}(\boldsymbol{x}-\boldsymbol{y}, t+s) = \int f_{\Delta X}(\boldsymbol{x}-\boldsymbol{z}, t)f_{\Delta X}(\boldsymbol{z}-\boldsymbol{y}, s)\,d\boldsymbol{z} \tag{7.34}$$

To show then that (7.34) takes the form of a convolution (7.31) it would be enough to change the variables according to the transformation $\boldsymbol{x}-\boldsymbol{y} \to \boldsymbol{x}$, e $\boldsymbol{z}-\boldsymbol{y} \to \boldsymbol{z}$

with Jacobian determinant equal to 1, to have

$$f_{\Delta X}(\boldsymbol{x}, t+s) = \int f_{\Delta X}(\boldsymbol{x}-\boldsymbol{z}, t) f_{\Delta X}(\boldsymbol{z}, s)\, d\boldsymbol{z}$$

The form (7.32) for the *chf*'s easily follows then from the convolution theorem. ■

Definition 7.13 We will say that a Markov process is **time homogeneous** when its transition *pdf* (7.25) only depends on the difference $\tau = t - s$; in this case the Chapman-Kolmogorov Markovianity condition takes the form (7.30).

Remark that the time homogeneity only requires the condition (7.24), while nothing is said about the other condition (7.23). As a consequence a time homogeneous Markov process is not in general a stationary process, not even in the wide sense.

Corollary 7.14 *Every process with independent and stationary increments is a time homogeneous Markov process, and—when an invariant distribution exist—it is also strict sense stationary if the initial distribution is chosen to be invariant.*

Proof When the independent increments are stationary the transition *pdf* (7.22) of the process only depends on Δt, and not on t, namely it is of the form (7.25) and hence the process is time homogeneous according to the Definition 7.13. If moreover also the one time distribution is invariant then both (7.23) and (7.24) hold and the process is strict sense stationary according to the Proposition 7.11. Remark however that in general the chosen initial distribution is not necessarily the invariant one, so that a time homogeneous Markov process may not be stationary (not even in wide sense). This may happen either because the invariant distribution does not exist at all (as for the Wiener and Poisson *sp*'s), or because the invariant *pdf* has not been chosen as the initial distribution for the Chapman-Kolmogorov equation (7.33). In these cases the process is not stationary according to the Definition 5.9, but it is only time homogeneous according to the Definition 7.13. ■

7.1.5 Distribution Ergodicity

Suppose we want to construct (the law of) a *stationary* Markov process starting with a given (at least in the retarded region $t > 0$) *time homogeneous* transition *pdf* $f(\boldsymbol{x}, t \mid \boldsymbol{y})$ such as (7.25). We should first check that the Chapman-Kolmogorov Markovianity condition in the form (7.30) is satisfied: if this is the case, according to the Proposition 7.7 we will be able to manufacture a whole family of processes by arbitrarily choosing the initial *pdf* $f_0(\boldsymbol{x})$. All these Markov processes will be time homogeneous by definition, but—retracing the discussion of the Corollary 7.14—they could all be non-stationary. The initial *pdf* $f_0(\boldsymbol{x})$ may indeed be non invariant, nay an invariant *pdf* could not exist at all for the given $f(\boldsymbol{x}, t \mid \boldsymbol{y})$. If this happens (either because an invariant law does not exist, or because we have chosen a non invariant initial law) the process will be time homogeneous, but not stationary and it will evolve according to

$$f(\boldsymbol{x}, t) = \int f(\boldsymbol{x}, t|\boldsymbol{y}) f_0(\boldsymbol{y})\, d\boldsymbol{y} \tag{7.35}$$

If instead $f_0(\boldsymbol{x})$ is chosen as the invariant *pdf* $f(\boldsymbol{x})$, then the Eq. (7.35) becomes (7.29) with $f(\boldsymbol{x}) = f_0(\boldsymbol{x})$, and the process is strict sense stationary. It is then especially important to provide a procedure to find—if it exists—an invariant distribution for a given time homogeneous transition *pdf*.

Going back then to the idea of ergodicity presented in the Sect. 5.5, take first an *ac* stationary Markov process $\boldsymbol{X}(t)$ with invariant *pdf* $f(\boldsymbol{x})$, and consider the problem of estimating its distribution

$$\boldsymbol{P}\{\boldsymbol{X}(t) \in B\} = \boldsymbol{E}\left[\chi_B\left(\boldsymbol{X}(t)\right)\right] = \int_B f(\boldsymbol{x})\, d\boldsymbol{x} \qquad B \in \mathcal{B}\left(\boldsymbol{R}^M\right) \tag{7.36}$$

(here again $\chi_B(\boldsymbol{x})$ is the indicator of the set B) by means of a time average on a fairly long time interval $[-T, T]$. To do that start by defining the process $Y(t) = \chi_B\left(\boldsymbol{X}(t)\right)$, so that from (7.36) it is

$$\boldsymbol{E}\left[Y(t)\right] = \int_B f(\boldsymbol{x})\, d\boldsymbol{x} \tag{7.37}$$

and then take the *rv*

$$\overline{Y}_T = \frac{1}{2T}\int_{-T}^{T} Y(t)\, dt = \frac{1}{2T}\int_{-T}^{T} \chi_B\left(\boldsymbol{X}(t)\right)\, dt$$

representing the time fraction of $[-T, T]$ during which $\boldsymbol{X}(t)$ sojourns in B (that is an estimate of the *relative frequency* of its findings in B).

Theorem 7.15 *Take the family of time homogeneous ac Markov processes associated (according to the Proposition 7.7) to a Markovian homogeneous transition pdf* $f(\boldsymbol{x}, t|\,\boldsymbol{y})$*: if it exists an asymptotic pdf of* $f(\boldsymbol{x}, t|\,\boldsymbol{y})$*, to wit an* $\bar{f}(\boldsymbol{x})$ *such that*

$$\lim_{t\to+\infty} f(\boldsymbol{x}, t\,|\,\boldsymbol{y}) = \bar{f}(\boldsymbol{x}) \qquad \forall\, \boldsymbol{y} \in \boldsymbol{R}^M \tag{7.38}$$

then $\bar{f}(\boldsymbol{x})$ *is both an invariant pdf, and the asymptotic pdf of every other time homogeneous non stationary process: namely, starting with an arbitrary non invariant initial pdf* $f_0(\boldsymbol{x})$*, we will always have*

$$\lim_{t\to+\infty} f(\boldsymbol{x}, t) = \bar{f}(\boldsymbol{x}) \tag{7.39}$$

Moreover, if $\boldsymbol{X}(t)$ *is stationary with invariant pdf* $\bar{f}(\boldsymbol{x})$*, and if the autocovariance* $C_Y(\tau)$ *of* $Y(t)$ *meets the condition*

$$\int_0^{+\infty} |C_Y(\tau)|\, d\tau < +\infty \tag{7.40}$$

than we will have

$$\operatorname*{lim\text{-}ms}_{T\to\infty} \overline{Y}_T = \int_B \bar{f}(\boldsymbol{x})\, d\boldsymbol{x} = \boldsymbol{P}\{\boldsymbol{X}(t) \in B\} \tag{7.41}$$

and $\boldsymbol{X}(t)$ will be said to be ***distribution ergodic****.*

Proof Taking for granted that we can exchange limits and integrals, from (7.38) an from the Chapman-Kolmogorov equation (7.30) we have first of all

$$\begin{aligned}\int f(\boldsymbol{x}, t \mid \boldsymbol{y})\bar{f}(\boldsymbol{y})\, d\boldsymbol{y} &= \int f(\boldsymbol{x}, t \mid \boldsymbol{y}) \lim_{s\to+\infty} f(\boldsymbol{y}, s \mid \boldsymbol{z})\, d\boldsymbol{y} \\ &= \lim_{s\to+\infty} \int f(\boldsymbol{x}, t \mid \boldsymbol{y}) f(\boldsymbol{y}, s \mid \boldsymbol{z})\, d\boldsymbol{y} = \lim_{s\to+\infty} f(\boldsymbol{x}, t+s \mid \boldsymbol{z}) = \bar{f}(\boldsymbol{x})\end{aligned}$$

so that the limit *pdf* turns out to be also the invariant *pdf*. If then we take an arbitrary (non invariant) initial distribution $f_0(\boldsymbol{x})$, from the Chapman-Kolmogorov equation (7.35) and from (7.38) we also find the result (7.39):

$$\lim_{t\to+\infty} f(\boldsymbol{x}, t) = \lim_{t\to+\infty} \int f(\boldsymbol{x}, t|\boldsymbol{y}) f_0(\boldsymbol{y})\, d\boldsymbol{y} = \bar{f}(\boldsymbol{x}) \int f_0(\boldsymbol{y})\, d\boldsymbol{y} = \bar{f}(\boldsymbol{x})$$

Within our notations, finally, the *ms*-convergence (7.41) amounts to require that the process $Y(t) = \chi_B(\boldsymbol{X}(t))$ be expectation ergodic in the sense of the Theorem 5.12, and this is assured, according to the Corollary 5.13, if the sufficient condition (7.40) is satisfied. In this regard it is finally useful to add that, being $\boldsymbol{X}(t)$ a stationary process, we have

$$\begin{aligned}C_Y(\tau) &= \boldsymbol{E}[Y(t+\tau)Y(t)] - \boldsymbol{E}[Y(t+\tau)]\,\boldsymbol{E}[Y(t)] \\ &= \boldsymbol{E}[Y(t+\tau)Y(t)] - \boldsymbol{E}[Y(t)]^2 \\ &= \int\!\!\int \chi_B(\boldsymbol{x})\chi_B(\boldsymbol{y}) f(\boldsymbol{x}, t+\tau; \boldsymbol{y}, t)\, d\boldsymbol{x} d\boldsymbol{y} - \left(\int_B \bar{f}(\boldsymbol{x})\, d\boldsymbol{x}\right)^2 \\ &= \int_B\!\int_B f(\boldsymbol{x}, \tau \mid \boldsymbol{y})\bar{f}(\boldsymbol{y})\, d\boldsymbol{x} d\boldsymbol{y} - \int_B\!\int_B \bar{f}(\boldsymbol{x})\bar{f}(\boldsymbol{y})\, d\boldsymbol{x} d\boldsymbol{y} \\ &= \int_B\!\int_B \left[f(\boldsymbol{x}, \tau \mid \boldsymbol{y}) - \bar{f}(\boldsymbol{x})\right] \bar{f}(\boldsymbol{y})\, d\boldsymbol{x} d\boldsymbol{y}\end{aligned}$$

so that the condition (7.38) entails first that $C_Y(\tau)$ is infinitesimal for $\tau \to +\infty$, an second that the condition (7.40) is satisfied too if the said convergence of (7.38) is fast enough: in this case, according to the Corollary 5.13, $Y(t)$ is expectation ergodic, and $\boldsymbol{X}(t)$ is distribution ergodic. ■

In conclusion, given a time homogeneous and ergodic Markov process with an asymptotic *pdf* $\bar{f}(\boldsymbol{x})$ in the sense of (7.38), all the initial *pdf's* $f_0(\boldsymbol{x})$ follow time evolutions $f(\boldsymbol{x}, t)$ tending toward the same invariant *pdf* $\bar{f}(\boldsymbol{x})$. As a matter of fact the process gradually loses memory of the initial distribution and tend toward a limit law that coincides with the invariant *pdf* $\bar{f}(\boldsymbol{x})$. Such a process, albeit non stationary in a proper sense, is asymptotically stationary in the sense that its limit law is invariant.

Remark however that under particular conditions (when for instance the available space is partitioned in separated, non communicating regions) there could be more than one invariant or asymptotic *pdf*. In this case we must pay attention to identify the invariant *pdf* of interest and its relation to the asymptotic *pdf*: we will neglect however to elaborate further on this point.

7.1.6 Lévy Processes

Definition 7.16 A *sp* $\boldsymbol{X}(t)$ is a **Lévy process** if

1. $\boldsymbol{X}(0) = 0 \qquad \boldsymbol{P}$-a.s.
2. it has independent and stationary increments
3. it is stochastically continuous according to the Definition 5.5, that is (keeping into account also 1. and 2.) if for every $\epsilon > 0$

$$\lim_{t \to 0^+} \boldsymbol{P}\{|\boldsymbol{X}(t)| > \epsilon\} = 0 \tag{7.42}$$

From what it has been stated in the previous sections, it follows that every Lévy process is a homogeneous Markov process, and that the laws of its increments must be infinitely divisible. Therefore the global law of a Lévy process is completely determined by the infinitely divisible, stationary laws of its increments according to the following result.

Proposition 7.17 *If* $\boldsymbol{X}(t)$ *is a Lévy process, then it exists an infinitely divisible chf* $\varphi(\boldsymbol{u})$ *and a time scale* $T > 0$ *such that the chf* $\varphi_{\Delta X}(\boldsymbol{u}, t)$ *of the increments of width* t *is*

$$\varphi_{\Delta X}(\boldsymbol{u}, t) = [\varphi(\boldsymbol{u})]^{t/T} \qquad t \geq 0 \tag{7.43}$$

Conversely, taken an arbitrary infinitely divisible chf $\varphi(\boldsymbol{u})$ *and a time scale* $T > 0$, *the (7.43) will always be the chf of the increments of a suitable Lévy process.*

Proof Without going into the details of a discussion that would exceed the boundaries of our presentation (for further details see [1, 2]), we will just remark about the *reverse* statement first that the infinite divisibility of the given *chf* entails that (7.43) continues to be an infinitely divisible *chf* for every $t \geq 0$, and then that the formula (7.43) or the increment *chf* is the simplest way to meet the Chapman-Kolmogorov Markovianity conditions (7.32) of the Corollary 7.12. ■

Example 7.18 According to the previous proposition, to assemble a Lévy process we must start by choosing an infinitely divisible law with *chf* $\varphi(\boldsymbol{u})$, and then we must define the process law by adopting the independent and stationary increment *chf* (7.43). The explicit form of the increment *pdf* can then be calculated—if possible—by inverting the *chf*. We have already seen at least two examples of Lévy processes: the **simple Poisson process** $N(t)$ is stochastically continuous, and the *chf* of its independent increments (6.9) follows from (7.43) just by taking an infinitely divisible Poisson law $\mathfrak{P}(\alpha)$ with *chf* $\varphi(u) = e^{\alpha(e^{iu}-1)}$, and then $\lambda = \alpha/T$ into (6.9). In the same way the **Wiener process** $W(t)$ is stochastically continuous, and the *chf* of its independent increments (6.44) follows from (7.43) taking the infinitely divisible law $\mathfrak{N}(0, \alpha^2)$ with *chf* $\varphi(u) = e^{-\alpha^2 u^2/2}$ and $D = \alpha^2/T$ into (6.44). A third example, the **Cauchy process**, will be presented in the Definition 7.25, and again the *chf* of its independent increments (7.53) will come from (7.43) starting from the infinitely divisible law $\mathfrak{C}(\alpha)$ with *chf* $\varphi(u) = e^{-\alpha|u|}$ then taking $a = \alpha/T$.

7.1.7 Continuity and Jumps

Different kinds of continuity have been presented in the Definition 5.5, and the conditions for the *ms*-continuity (and hence for the stochastic continuity too) have been discussed in the Proposition 5.6. We will look now into the conditions for the ***sample continuity of a Markov process*** $\boldsymbol{X}(t)$.

Theorem 7.19 *A Markov process $\boldsymbol{X}(t)$ is sample continuous iff the following* ***Lindeberg conditions*** *are met*

$$\sup_{\boldsymbol{y},t} \boldsymbol{P}\{|\Delta\boldsymbol{X}(t)| > \epsilon \mid \boldsymbol{X}(t) = \boldsymbol{y}\} = o(\Delta t) \qquad \forall\epsilon > 0 \quad \Delta t \to 0 \tag{7.44}$$

If the process also has stationary and independent increments such conditions become

$$\boldsymbol{P}\{|\Delta\boldsymbol{X}(t)| > \epsilon\} = o(\Delta t) \qquad \forall\epsilon > 0 \quad \Delta t \to 0 \tag{7.45}$$

Proof Omitted: see [3], vol. II, p. 333, and [4], p. 46. ■

According to this result the sample continuity of a Markov process (to wit that all its sample trajectories are continuous for every t, but for a subset of zero probability) is ensured when (uniformly in $\boldsymbol{y}$ and t) the probability that a Δt-increment exceeds in absolute value an arbitrary threshold $\epsilon > 0$ is infinitesimal of order larger than $\Delta t \to 0$. In other words the Lindeberg condition is an apparent requirement on the vanishing rate of $|\Delta\boldsymbol{X}(t)|$ when $\Delta t \to 0$. Remark finally that if the process is *ac* the condition (7.44) becomes

$$\lim_{\Delta t \to 0} \frac{1}{\Delta t} \sup_{\boldsymbol{y},t} \int_{|\boldsymbol{x}-\boldsymbol{y}|>\epsilon} f(\boldsymbol{x}, t + \Delta t \mid \boldsymbol{y}, t)\, d\boldsymbol{x} = 0 \qquad \forall\epsilon > 0 \tag{7.46}$$

that is

$$\lim_{\Delta t \to 0} \frac{1}{\Delta t} \int_{|x-y|>\epsilon} f(x, t + \Delta t \mid y, t)\, dx = 0 \qquad \forall \epsilon > 0 \tag{7.47}$$

uniformly in y and t. If moreover the process also has independent and stationary increments, taking $x - y \to x$, from (7.26) the Lindeberg condition becomes

$$\lim_{\Delta t \to 0} \frac{1}{\Delta t} \int_{|x|>\epsilon} f_{\Delta X}(x, \Delta t)\, dx = 0 \qquad \forall \epsilon > 0 \tag{7.48}$$

Remark that there is a sort of competition between ***Markovianity and regularity*** of the process paths according to the chosen ***time scale*** of the observations. We could say that the shorter the observation times, the more continuous a process, but at the same time the less Markovian. In a sense this depends on the fact that a continuous description requires more information on the past of the trajectory and therefore comes into conflict with Markovianity. For instance a model for the molecular movement in a gas based on hard spheres and instantaneous collisions provides piecewise rectilinear trajectories with sudden velocity changes: in this case the velocity process will be discontinuous, while the position is continuous and Markovian (the position after a collision will depend on the starting point, but not on its previous path). If however we go to shorter times with a more detailed physical description (elasticity, deformation ...) the velocity may become continuous too, but the position is less Markovian because a longer section of its history will be needed to predict the future even only probabilistically. Look in this regard also to the difference between *random* and *stochastic* differential equations sketched in the forthcoming Sect. 8.4.2.

Of course, a process may be stochastically continuous even without being sample continuous, but in this case the trajectories may present ***discontinuities (jumps)***: this is not unusual, for example, among the Lévy processes. Typically the jumping times of our Markov processes are random as well as their jump sizes (think to a compound Poisson process); in general, however, they are *first kind discontinuities* and anyhow the trajectories will be *cadlag*, namely right continuous and admitting a left limit in every instant. The study of both the discontinuities and their distributions is a major topic in the investigations about the Lévy processes that, however, we will be obliged to neglect.

7.1.8 Poisson, Wiener and Cauchy Processes

In the next two sections we will deduce the ***univariate distributions*** of the Markov processes of our interest starting from their Markovian transition laws, and we will exploit them in order to scrutinize the process properties. In two cases (Poisson and Wiener processes) the processes have already been heuristically introduced working up their trajectories and then deducing their basic probabilistic attributes: here we will dwell in particular on their possible *sample continuity*. In other two cases (Cauchy and Ornstein-Uhlenbeck processes) the processes will be introduced here for the first time from their transition distributions moving on then to obtain all their other properties.

7.1.8.1 Poisson Process

Definition 7.20 We say that $N(t)$ is a **simple Poisson process** of intensity λ if it takes integer values, and has stationary and independent increments, with the homogeneous Markovian transition probability (in the retarded region $\Delta t > 0$)

$$p_N(n, t + \Delta t | m, t) = e^{-\lambda \Delta t} \frac{(\lambda \Delta t)^{n-m}}{(n-m)!} \qquad n = 0, 1, 2, \ldots; \; 0 \leq m \leq n \quad (7.49)$$

consistent with the distribution (6.8) of the increments of width $\Delta t > 0$. Since we know from the Proposition 6.7 that a Poisson process is ms-, and hence stochastically continuous, when moreover $N(0) = 0$, $\boldsymbol{P}$-a.s., then $N(t)$ is also a Lévy process.

Recalling then that from (6.9) the increment *chf* is

$$\varphi_{\Delta N}(u, t) = e^{\lambda t (e^{iu} - 1)} \quad (7.50)$$

the Lévy Poisson process can also be produced according to the Proposition 7.17 starting from the infinitely divisible Poisson *chf*

$$\varphi(u) = e^{\lambda T (e^{iu} - 1)}$$

where $T > 0$ is the usual time constant. Yet, given the jumping character of its trajectories discussed in the Sect. 6.1.2, it easy to understand that $N(t)$ can not be sample continuous, as it is also upheld by the following result that exploits the Lindeberg conditions.

Proposition 7.21 *A Poisson process $N(t)$ does not satisfy the Lindeberg conditions (7.45) and hence it is not sample continuous.*

Proof We see indeed for $0 < \epsilon < 1$ and $t \to 0$ that the probability

$$\begin{aligned} \boldsymbol{P}\{|\Delta N| > \epsilon\} &= 1 - \boldsymbol{P}\{|\Delta N| \leq \epsilon\} = 1 - \boldsymbol{P}\{|\Delta N| = 0\} = 1 - e^{-\lambda t} \\ &= 1 - \left(1 - \lambda t + \cdots + (-1)^n \frac{\lambda^n t^n}{n!} + \cdots\right) = \lambda t + o(t) \end{aligned}$$

is of the first order in t, and hence the Lindeberg conditions are not respected. ■

We also recall that $N(t)$ is homogeneous, but it is not stationary (not even in the wide sense because its expectation is not constant), and that for $t \to +\infty$ its transition probabilities do not converge toward a probability distribution: they rather flatten to zero for every n. As a consequence there is no invariant law for the Poisson transition distributions, and $N(t)$ is not ergodic.

7.1.8.2 Wiener Process

Definition 7.22 We say that $W(t)$ is a **Wiener process** with diffusion coefficient D if it is *ac*, and has stationary and independent increments, with the homogeneous Markovian transition *pdf* (in the retarded region)

$$f_W(x, t + \Delta t | y, t) = \frac{e^{-(x-y)^2/2D\Delta t}}{\sqrt{2\pi D\Delta t}} \qquad \Delta t > 0 \tag{7.51}$$

consistent with the distribution (6.43) of the increments of width $\Delta t > 0$. Since we know from the Proposition 6.18 that a Wiener process is ms-, and hence stochastically continuous, when moreover $W(0) = 0$, $\boldsymbol{P}$-a.s., then $W(t)$ is also a Lévy process.

Recalling then that from (6.44) the increment *chf* is

$$\varphi_{\Delta W}(u, t) = e^{-Dt\,u^2/2} \tag{7.52}$$

the Lévy Wiener process can also be produced according to the Proposition 7.17 starting from the infinitely divisible, centered Gaussian *chf*

$$\varphi(u) = e^{-DT\,u^2/2}$$

where $T > 0$ is the usual time constant. At variance with the Poisson process, however, $W(t)$ is also sample continuous, as anticipated in the Proposition 6.18 and upheld by the following result that again exploits the Lindeberg conditions.

Proposition 7.23 *A Wiener process $W(t)$ is Gaussian if $W(0)$ is Gaussian; it moreover meets the Lindeberg conditions (7.45) and hence it is sample continuous.*

Proof The Gaussianity follows by direct inspection of the joint *pdf*'s of the *r-vec* $W(t_1), \ldots, W(t_n)$ (with arbitrary instants) that result from the chain rule (7.14). As for the sample continuity, with $t \to 0^+$ we have indeed from (7.51)

$$\boldsymbol{P}\{|\Delta W| > \epsilon\} = \int_{|x|>\epsilon} f_{\Delta W}(x, t)\, dx = 2\left[1 - \Phi\left(\frac{\epsilon}{\sqrt{Dt}}\right)\right] \longrightarrow 0$$

where we adopted the notation

$$\Phi(x) = \frac{1}{\sqrt{2\pi}} \int_{-\infty}^{x} e^{-y^2/2} dy \qquad \Phi'(x) = \frac{e^{-x^2/2}}{\sqrt{2\pi}}$$

Then by using l'Hôpital's rule, and by taking $\alpha = \epsilon/\sqrt{Dt}$ we have

$$\lim_{t\downarrow 0} \frac{1}{t}\left[1 - \Phi\left(\frac{\epsilon}{\sqrt{Dt}}\right)\right] = \lim_{t\downarrow 0} \frac{\epsilon}{2t} \frac{e^{-\epsilon^2/2Dt}}{\sqrt{2\pi Dt}} = \lim_{\alpha\to+\infty} \frac{D\alpha^3}{\epsilon^2} \frac{e^{-\alpha^2/2}}{\sqrt{2\pi}} = 0$$

so that the Lindeberg conditions are obeyed and $W(t)$ is sample continuous. ■

We already know that $W(t)$ is homogeneous but not stationary, and that for $t \to +\infty$ its transition *pdf*'s do not converge to some other *pdf* (they rather flatten to the uniform Lebesgue measure on $\boldsymbol{R}$ that however is not a *pdf*): then there is no invariant distribution for the Wiener transition *pdf*'s and hence $W(t)$ is not ergodic.

7.1.8.3 Cauchy Process

Proposition 7.24 *The stationary increments* $\Delta X(t) \sim \mathfrak{C}(a\Delta t)$ *for* $a > 0$, $\Delta t > 0$, *with a chf*

$$\varphi_{\Delta X}(u, \Delta t) = e^{-a\Delta t|u|} \tag{7.53}$$

and a pdf

$$f_{\Delta X}(x, \Delta t) = \frac{1}{\pi} \frac{a\Delta t}{x^2 + a^2\Delta t^2} \tag{7.54}$$

are Markovian, and thus enable us to define an entire family of independent increments Markov processes.

Proof The *chf*'s (7.53) trivially comply with the conditions required in the Corollary 7.12 because for increments $s > 0$ and $t > 0$ we find

$$\varphi_{\Delta X}(u, t+s) = e^{-a(t+s)|u|} = e^{-at|u|}e^{-as|u|} = \varphi_{\Delta X}(u, t)\varphi_{\Delta X}(u, s)$$

As a consequence, according to the Proposition 7.7 we can consistently work out (the laws of) an entire family of independent and stationary increment processes $X(t)$ from the increment distribution (7.54). ■

Definition 7.25 We say that $X(t)$ is a **Cauchy process** with parameter a if it is *ac*, and has stationary and independent increments, with the homogeneous Markovian transition *pdf* (in the retarded region)

$$f_X(x, t+\Delta t \mid y, t) = \frac{1}{\pi} \frac{a\Delta t}{(x-y)^2 + (a\Delta t)^2} \qquad \Delta t > 0 \tag{7.55}$$

consistent with the distribution (7.54) of the increments of width $\Delta t > 0$.

Remark that, because of the properties of the Cauchy distributions, a Cauchy process $X(t)$ is not provided with expectation, variance and autocorrelation: we will always be able to calculate probabilities, medians or quantiles of every order, but we will be obliged to do without the more familiar moments of order larger or equal to 1.

Proposition 7.26 *A Cauchy process $X(t)$ is stochastically continuous, but it is not sample continuous. As a consequence, if we also take $X(0) = 0$ $\boldsymbol{P}$-a.s., then $X(t)$ is a Lévy process.*

Proof The process $X(t)$ is stochastically continuous (and hence is a Lévy process if $X(0) = 0$, $\boldsymbol{P}$-a.s.) because from its distribution

$$f_X(x,t) = \frac{1}{\pi}\frac{at}{x^2 + a^2t^2} \qquad\qquad \varphi_X(u,t) = e^{-at|u|} \tag{7.56}$$

it is easy to see that for $\epsilon > 0$

$$\lim_{t\downarrow 0} \boldsymbol{P}\{|X(t)| > \epsilon\} = \lim_{t\downarrow 0} \int_{|x|>\epsilon} f_X(x,t)\,dx = \lim_{t\downarrow 0}\left(1 - \frac{2}{\pi}\arctan\frac{\epsilon}{at}\right) = 0$$

namely that the requirement (7.42) is met. On the other hand from (7.54) we again have

$$\lim_{\Delta t\downarrow 0} \boldsymbol{P}\{|\Delta X(t)| > \epsilon\} = \lim_{\Delta t\downarrow 0} \int_{|x|>\epsilon} f_{\Delta X}(x,\Delta t)\,dx = \lim_{\Delta t\downarrow 0}\left(1 - \frac{2}{\pi}\arctan\frac{\epsilon}{a\Delta t}\right) = 0$$

but then from l'Hôpital's rule we see that

$$\lim_{\Delta t\downarrow 0}\frac{1}{\Delta t}\left(1 - \frac{2}{\pi}\arctan\frac{\epsilon}{a\Delta t}\right) = \frac{2\epsilon}{a\pi}\lim_{\Delta t\downarrow 0}\frac{a^2}{\epsilon^2 + a^2\Delta t^2} = \frac{2a}{\epsilon\pi} > 0$$

namely that the Cauchy process does not comply with the Lindeberg condition (7.45) so that it is not sample continuous. ■

Because of the previous result the Cauchy process makes jumps even if—at variance with the other jumping process, the Poisson one—it takes continuous values in $\boldsymbol{R}$ (it is indeed *ac*). It could also be shown that the lengths of the Cauchy jumps actually cluster around infinitesimal values, but this is not inconsistent with the existence of finite size discontinuities too. Finally the Cauchy process is apparently homogeneous, but it is neither stationary nor ergodic because its transition *pdf*'s do not converge toward a limit *pdf* for $t \to \infty$.

7.1.9 Ornstein-Uhlenbeck Processes

Proposition 7.27 *The homogeneous transition pdf* ($\alpha > 0,\ \beta > 0,\ \Delta t > 0$)

$$f(x, t + \Delta t \mid y, t) = f(x, \Delta t \mid y) = \frac{e^{-(x-ye^{-\alpha\Delta t})^2/2\beta^2(1-e^{-2\alpha\Delta t})}}{\sqrt{2\pi\beta^2(1-e^{-2\alpha\Delta t})}} \tag{7.57}$$

of the Gaussian law $\mathfrak{N}(ye^{-\alpha\Delta t}, \beta^2(1-e^{-2\alpha\Delta t}))$, *is Markovian and ergodic with invariant pdf*

$$f(x) = \frac{e^{-x^2/2\beta^2}}{\sqrt{2\pi\beta^2}} \tag{7.58}$$

namely that of the Gaussian law $\mathfrak{N}(0, \beta^2)$.

Proof Since (7.57) is time homogeneous, to prove the Markovianity we should check the second Chapman-Kolmogorov equation in the form (7.30): to this end we remark that by taking $v = ze^{-\alpha t}$, and by keeping into account the reproductive properties (3.67) of the normal laws, we find

$$\begin{aligned}
\int f(x, t|z) f(z, s|y)\, dz &= \int \frac{e^{-(x-ze^{-\alpha t})^2/2\beta^2(1-e^{-2\alpha t})}}{\sqrt{2\pi\beta^2(1-e^{-2\alpha t})}} \frac{e^{-(z-ye^{-\alpha s})^2/2\beta^2(1-e^{-2\alpha s})}}{\sqrt{2\pi\beta^2(1-e^{-2\alpha s})}}\, dz \\
&= \int \frac{e^{-(x-v)^2/2\beta^2(1-e^{-2\alpha t})}}{\sqrt{2\pi\beta^2(1-e^{-2\alpha t})}} \frac{e^{-(v-ye^{-\alpha(t+s)})^2/2\beta^2 e^{-2\alpha t}(1-e^{-2\alpha s})}}{\sqrt{2\pi\beta^2 e^{-2\alpha t}(1-e^{-2\alpha s})}}\, dv \\
&= \mathfrak{N}(0, \beta^2(1-e^{-2\alpha t})) * \mathfrak{N}(ye^{-\alpha(t+s)}, \beta^2 e^{-2\alpha t}(1-e^{-2\alpha s})) \\
&= \mathfrak{N}(ye^{-\alpha(t+s)}, \beta^2(1-e^{-2\alpha(t+s)})) = f(x, t+s|y)
\end{aligned}$$

as required for the Markovianity. The ergodicity follows then from the fact that

$$\mathfrak{N}(ye^{-\alpha\tau}, \beta^2(1-e^{-2\alpha\tau})) \longrightarrow \mathfrak{N}(0, \beta^2) \qquad \tau \to +\infty$$

with the limit *pdf* (7.58), and we can check by direct calculation that this limit law is also invariant: taking indeed $v = ye^{-\alpha t}$ as before, it is

$$\begin{aligned}
\int f(x, t|y) f(y)\, dy &= \int \frac{e^{-(x-ye^{-\alpha t})^2/2\beta^2(1-e^{-2\alpha t})}}{\sqrt{2\pi\beta^2(1-e^{-2\alpha t})}} \frac{e^{-y^2/2\beta^2}}{\sqrt{2\pi\beta^2}}\, dy \\
&= \int \frac{e^{-(x-v)^2/2\beta^2(1-e^{-2\alpha t})}}{\sqrt{2\pi\beta^2(1-e^{-2\alpha t})}} \frac{e^{-v^2/2\beta^2 e^{-2\alpha t}}}{\sqrt{2\pi\beta^2 e^{-2\alpha t}}}\, dv \\
&= \mathfrak{N}(0, \beta^2(1-e^{-2\alpha t})) * \mathfrak{N}(0, \beta^2 e^{-2\alpha t}) = \mathfrak{N}(0, \beta^2) = f(x)
\end{aligned}$$

where we used again the reproductive properties (3.67). ■

Definition 7.28 We will say that $X(t)$ is an **Ornstein-Uhlenbeck process** if it is a homogeneous and ergodic Markov process with transition *pdf* (7.57); its invariant limit *pdf* (7.58) apparently selects the **stationary** Ornstein-Uhlenbeck process when it is taken as the initial *pdf*.

Proposition 7.29 *An Ornstein-Uhlenbeck process $X(t)$ is sample continuous, and it is Gaussian if $X(0)$ is Gaussian. In particular the stationary process (that starting from (7.58) is Gaussian with $\mathbf{E}[X(t)] = 0$, $\mathbf{V}[X(t)] = \beta^2$ and*

$$R(\tau) = C(\tau) = \beta^2 e^{-\alpha|\tau|} \qquad \rho(\tau) = e^{-\alpha|\tau|} \qquad S(\varpi) = \frac{\beta^2}{\pi}\frac{\alpha}{\alpha^2 + \varpi^2} \tag{7.59}$$

while, for every $t_1, \dots, t_n$, it is $(X(t_1), \dots, X(t_n)) \sim \mathfrak{N}(0, \mathbb{A})$ with covariance matrix

$$\mathbb{A} = \beta^2 \begin{pmatrix} 1 & e^{-\alpha|\tau_{12}|} & \dots & e^{-\alpha|\tau_{1n}|} \\ e^{-\alpha|\tau_{21}|} & 1 & & e^{-\alpha|\tau_{2n}|} \\ \vdots & & \ddots & \vdots \\ e^{-\alpha|\tau_{n1}|} & e^{-\alpha|\tau_{n2}|} & \dots & 1 \end{pmatrix} \qquad \tau_{jk} = t_j - t_k \tag{7.60}$$

The increments $\Delta X = X(t) - X(s)$ of an Ornstein-Uhlenbeck process are not independent, and therefore it can never be a Lévy process.

Proof Postponing to the Sect. 7.40 the proof of the sample continuity, and to the Sect. 8.5.4 that of the Gaussianity for arbitrary Gaussian initial conditions $X(0)$, we will confine the present discussion first to the Gaussianity of the stationary process: from (7.57) and (7.58) we have indeed for two times with $\tau = t - s > 0$

$$\begin{aligned} f(x, t; y, s) &= f(x, t \mid y, s) f(y, s) = f(x, \tau \mid y) f(y) \\ &= \frac{e^{-(x-ye^{-\alpha\tau})^2/2\beta^2(1-e^{-2\alpha\tau})}}{\sqrt{2\pi\beta^2(1-e^{-2\alpha\tau})}} \frac{e^{-y^2/2\beta^2}}{\sqrt{2\pi\beta^2}} = \frac{e^{-(x^2+y^2-2xye^{-\alpha\tau})/2\beta^2(1-e^{-2\alpha\tau})}}{\sqrt{2\pi\beta^2}\sqrt{2\pi\beta^2(1-e^{-2\alpha\tau})}} \end{aligned}$$

and juxtaposing it to the general form (2.24) of a bivariate normal *pdf* we deduce that the *r-vec* $(X(s), X(t))$ is a jointly bivariate Gaussian $\mathfrak{N}(0, \mathbb{A})$ with

$$\mathbb{A} = \beta^2 \begin{pmatrix} 1 & e^{-\alpha\tau} \\ e^{-\alpha\tau} & 1 \end{pmatrix}$$

The results (7.59) easily follow then from this bivariate normal distribution, while the generalization to n time instants is achieved by means of a rather tedious iterative procedure.

As for the non independence of the increments it will be discussed here only for the stationary process too: in this case we will take advantage of the previous bivariate law $\mathfrak{N}(0, \mathbb{A})$ to show that two increments on non overlapping intervals are correlated, and hence they are not independent. With $s_1 < s_2 \le t_1 < t_2$ we have indeed from the previous results

$$\begin{aligned}
&\boldsymbol{E}\left[\big(X(t_2) - X(t_1)\big)\big(X(s_2) - X(s_1)\big)\right] \\
&\quad = \boldsymbol{E}\left[X(t_2)X(s_2)\right] + \boldsymbol{E}\left[X(t_1)X(s_1)\right] - \boldsymbol{E}\left[X(t_2)X(s_1)\right] - \boldsymbol{E}\left[X(t_1)X(s_2)\right] \\
&\quad = \beta^2\left[e^{-\alpha(t_2-s_2)} + e^{-\alpha(t_1-s_1)} - e^{-\alpha(t_2-s_1)} - e^{-\alpha(t_1-s_2)}\right] \\
&\quad = \beta^2\left(e^{-\alpha t_2} - e^{-\alpha t_1}\right)\left(e^{\alpha s_2} - e^{\alpha s_1}\right)
\end{aligned}$$

that in general is a non vanishing quantity, so that the increments correlation is non zero and their independence is ruled out. ■

7.1.10 Non Markovian, Gaussian Processes

Non Markovian stochastic processes do in fact exist, and of course they are bereft of most of the properties hitherto exposed. In particular we will no longer be able to find the global law of the process just from the transition *pdf* by means of the usual chain rule: these simplifications are lost, and the hierarchy of the finite joint laws needed to define the overall distribution must be given otherwise.

Example 7.30 Non Markovian transition distributions: Take first the conditional *pdf*'s **uniform** in $[y - \alpha(t-s)\,,\; y + \alpha(t-s)]$

$$f(x, t \,|\, y, s) = \begin{cases} \frac{1}{2\alpha(t-s)} & if\ |x - y| \le \alpha(t-s) \\ 0 & if\ |x - y| > \alpha(t-s) \end{cases} \tag{7.61}$$

It is easy to see then that, with the notation $\big|[a, b]\big| = |b - a|$, it is

$$\begin{aligned}
&\int f(x, t \,|\, z, r) f(z, r \,|\, y, s)\, dz \\
&\qquad = \frac{\big|[x - \alpha(t-r), x + \alpha(t-r)] \cap [y - \alpha(r-s), y + \alpha(r-s)]\big|}{4\alpha^2(t-r)(r-s)} \\
&\qquad \neq f(x, t \,|\, y, s)
\end{aligned}$$

so that the Chapman-Kolmogorov condition (7.17) is not satisfied, although (7.61) is a legitimate conditional *pdf*: hence in no way a conditional uniform distribution can be taken as the starting point to define the laws of a Markov process.

A second example is the family of the conditional **Student** *pdf*'s ($a > 0, \nu > 0$)

$$f(x, t + \Delta t \,|\, y, t) = \frac{1}{a \Delta t\, B\left(\frac{1}{2}, \frac{\nu}{2}\right)} \left(\frac{a^2 \Delta t^2}{(x-y)^2 + a^2 \Delta t^2}\right)^{\frac{\nu+1}{2}} \tag{7.62}$$

where $B(u, v)$ is the Riemann beta function defined as

$$B(u, v) = \frac{\Gamma(u)\Gamma(v)}{\Gamma(u+v)} \tag{7.63}$$

taking advantage of the gamma function (3.68). It is easy to see that the transition *pdf*'s (7.62) are a generalization of the Cauchy *pdf* (7.55) that is recovered for $\nu = 1$. Albeit a legitimate transition *pdf*, a long calculation that we will neglect here would prove that the (7.62) too does not meet the second Chapman-Kolmogorov equation (7.17), but for the unique particular Cauchy case with $\nu = 1$ that, as we already know, is Markovian. As a matter of fact a Student Lévy (and hence Markov) process does exist, but its transition *pdf*'s have not the form (7.62), and are not even explicitly known with the exception of the Cauchy case.

It is important then to emphasize that there is another relevant family of processes (that in general are not Markovian) whose global distributions can still be provided in an elementary way, viz. the *Gaussian processes* of the Definition 5.3. Their main simplification comes from the properties of the multivariate, joint Gaussian laws that are wholly specified by means of a covariance matrix and a mean vector. If indeed $X(t)$ (with just one component for short) is a Gaussian process, from the Definition 4.17 we find that all its joint laws will be completely specified through the functions (5.1) and (5.3)

$$m(t) = \boldsymbol{E}[X(t)] \qquad C(t, s) = \boldsymbol{E}[X(t)X(s)] - m(t)m(s)$$

If in fact we have an arbitrary $m(t)$ and a symmetric, non negative definite $C(t, s)$, then for every choice of $t_1, \ldots, t_n$ the *r-vec* $(X(t_1), \ldots, X(t_n))$ will be distributed according to the law $\mathfrak{N}(\boldsymbol{b}, \mathbb{A})$ with

$$b_j = m(t_j) \qquad a_{jk} = C(t_j, t_k)$$

In other words the *chf* of an arbitrary finite joint distribution of $X(t)$ takes the form

$$\varphi(u_1, t_1; \ldots; u_n, t_n) = e^{i\sum_j m(t_j)u_j - \frac{1}{2}\sum_{jk} C(t_j, t_k)u_j u_k} \tag{7.64}$$

so that the law of a Gaussian process $X(t)$ is completely specified when $m(t)$ and $C(t, s)$ (non negative defined) are given. We already met a few instances of Gaussian processes that were also Markov processes: the Wiener process $W(t)$ with $W(0) = 0$ $\boldsymbol{P}$-a.s., about which, from the Propositions 6.17 and 6.18, we know that

$$m_W(t) = 0 \qquad C_W(t, s) = D \min\{s, t\}$$

and the stationary Ornstein-Uhlenbeck process $X(t)$ about which, from the Proposition 7.29, we have that

$$m_X(t) = 0 \qquad C_X(t, s) = \beta^2 e^{-\alpha|t-s|}$$

Several other Gaussian processes can be defined in this way: a notable example of a Gaussian, *non Markovian* process, the ***fractional Brownian motion***, is briefly presented in the Appendix J.

7.2 Jump-Diffusion Processes

The second Chapman-Kolmogorov equation (7.17) is a major compatibility condition for the Markovian transition *pdf*'s, but—since it is a non linear, integral equation—it would be troublesome to actually use it to find the transition distributions of a Markov process. It is crucial then to show that, for a wide class of Markov processes, (7.17) can be put in a more tractable layout. Remark moreover that this new form will turn out to be nothing else than a generalization of the Einstein diffusion equations put forward in 1905, and therefore it stands within the first of the two lines of thought mentioned in the Sect. 6.4: it will be indeed an equation for the distributions, not for the trajectories of the process. For short in the following we will generally confine our presentation to the case of *ac* processes endowed with *pdf*'s.

Definition 7.31 We will say that a Markov process $\boldsymbol{X}(t) = (X_1(t), \dots, X_M(t))$, with a transition *pdf* $f(\boldsymbol{x}, t \mid \boldsymbol{y}, s)$ given at least in the retarded region $t > s$, is a **jump-diffusion** when it conforms to the following conditions (see also [3], vol. II, p. 333, and [4], p. 47):

1. it exists $\ell(\boldsymbol{x}|\boldsymbol{z}, t) \geq 0$ called **Lévy density** such that, for every $\epsilon > 0$, and uniformly in $\boldsymbol{x}, \boldsymbol{z}, t$, it is

$$\lim_{\Delta t \downarrow 0} \frac{1}{\Delta t} f(\boldsymbol{x}, t + \Delta t \mid \boldsymbol{z}, t) = \ell(\boldsymbol{x}|\boldsymbol{z}, t) \qquad \text{for } |\boldsymbol{x} - \boldsymbol{z}| > \epsilon \tag{7.65}$$

   and if in particular $\ell(\boldsymbol{x}|\boldsymbol{z}, t) = 0$ the process is simply called a **diffusion**
2. it exists $\boldsymbol{A}(\boldsymbol{z}, t)$ called **drift vector** such that, for every $\epsilon > 0$, and uniformly in $\boldsymbol{z}, t$, it is

$$\lim_{\Delta t \downarrow 0} \frac{1}{\Delta t} \int_{|\boldsymbol{x}-\boldsymbol{z}|<\epsilon} (x_i - z_i) f(\boldsymbol{x}, t + \Delta t \mid \boldsymbol{z}, t)\, d\boldsymbol{x} = A_i(\boldsymbol{z}, t) + O(\epsilon) \tag{7.66}$$

   which is equivalent to say that

$$A_i(\boldsymbol{z}, t) = \lim_{\epsilon \downarrow 0} \lim_{\Delta t \downarrow 0} \boldsymbol{E}\left[\frac{\Delta X_i(t)}{\Delta t} \chi_{[0,\epsilon)}\big(|\Delta \boldsymbol{X}(t)|\big) \,\Big|\, \boldsymbol{X}(t) = \boldsymbol{z} \right] \tag{7.67}$$

   where $\chi_B(\cdot)$ is the indicator of a set B
3. it exists $\mathbb{B}(\boldsymbol{z}, t)$ called **diffusion matrix** such that, for every $\epsilon > 0$, and uniformly in $\boldsymbol{z}, t$, it is

$$\lim_{\Delta t \downarrow 0} \frac{1}{\Delta t} \int_{|x-z|<\epsilon} (x_i - z_i)(x_j - z_j) f(\boldsymbol{x}, t + \Delta t \mid \boldsymbol{z}, t)\, d\boldsymbol{x} = B_{ij}(\boldsymbol{z}, t) + O(\epsilon) \tag{7.68}$$

which is equivalent to say that

$$B_{ij}(\boldsymbol{z}, t) = \lim_{\epsilon \downarrow 0} \lim_{\Delta t \downarrow 0} \boldsymbol{E}\left[\frac{\Delta X_i(t) \Delta X_j(t)}{\Delta t} \chi_{[0,\epsilon)}\big(|\Delta \boldsymbol{X}(t)|\big) \,\Big|\, \boldsymbol{X}(t) = \boldsymbol{z} \right] \tag{7.69}$$

It is possible to show that higher order terms of the type (7.66) and (7.68) would vanish, and therefore they will be simply neglected. Remark that, since $\ell(\boldsymbol{x}|\boldsymbol{z}, t) = 0$ for $\boldsymbol{x} \neq \boldsymbol{z}$ entails that the Lindeberg condition (7.47) is apparently satisfied, a diffusion process will always be sample continuous. A non vanishing $\ell(\boldsymbol{x}|\boldsymbol{z}, t)$ would point instead to the existence of discontinuous trajectories, that is to the jumping behavior of a generic jump-diffusion.

7.2.1 Forward Equations

Theorem 7.32 *The pdf $f(\boldsymbol{x}, t)$ of a jump-diffusion $\boldsymbol{X}(t)$ with $\boldsymbol{X}(0) = \boldsymbol{X}_0$, $\boldsymbol{P}$-a.s., is a solution of the so-called* ***forward equation***

$$\partial_t f(\boldsymbol{x}, t) = -\sum_i \partial_i [A_i(\boldsymbol{x}, t) f(\boldsymbol{x}, t)] + \frac{1}{2} \sum_{i,j} \partial_i \partial_j [B_{ij}(\boldsymbol{x}, t) f(\boldsymbol{x}, t)]$$
$$+ \int_{\boldsymbol{z} \neq \boldsymbol{x}} [\ell(\boldsymbol{x}|\boldsymbol{z}, t)\, f(\boldsymbol{z}, t) - \ell(\boldsymbol{z}|\boldsymbol{x}, t)\, f(\boldsymbol{x}, t)]\, d\boldsymbol{z} \tag{7.70}$$

with the initial condition

$$f(\boldsymbol{x}, 0^+) = f_0(\boldsymbol{x}) \tag{7.71}$$

where $f_0(\boldsymbol{x})$ is the $\boldsymbol{X}_0$ pdf. Moreover its transition pdf $f(\boldsymbol{x}, t \mid \boldsymbol{y}, s)$ in the retarded region $t > s$, with $\boldsymbol{X}(s) = \boldsymbol{y}$, is a solution (7.70) with the initial condition

$$f(\boldsymbol{x}, s^+) = \delta(\boldsymbol{x} - \boldsymbol{y}) \tag{7.72}$$

Proof Remembering that the uniformity of the convergence in the Definition 7.31 will enable us below to exchange limits and integrals, we will prove first the second statement to the effect that the transition *pdf* $f(\boldsymbol{x}, t \mid \boldsymbol{y}, s)$ of a jump-diffusion in the retarded region $t \geq s$ is solution of the forward equation (7.70). Take indeed a function $h(\boldsymbol{x})$ at least twice differentiable (in order to be able to implement the Taylor formula up to the second order) and, recalling that a derivative—if it exists at all—coincides with the *right* derivative, we will have

$$\partial_t \boldsymbol{E}\left[h\left(\boldsymbol{X}(t)\right) \mid \boldsymbol{X}(s) = \boldsymbol{y}\right] = \partial_t \int h(\boldsymbol{x}) f(\boldsymbol{x}, t \mid \boldsymbol{y}, s)\, d\boldsymbol{x} = \int h(\boldsymbol{x})\, \partial_t f(\boldsymbol{x}, t \mid \boldsymbol{y}, s)\, d\boldsymbol{x}$$
$$= \lim_{\Delta t \downarrow 0} \int h(\boldsymbol{x}) \frac{f(\boldsymbol{x}, t + \Delta t \mid \boldsymbol{y}, s) - f(\boldsymbol{x}, t \mid \boldsymbol{y}, s)}{\Delta t}\, d\boldsymbol{x}$$

For $\Delta t > 0$ and renaming the variables wherever needed, from the Chapman-Kolmogorov equation (7.17) and a simple normalization we can write

$$\int h(\boldsymbol{x}) f(\boldsymbol{x}, t + \Delta t \mid \boldsymbol{y}, s)\, d\boldsymbol{x} = \iint h(\boldsymbol{x}) f(\boldsymbol{x}, t + \Delta t \mid \boldsymbol{z}, t) f(\boldsymbol{z}, t \mid \boldsymbol{y}, s)\, d\boldsymbol{x} d\boldsymbol{z}$$
$$\int h(\boldsymbol{x}) f(\boldsymbol{x}, t \mid \boldsymbol{y}, s)\, d\boldsymbol{x} = \int h(\boldsymbol{z}) f(\boldsymbol{z}, t \mid \boldsymbol{y}, s)\, d\boldsymbol{z}$$
$$= \iint h(\boldsymbol{z}) f(\boldsymbol{x}, t + \Delta t \mid \boldsymbol{z}, t) f(\boldsymbol{z}, t \mid \boldsymbol{y}, s)\, d\boldsymbol{x} d\boldsymbol{z}$$

so that, decomposing the integration domain in $|\boldsymbol{x} - \boldsymbol{z}| < \epsilon$ and $|\boldsymbol{x} - \boldsymbol{z}| \geq \epsilon$ by means of an arbitrary $\epsilon > 0$, we will have

$$\int h(\boldsymbol{x})\, \partial_t f(\boldsymbol{x}, t \mid \boldsymbol{y}, s)\, d\boldsymbol{x}$$
$$= \lim_{\Delta t \downarrow 0} \iint [h(\boldsymbol{x}) - h(\boldsymbol{z})] \frac{f(\boldsymbol{x}, t + \Delta t \mid \boldsymbol{z}, t)}{\Delta t} f(\boldsymbol{z}, t \mid \boldsymbol{y}, s)\, d\boldsymbol{x} d\boldsymbol{z}$$
$$= \lim_{\epsilon \downarrow 0} \lim_{\Delta t \downarrow 0} \left[\iint_{|\boldsymbol{x} - \boldsymbol{z}| < \epsilon} [h(\boldsymbol{x}) - h(\boldsymbol{z})] \frac{f(\boldsymbol{x}, t + \Delta t \mid \boldsymbol{z}, t)}{\Delta t} f(\boldsymbol{z}, t \mid \boldsymbol{y}, s)\, d\boldsymbol{x} d\boldsymbol{z} \right.$$
$$\left. + \iint_{|\boldsymbol{x} - \boldsymbol{z}| \geq \epsilon} [h(\boldsymbol{x}) - h(\boldsymbol{z})] \frac{f(\boldsymbol{x}, t + \Delta t \mid \boldsymbol{z}, t)}{\Delta t} f(\boldsymbol{z}, t \mid \boldsymbol{y}, s)\, d\boldsymbol{x} d\boldsymbol{z} \right]$$

Take first the integration on the domain $|\boldsymbol{x} - \boldsymbol{z}| < \epsilon$: for $\epsilon \to 0$ we can use for $h(\boldsymbol{x})$ the Taylor formula up to the second order in a neighborhood of $\boldsymbol{z}$

$$h(\boldsymbol{x}) = h(\boldsymbol{z}) + \sum_i (x_i - z_i) \partial_i h(\boldsymbol{z})$$
$$+ \frac{1}{2} \sum_{i,j} (x_i - z_i)(x_j - z_j) \partial_i \partial_j h(\boldsymbol{z}) + |\boldsymbol{x} - \boldsymbol{z}|^2 R(\boldsymbol{x}, \boldsymbol{z})$$

where for the remainder it is understood that $R(\boldsymbol{x}, \boldsymbol{z}) \to 0$ when $|\boldsymbol{x} - \boldsymbol{z}| \to 0$. We then have for the integral on $|\boldsymbol{x} - \boldsymbol{z}| < \epsilon$

$$\iint_{|x-z|<\epsilon} [h(x) - h(z)] \frac{f(x, t + \Delta t \mid z, t)}{\Delta t} f(z, t|y, s)\, dx dz$$
$$= \iint_{|x-z|<\epsilon} \left[\sum_i (x_i - z_i)\partial_i h(z) \right.$$
$$\left. + \frac{1}{2} \sum_{i,j} (x_i - z_i)(x_j - z_j)\partial_i \partial_j h(z) \right] \frac{f(x, t + \Delta t \mid z, t)}{\Delta t} f(z, t|y, s)\, dx dz$$
$$+ \iint_{|x-z|<\epsilon} |x - z|^2 R(x, z) \frac{f(x, t + \Delta t \mid z, t)}{\Delta t} f(z, t|y, s)\, dx dz$$

that, in the limits $\Delta t \to 0$ and $\epsilon \to 0$, because of (7.66) and (7.68) and with a few integrations by parts, becomes

$$\int \left[\sum_i A_i(z, t)\partial_i h(z) + \sum_{i,j} \frac{1}{2} B_{ij}(z, t)\partial_i \partial_j h(z) \right] f(z, t|y, s)\, dz$$
$$= \int h(z) \left\{ - \sum_i \partial_{z_i} \left[A_i(z, t) f(z, t \mid y, s) \right] + \frac{1}{2} \sum_{i,j} \partial_{z_i} \partial_{z_j} \left[B_{ij}(z, t) f(z, t \mid y, s) \right] \right\} dz$$

In the external domain $|x - z| \geq \epsilon$ instead—decomposing the integral in two terms and exchanging the names of the integration variables x and z in the first addend—we have

$$\iint_{|x-z|\geq\epsilon} [h(x) - h(z)] \frac{f(x, t + \Delta t \mid z, t)}{\Delta t} f(z, t|y, s)\, dx dz$$
$$= \iint_{|x-z|\geq\epsilon} h(z) \left[\frac{f(z, t + \Delta t \mid x, t)}{\Delta t} f(x, t|y, s) \right.$$
$$\left. - \frac{f(x, t + \Delta t \mid z, t)}{\Delta t} f(z, t|y, s) \right] dx dz$$

Then from (7.65), in the limit $\Delta t \to 0$ we first get

$$\iint_{|x-z|\geq\epsilon} h(z) \big[\ell(z|x, t)\, f(x, t|y, s) - \ell(x|z, t)\, f(z, t|y, s) \big]\, dx dz$$

and subsequently for $\epsilon \to 0$

$$\int h(z) \left\{ \int_{x \neq z} \big[\ell(z|x, t)\, f(x, t|y, s) - \ell(x|z, t)\, f(z, t|y, s) \big]\, dx \right\} dz$$

where we adopted the shorthand notation

$$\int_{x \neq z} \dots dx = \lim_{\epsilon \to 0} \int_{|x-z| \geq \epsilon} \dots dx$$

Collecting all the results we then have

$$\int h(z)\, \partial_t f(z, t \mid y, s)\, dz$$

$$= \int h(z) \left\{ - \sum_i \partial_{z_i} [A_i(z, t) f(z, t \mid y, s)] + \frac{1}{2} \sum_{i,j} \partial_{z_i} \partial_{z_j} [B_{ij}(z, t) f(z, t \mid y, s)] \right.$$

$$\left. + \int_{x \neq z} \big[\ell(z|x, t)\, f(x, t|y, s) - \ell(x|z, t)\, f(z, t|y, s) \big]\, dx \right\} dz$$

and since $h(x)$ is arbitrary we will be finally able to write (exchanging for convenience z and x)

$$\partial_t f(x, t \mid y, s) = - \sum_i \partial_{x_i} [A_i(x, t) f(x, t \mid y, s)] + \frac{1}{2} \sum_{i,j} \partial_{x_i} \partial_{x_j} [B_{ij}(x, t) f(x, t \mid y, s)]$$

$$+ \int_{z \neq x} [\ell(x|z, t)\, f(z, t|y, s) - \ell(z|x, t)\, f(x, t|y, s)]\, dz \qquad (7.73)$$

that is the form of our integro-differential *forward equation* (7.70) adapted to the case of the transition *pdf*'s. To have that in the initial form (7.70) for $f(x, t)$ it will be enough to multiply (7.73) by $f(y, s)$ and to integrate it in dy: the first Chapman-Kolmogorov equation (7.16) will then entail that also $f(x, t)$ is a solution of (7.70), in particular for $s = 0$. ■

Theorem 7.33 *Let us take*

1. *a non negative Lévy density* $\ell(x|y, t)$
2. *a drift vector* $A(x, t)$
3. *a definite nonnegative covariance matrix* $\mathbb{B}(x, t)$

then it exists a unique non negative and normalized solution $f(x, t \mid y, s)$ $(t > s)$ *of the forward equation (7.70) with degenerate initial conditions (7.72) that satisfies the Chapman-Kolmogorov equation (7.17): this Markovian transition pdf in the retarded region fulfills the requirements of the Definition 7.31, and hence, according to the Proposition 7.7, it selects a family of jump-diffusions, one for every initial condition.*

Proof Omitted: see [4], p. 51 and [5], vol. II, Chap. 1, §1. ■

We have derived the previous results by supposing that the process was endowed with a *pdf* f: it is apparent then that these statements must be suitable modified in order to encompass also the cases of processes bereft of a *pdf*. Confining ourselves to the simplest instances, let us take a process with integer values (as the simple Poisson process $N(t)$) and revise first the conditions (7.65), (7.66) and (7.68). To this end remark that for an integer values process the requirements $|x - z| > \epsilon$ and

$|\boldsymbol{x} - \boldsymbol{z}| \leq \epsilon$ with arbitrary $\epsilon > 0$, just mean $\boldsymbol{n} \neq \boldsymbol{m}$ and $\boldsymbol{n} = \boldsymbol{m}$. As a consequence the conditions (7.66) and (7.68) are trivially reduced to $\boldsymbol{A} = 0$ and $\mathbb{B} = 0$. The first condition (7.65) is instead to be replaced with its discrete version

$$\lim_{\Delta t \downarrow 0} \frac{1}{\Delta t} p(\boldsymbol{n}, t + \Delta t \mid \boldsymbol{m}, t) = \ell(\boldsymbol{n}|\boldsymbol{m}, t) \qquad \text{with } \boldsymbol{n} \neq \boldsymbol{m} \tag{7.74}$$

uniformly in $\boldsymbol{n}, \boldsymbol{m}, t$. Under these new conditions it is possible to show then that the forward equation (7.73) is replaced by the ***master equation*** (see later Sect. 7.2.3 for further details)

$$\partial_t p(\boldsymbol{n}, t) = \sum_{\boldsymbol{k}} \big[\ell(\boldsymbol{n}|\boldsymbol{k}, t) p(\boldsymbol{k}, t) - \ell(\boldsymbol{k}|\boldsymbol{n}, t) p(\boldsymbol{n}, t)\big] \tag{7.75}$$

whose solution is the transition probability $p(\boldsymbol{n}, t \mid \boldsymbol{m}, s)$ if the initial condition is $p(\boldsymbol{n}, s^+) = \delta_{\boldsymbol{nm}}$. Of course for such a kind of processes with integer values their jumping behavior is a foregone conclusion, and this accounts for the prominent role played by $\ell(\boldsymbol{n}|\boldsymbol{m}, t)$. But it would be wrong to suppose in reverse that $\ell(\boldsymbol{x}|\boldsymbol{y}, t)$ should vanish only because a process takes continuous values. We already remarked indeed at the end of the Sect. 7.1.1 that even these processes can have discontinuous, jumping trajectories as for instance the Cauchy process that will be further elaborated later on in the present chapter.

7.2.2 *Backward Equations*

Equation (7.73) is known as *forward equation* because it—understood as an equation for the transition *pdf*'s in the retarded region—involves operations on the *final* variables $\boldsymbol{x}, t$ of $f(\boldsymbol{x}, t \mid \boldsymbol{y}, s)$ imposing *initial conditions* at the time $s < t$. It admits however also another formulation called *backward equation*: in this second case the integro-differential operations are performed on the initial variables $\boldsymbol{y}, s$, while the solutions must satisfy appropriate *final conditions* at the time $t > s$. It is possible to prove indeed that the transition *pdf*'s of a jump-diffusion process are also solutions of the following **backward equation**

$$\begin{aligned}\partial_s f(\boldsymbol{x}, t \mid \boldsymbol{y}, s) = &- \sum_i A_i(\boldsymbol{y}, s) \partial_{y_i} f(\boldsymbol{x}, t \mid \boldsymbol{y}, s) \\ &- \frac{1}{2} \sum_{i,j} B_{ij}(\boldsymbol{y}, s) \partial_{y_i} \partial_{y_j} f(\boldsymbol{x}, t \mid \boldsymbol{y}, s) \\ &+ \int_{\boldsymbol{z} \neq \boldsymbol{y}} \ell(\boldsymbol{z}|\boldsymbol{y}, s)\big[f(\boldsymbol{x}, t|\boldsymbol{y}, s) - f(\boldsymbol{x}, t|\boldsymbol{z}, s)\big]\, d\boldsymbol{z}\end{aligned} \tag{7.76}$$

with ***final conditions*** $f(\boldsymbol{x}, t \mid \boldsymbol{y}, t^-) = \delta(\boldsymbol{x} - \boldsymbol{y})$. These two formulations. forward (7.73) and backward (7.76), are in fact equivalent: a retarded *pdf*'s $f(\boldsymbol{x}, t \mid \boldsymbol{y}, s)$ solution of the forward equations in $\boldsymbol{x}, t$ with initial conditions $\boldsymbol{y}, s$, is also a solution of the backward equation in $\boldsymbol{y}, s$ with the same coefficients and final conditions $\boldsymbol{x}, t$. Both these formulations can be adopted according to the needs: the forward equations are more popular in the physical applications, but also the backward ones are employed in several problems, as for instance that of the first passage time.

From a mathematical standpoint, however, the backward equation (7.76) is more desirable precisely because it operate on the conditioning (initial) variables $\boldsymbol{y}, s$. To write the forward equation in the form (7.73) we need in fact the existence of the *pdf* $f(\boldsymbol{x}, t \mid \boldsymbol{y}, s)$ because its integro-differential operations are performed on the final variables $\boldsymbol{x}, t$. On the other hand this requirement is apparently not always met, and hence (as for the integer values processes) a different formulation is needed for the cases without a *pdf*. The conditioning variables $\boldsymbol{y}, s$ are conversely always explicitly spelled in every distribution or expectation conditioned by $\{\boldsymbol{X}(s) = \boldsymbol{y}\}$. This enables us to find a general form of the evolution equations without being obliged to tell apart the *ac* cases with *pdf* from the discrete ones.

Without going into needless details and neglecting for simplicity the jump (integral) terms, let us take just the backward equations (7.76) in its diffusive form

$$\begin{aligned}\partial_s f(\boldsymbol{x}, t \mid \boldsymbol{y}, s) = &-\sum_i A_i(\boldsymbol{y}, s)\partial_{y_i} f(\boldsymbol{x}, t \mid \boldsymbol{y}, s)\\ &-\frac{1}{2}\sum_{i,j} B_{ij}(\boldsymbol{y}, s)\partial_{y_i}\partial_{y_j} f(\boldsymbol{x}, t \mid \boldsymbol{y}, s)\end{aligned} \tag{7.77}$$

and an arbitrary function $h(\boldsymbol{x})$: if we now define

$$g(\boldsymbol{y}, s) = \boldsymbol{E}\left[h(\boldsymbol{X}(t)) \mid \boldsymbol{X}(s) = \boldsymbol{y}\right] = \int h(\boldsymbol{x}) f(\boldsymbol{x}, t \mid \boldsymbol{y}, s)\, d\boldsymbol{x}$$

a multiplication of (7.77) by $h(\boldsymbol{x})$ and a subsequent $\boldsymbol{x}$-integration yield

$$\partial_s g(\boldsymbol{y}, s) = -\sum_i A_i(\boldsymbol{y}, s)\partial_i g(\boldsymbol{y}, s) - \frac{1}{2}\sum_{i,j} B_{ij}(\boldsymbol{y}, s)\partial_i\partial_j g(\boldsymbol{y}, s) \tag{7.78}$$

with final condition $g(\boldsymbol{y}, t^-) = h(\boldsymbol{y})$. Equation (7.78), that no longer makes an explicit reference to a *pdf*, is known in the literature as a particular case of ***Kolmogorov equation***.

7.2.3 Main Classes of Jump-Diffusions

7.2.3.1 Pure Jump Processes: Master Equation

Consider first the case $\boldsymbol{A} = \mathbb{B} = 0$, while only $\ell \neq 0$: the forward equation (7.70) then becomes

$$\partial_t f(\boldsymbol{x}, t) = \int_{z \neq x} [\ell(\boldsymbol{x}|z, t) f(z, t) - \ell(z|\boldsymbol{x}, t) f(\boldsymbol{x}, t)] \, dz \qquad (7.79)$$

When in particular the process takes only integer values (like the simple Poisson process) the Eq. (7.79) appears in the discrete form (7.75). Equations of this kind are called ***master equations***, and the processes $\boldsymbol{X}(t)$ ruled by them (even if they take continuous values) are known as ***pure jump processes*** because they lack both a drift and a diffusive component.

Proposition 7.34 *If $\boldsymbol{X}(t)$ is a pure jump process taking continuous values, the probability of performing a finite size jump in the time interval $[t, t + dt]$ is*

$$\boldsymbol{P}\{\boldsymbol{X}(t + dt) \neq \boldsymbol{X}(t) \,|\, \boldsymbol{X}(t) = \boldsymbol{y}\} = dt \int_{x \neq y} \ell(\boldsymbol{x}|\boldsymbol{y}, t) \, d\boldsymbol{x} + o\,(dt) \qquad (7.80)$$

Proof The retarded transition *pdf* $f(\boldsymbol{x}, t|\boldsymbol{y}, s)$, with $t > s$ and initial condition $f(\boldsymbol{x}, s^+|\boldsymbol{y}, s) = \delta(\boldsymbol{x} - \boldsymbol{y})$, obeys to the master equation (7.79) in the form

$$\partial_t f(\boldsymbol{x}, t|\boldsymbol{y}, s) = \int_{z \neq x} \big[\ell(\boldsymbol{x}|z, t) f(z, t|\boldsymbol{y}, s) - \ell(z|\boldsymbol{x}, t) f(\boldsymbol{x}, t|\boldsymbol{y}, s)\big] \, dz$$

that for $s = t$ can also be symbolically written as

$$\frac{f(\boldsymbol{x}, t + dt|\boldsymbol{y}, t) - \delta(\boldsymbol{x} - \boldsymbol{y})}{dt} = \int_{z \neq x} \big[\ell(\boldsymbol{x}|z, t)\delta(z - \boldsymbol{y}) - \ell(z|\boldsymbol{x}, t)\delta(\boldsymbol{x} - \boldsymbol{y})\big] \, dz$$

namely (neglecting for short the higher order infinitesimals) even as

$$f(\boldsymbol{x}, t + dt|\boldsymbol{y}, t) = \left[1 - dt \int_{z \neq x} \ell(z|\boldsymbol{x}, t) \, dz\right] \delta(\boldsymbol{x} - \boldsymbol{y}) + dt \int_{z \neq x} \ell(\boldsymbol{x}|z, t)\delta(z - \boldsymbol{y}) \, dz$$

We then have

$$\begin{aligned}
\boldsymbol{P}\{\boldsymbol{X}(t+dt) \neq \boldsymbol{X}(t) \mid \boldsymbol{X}(t) = \boldsymbol{y}\} &= \int_{\boldsymbol{x}\neq\boldsymbol{y}} f(\boldsymbol{x}, t+dt|\boldsymbol{y}, t)\, d\boldsymbol{x} \\
&= \int_{\boldsymbol{x}\neq\boldsymbol{y}} d\boldsymbol{x} \left[1 - dt \int_{\boldsymbol{z}\neq\boldsymbol{x}} \ell(\boldsymbol{z}|\boldsymbol{x}, t)\, d\boldsymbol{z}\right] \delta(\boldsymbol{x}-\boldsymbol{y}) \\
&\quad + dt \int_{\boldsymbol{x}\neq\boldsymbol{y}} d\boldsymbol{x} \int_{\boldsymbol{z}\neq\boldsymbol{x}} \ell(\boldsymbol{x}|\boldsymbol{z}, t)\delta(\boldsymbol{z}-\boldsymbol{y})\, d\boldsymbol{z}
\end{aligned}$$

and since from the Dirac delta properties it is

$$\int_{\boldsymbol{x}\neq\boldsymbol{y}} d\boldsymbol{x} \left[1 - dt \int_{\boldsymbol{z}\neq\boldsymbol{x}} \ell(\boldsymbol{z}|\boldsymbol{x}, t)\, d\boldsymbol{z}\right] \delta(\boldsymbol{x}-\boldsymbol{y}) = 0$$

$$\int_{\boldsymbol{x}\neq\boldsymbol{y}} d\boldsymbol{x} \int_{\boldsymbol{z}\neq\boldsymbol{x}} \ell(\boldsymbol{x}|\boldsymbol{z}, t)\delta(\boldsymbol{z}-\boldsymbol{y})\, d\boldsymbol{z} = \int_{\boldsymbol{x}\neq\boldsymbol{y}} \ell(\boldsymbol{x}|\boldsymbol{y}, t)\, d\boldsymbol{x}$$

we finally get the result (7.80). ■

This result enables us to identify both the jumping character of the process and the meaning of the Lévy density $\ell(\boldsymbol{x}|\boldsymbol{y}, t)$: for an infinitesimal dt the term $\ell(\boldsymbol{x}|\boldsymbol{y}, t)\, dt$ plays the role of a density for the probability of *not* staying in the initial position $\boldsymbol{y}$ by performing a jump of finite size $\boldsymbol{x} - \boldsymbol{y}$ in a time dt, as pointed out also in (7.80). Keep in mind though that this interpretation cannot be pushed beyond a certain limit because generally speaking $\ell(\boldsymbol{x}|\boldsymbol{y}, t)$ itself is not normalizable and hence can not be considered as an authentic *pdf*: the previous integral of $\ell(\boldsymbol{x}|\boldsymbol{y}, t)$ can indeed approximate a probability only as an infinitesimal, that is only if multiplied by dt.

7.2.3.2 Diffusion Processes: Fokker-Planck Equation

Let us consider now the case $\ell = 0$, but $\mathbb{B} \neq 0$ (it is not relevant whether $\boldsymbol{A}$ vanishes or not): for such a *diffusion process* the Lindeberg criterion guarantees then that $\boldsymbol{X}(t)$ is sample continuous, while the forward equation becomes

$$\partial_t f(\boldsymbol{x}, t) = -\sum_i \partial_{x_i}[A_i(\boldsymbol{x}, t) f(\boldsymbol{x}, t)] + \frac{1}{2}\sum_{i,j} \partial_{x_i}\partial_{x_j}[B_{ij}(\boldsymbol{x}, t) f(\boldsymbol{x}, t)] \quad (7.81)$$

and takes the name of ***Fokker-Planck equation*** . This, at variance with (7.70), is moreover an exclusively partial differential equation without additional integral terms. Equation (6.53) previously found for the Wiener process with just one component is a particular case with $\boldsymbol{A} = 0$ and $\mathbb{B} = D$

Proposition 7.35 *If on a diffusion process we impose the condition $\boldsymbol{X}(t)=\boldsymbol{y}$, $\boldsymbol{P}$-a.s. at an arbitrary instant $t > 0$, then after an infinitesimal delay $dt > 0$ the process law becomes*

$$\boldsymbol{X}(t+dt) \sim \mathfrak{N}\big(\boldsymbol{y} + \boldsymbol{A}(\boldsymbol{y}, t)dt\,,\ \mathbb{B}(\boldsymbol{y}, t)dt\big) \quad (7.82)$$

Proof We know that the retarded transition *pdf* of our process $f(\boldsymbol{x}, t|\boldsymbol{y}, s)$, with $t > s$ and initial condition $f(\boldsymbol{x}, s^+|\boldsymbol{y}, s) = \delta(\boldsymbol{x} - \boldsymbol{y})$, is a solution of the Fokker-Planck equation (7.81) in the form

$$\partial_t f(\boldsymbol{x}, t|\boldsymbol{y}, s) = -\sum_i \partial_{x_i}[A_i(\boldsymbol{x}, t) f(\boldsymbol{x}, t|\boldsymbol{y}, s)] + \frac{1}{2}\sum_{i,j} \partial_{x_i}\partial_{x_j}[B_{ij}(\boldsymbol{x}, t) f(\boldsymbol{x}, t|\boldsymbol{y}, s)]$$

If the time interval $[s, t]$ is infinitesimal the transition *pdf* f at a time t very near to s will still be squeezed around its initial position $\boldsymbol{y}$, so that—near to $\boldsymbol{y}$ where it is not zero—f will exhibit very large spatial derivatives. We can then assume that the corresponding spatial derivatives of $\boldsymbol{A}$ and $\mathbb{B}$ will be negligible w.r.t. that of f, and hence that in a first approximation the functions $\boldsymbol{A}$ and $\mathbb{B}$ can reasonably be considered as constant and still coincident with their values in $\boldsymbol{y}$ at the time s. If this is so the previous Fokker-Planck equation is reduced to

$$\partial_t f(\boldsymbol{x}, t \mid \boldsymbol{y}, s) = -\sum_i A_i(\boldsymbol{y}, s)\partial_{x_i} f(\boldsymbol{x}, t \mid \boldsymbol{y}, s) + \frac{1}{2}\sum_{i,j} B_{ij}(\boldsymbol{y}, s)\partial_{x_i}\partial_{x_j} f(\boldsymbol{x}, t \mid \boldsymbol{y}, s)$$

where $\boldsymbol{A}$ and $\mathbb{B}$ depend now only on the variables $\boldsymbol{y}, s$ not involved in the differentiations. Remark that, despite their formal similarities, this equation differs from the backward equation without jumping terms (7.77): not only the sign of the diffusive term $\mathbb{B}$ is reversed, but in (7.77) the derivatives involve $\boldsymbol{y}, s$, not $\boldsymbol{x}, t$, so that the terms $\boldsymbol{A}$ and $\mathbb{B}$ can not be considered constant as we do here. Our approximated Fokker-Planck equation with constant coefficients can now be easily solved: it is possible to check indeed by direct calculation that, with $t - s = dt > 0$, the solution is

$$f(\boldsymbol{x}, t \mid \boldsymbol{y}, s) = \sqrt{\frac{|\mathbb{B}^{-1}(\boldsymbol{y}, s)|}{(2\pi)^M dt}}\, e^{-[\boldsymbol{x}-\boldsymbol{y}-\boldsymbol{A}(\boldsymbol{y},s)dt]\cdot\mathbb{B}^{-1}(\boldsymbol{y},s)[\boldsymbol{x}-\boldsymbol{y}-\boldsymbol{A}(\boldsymbol{y},s)dt]\,/\,2dt}$$

where $|\mathbb{B}^{-1}|$ is the determinant of the matrix $\mathbb{B}^{-1}$. It is obvious then that—if we adapt our notations to that of (7.82) with the replacements $s \to t$, and $t \to t + dt$—we are now able to state that, starting from $\boldsymbol{X}(t) = \boldsymbol{y}$, in an infinitesimal interval $[t, t + dt]$ a solution of (7.81) evolves exactly toward the law (7.82) of our proposition. This apparently elucidates both the role of drift velocity of $\boldsymbol{A}$, and that of diffusion coefficient of $\mathbb{B}$, and hence fully accounts for their names. ■

7.2.3.3 Degenerate Processes: Liouville Equation

Take finally $\ell = \mathbb{B} = 0$ and only $\boldsymbol{A} \neq 0$ so that the forward equation becomes

$$\partial_t f(\boldsymbol{x}, t) = -\sum_i \partial_i[A_i(\boldsymbol{x}, t) f(\boldsymbol{x}, t)] \tag{7.83}$$

also known as ***Liouville equation***. This is now a differential equation ruling the *pdf* evolution with neither jumps nor diffusion terms, and hence the process will predictably follow ***degenerate*** trajectories: we will show indeed that its solution with initial condition $f(\boldsymbol{x}, s^+) = \delta(\boldsymbol{x} - \boldsymbol{y})$ progresses without spreading, and remains concentrated in one point that follows a deterministic trajectory.

Proposition 7.36 *Take a dynamic system $\boldsymbol{x}(t)$ ruled by the equation*

$$\dot{\boldsymbol{x}}(t) = \boldsymbol{A}\big[\boldsymbol{x}(t), t\big] \qquad t \geq s \tag{7.84}$$

and its solution $\boldsymbol{x}(t|\boldsymbol{y}, s)$ labeled by its initial condition $\boldsymbol{x}(s) = \boldsymbol{y}$: then the solution of the Liouville equation (7.83) with initial condition $f(\boldsymbol{x}, s^+ \mid \boldsymbol{y}, s) = \delta(\boldsymbol{x} - \boldsymbol{y})$ is

$$f(\boldsymbol{x}, t \mid \boldsymbol{y}, s) = \delta\big[\boldsymbol{x} - \boldsymbol{x}(t|\boldsymbol{y}, s)\big] \tag{7.85}$$

namely the process is invariably degenerate in $\boldsymbol{x}(t|\boldsymbol{y}, s)$ and follows its trajectory without diffusion.

Proof We will first prove the following property of the δ distributions:

$$\partial_t \delta\big[\boldsymbol{x} - \boldsymbol{g}(t)\big] = -\sum_i \dot{g}_i(t)\, \partial_i \delta\big[\boldsymbol{x} - \boldsymbol{g}(t)\big] \tag{7.86}$$

Confining ourselves for simplicity to the one-dimensional case, we find indeed for an arbitrary test function $\varphi(x)$ that

$$\begin{aligned}
\int \varphi(x)\, \partial_t \delta\big[x - g(t)\big]\, dx &= \partial_t \int \varphi(x) \delta\big[x - g(t)\big]\, dx \\
&= \partial_t \big[\varphi(g(t))\big] = \dot{g}(t)\, \varphi'(g(t)) \\
-\int \varphi(x) \dot{g}(t)\, \partial_x \delta\big[x - g(t)\big]\, dx = \dot{g}(t) &\int \varphi'(x)\, \delta\big[x - g(t)\big]\, dx = \dot{g}(t)\, \varphi'(g(t))
\end{aligned}$$

and hence the two sides of (7.86) coincide. Keeping then into account also (7.84) we find

$$\begin{aligned}
-\sum_i \partial_i \big[A_i(\boldsymbol{x}, t)\, \delta\big(\boldsymbol{x} - \boldsymbol{x}(t \mid \boldsymbol{y}, s)\big)\big] &= -\sum_i \partial_i \big[A_i\big(\boldsymbol{x}(t \mid \boldsymbol{y}, s), t\big)\, \delta\big(\boldsymbol{x} - \boldsymbol{x}(t \mid \boldsymbol{y}, s)\big)\big] \\
&= -\sum_i A_i\big(\boldsymbol{x}(t \mid \boldsymbol{y}, s), t\big)\, \partial_i \delta\big(\boldsymbol{x} - \boldsymbol{x}(t \mid \boldsymbol{y}, s)\big) \\
&= -\sum_i \dot{x}_i(t \mid \boldsymbol{y}, s)\, \partial_i \delta\big(\boldsymbol{x} - \boldsymbol{x}(t \mid \boldsymbol{y}, s)\big) \\
&= \partial_t \delta\big(\boldsymbol{x} - \boldsymbol{x}(t \mid \boldsymbol{y}, s)\big)
\end{aligned}$$

showing in this way that (7.85) is a solution of the Liouville equation (7.83). ■

7.2.4 Notable Jump-Diffusion Processes

We will analyze now a few typical examples of *forward equations*: for simplicity in the present section we will confine the discussion to the one-component processes so that in the *ac* case with *pdf* f the Eq. (7.70) becomes

$$\partial_t f(x,t) = -\partial_x[A(x,t)f(x,t)] + \frac{1}{2}\partial_x^2[B(x,t)f(x,t)] + \int_{z\neq x}[\ell(x|z,t)\,f(z,t) - \ell(z|x,t)\,f(x,t)]\,dz \quad (7.87)$$

while for the discrete processes taking integer values the master equation (7.75) becomes

$$\partial_t p(n,t) = \sum_k \big[\ell(n|k,t)p(k,t) - \ell(k|n,t)p(n,t)\big] \quad (7.88)$$

In the following we will explicitly find the coefficients A, B and ℓ of some notable *forward equation* taking advantage of the Markovian transition *pdf*'s that select the families of jump-diffusions already defined in the previous sections; we will also give a few hints about the solution methods for these equations.

Proposition 7.37 *The distributions and the transition probabilities (7.49) of a* ***simple Poisson process*** $N(t)$ *satisfy the master equation*

$$\partial_t p(n,t) = -\lambda\big[p(n,t) - p(n-1,t)\big] \quad (7.89)$$

Proof Since $N(t)$ is a counting process it only takes integer values and hence, as already suggested at the end of the Sect. 7.2.1, we first of all have $A = B = 0$, so that its forward equation will be a master equation of the type (7.88). To find then $\ell(n|m,t)$ we will use the one-dimensional version of (7.74): taking indeed the transition probability (7.49) with $n \neq m$, in the limit $\Delta t \to 0$ we find

$$\frac{1}{\Delta t}\, p_N(n, t+\Delta t\,|\,m,t) = \begin{cases} \lambda e^{-\lambda\Delta t} \to \lambda & \text{if } n = m+1 \\ O(\Delta t^{n-m}) \to 0 & \text{if } n \geq m+2 \end{cases}$$

By summarizing we can then say that for a simple Poisson process it is

$$\ell(n|m,t) = \lambda\delta_{n,\,m+1} \quad (7.90)$$

that plugged into (7.88) gives the master equation (7.89). Scanning then through the possible initial conditions we will find out all the simple Poisson processes of intensity λ of the Definition 7.20: the transition probability (7.49) is in particular associated to the degenerate initial condition $p(n,t) = \delta_{nm}$. ■

If conversely a master equation is given, we will face the problem of solving it with an appropriate initial condition, to find the law of the corresponding discrete Markov process $N(t)$. A well known method takes advantage of the so called ***generating function***: for example, in the case of the master equation (7.89) with degenerate initial conditions $p(n, 0) = \delta_{n0}$, we first check—by taking (7.89) into account—that the generating function of $N(t)$ defined as

$$\gamma(u, t) = \boldsymbol{E}\left[u^{N(t)}\right] = \sum_n u^n p(n, t) \tag{7.91}$$

satisfies the transformed equation

$$\partial_t \gamma(u, t) = \lambda(u - 1)\gamma(u, t) \qquad\qquad \gamma(u, 0) = 1$$

then we find that its solution apparently is

$$\gamma(u, t) = e^{\lambda(u-1)t}$$

and finally, comparing its Taylor expansion around $u = 0$

$$\gamma(u, t) = e^{-\lambda t} \sum_n u^n \frac{(\lambda t)^n}{n!}$$

with its definition (7.91), we finally get

$$p(n, t) = e^{-\lambda t} \frac{(\lambda t)^n}{n!}$$

This result just corroborates our initial suggestion that the solution of the master equation (7.89) with a degenerate initial condition provides the law of a simple Poisson process

Proposition 7.38 *The distributions and the transition pdf (7.51) of a* ***Wiener process*** *$W(t)$ satisfy the Fokker-Planck equation*

$$\partial_t f(x, t) = \frac{D}{2}\,\partial_x^2 f(x, t) \tag{7.92}$$

Proof Instead of a direct check, we first find from (7.51) and l'Hôpital's rule that for $x \neq y$

$$\frac{1}{\Delta t}\, f_W(x, t + \Delta t \mid y, t) = \frac{e^{-(x-y)^2/2D\Delta t}}{\Delta t\sqrt{2\pi D\Delta t}} \underset{\Delta t \to 0}{\longrightarrow} 0$$

namely $\ell(x|y, t) = 0$, a result consistent with that of the Proposition 7.23 stating that a Wiener process is sample continuous. We will moreover find $A = 0$ by symmetry, while for the diffusion coefficient B, taking $y = (x - z)/\sqrt{D\Delta t}$, we will have

$$\frac{1}{\Delta t}\int_{z-\epsilon}^{z+\epsilon}(x-z)^2 f_W(x,t+\Delta t\,|\,z,t)\,dx = D\int_{-\epsilon/\sqrt{D\Delta t}}^{+\epsilon/\sqrt{D\Delta t}}\frac{y^2e^{-y^2/2}}{\sqrt{2\pi}}\,dy \underset{\Delta t\to 0}{\longrightarrow} D$$

Collecting finally all these remarks we find that the *pdf* of a Wiener process satisfies the Fokker-Planck equation (7.92)—coincident with the (6.74) first derived by Einstein—and that its transition *pdf* (7.51) is the solution selected by the initial condition $f(x, t^+) = \delta(x - y)$. ■

If conversely the following Fokker-Planck equation with degenerate initial condition is given

$$\partial_t f(x,t) = \frac{D}{2}\,\partial_x^2 f(x,t) \qquad f(x,s^+) = \delta(x-y) \tag{7.93}$$

by solving it we find the transition *pdf* of the associated wiener process. Here too a well known solution method is that of the ***Fourier transform***: we have indeed that the *chf*

$$\varphi(u,t) = \int_{-\infty}^{+\infty} e^{iux} f(x,t)\,dx$$

turns out to abide by the transformed equation

$$\partial_t\varphi(u,t) = -\frac{Du^2}{2}\,\varphi(u,t) \qquad \varphi(u,s) = e^{iuy}$$

whose well known solution is

$$\varphi(u,t) = e^{iuy}e^{-Du^2(t-s)/2}$$

The straightforward inversion of this *chf* provides then a transition *pdf*

$$f(x,t\,|\,y,s) = \frac{e^{-(x-y)^2/2D(t-s)}}{\sqrt{2\pi\,D(t-s)}}$$

consistent with that of the Definition 7.22.

Proposition 7.39 *The distributions and the transition pdf (7.55) of a* ***Cauchy process*** *$X(t)$ satisfy the master equation*

$$\partial_t f(x,t) = \frac{a}{\pi}\int_{z\neq x}\frac{f(z,t)-f(x,t)}{(x-z)^2}\,dz \tag{7.94}$$

Proof First of all we have indeed

$$\frac{1}{\Delta t}\,f_X(x,t+\Delta t\,|\,y,t) = \frac{a}{\pi}\,\frac{1}{(x-y)^2+(a\Delta t)^2} \underset{\Delta t\to 0}{\longrightarrow} \ell(x|y,t) = \frac{a}{\pi(x-y)^2}$$

so that the process is not sample continuous and its trajectories will make jumps. Moreover it is $A = 0$ by symmetry, while for the diffusion coefficient we have with $y = (x - z)/a\Delta t$ that

$$\frac{1}{\Delta t}\int_{z-\epsilon}^{z+\epsilon}(x-z)^2 f_X(x, t+\Delta t\,|\,z, t)\,dx = \frac{a^2\Delta t}{\pi}\int_{-\epsilon/a\Delta t}^{+\epsilon/a\Delta t}\frac{y^2}{1+y^2}\,dy$$
$$= \frac{2a^2\Delta t}{\pi}\left(\frac{\epsilon}{a\Delta t} - \arctan\frac{\epsilon}{a\Delta t}\right) \underset{\Delta t\to 0}{\longrightarrow} \frac{2a\epsilon}{\pi}$$

and hence that $B = 0$ in the limit $\epsilon \to 0$. The Cauchy process is therefore a pure jump process and the equation for its *pdf* is the master equation (7.94). The transition *pdf* (7.55) is of course the solution selected with the degenerate initial condition $f(x, t) = \delta(x - y)$. ■

At variance with the previous examples the Eq. (7.94) had not been previously mentioned among our heuristic considerations: it is in fact only derivable in the present framework of a discussion about Markovian jump-diffusions. Even in this instance, of course, it is in principle possible to take the reverse standpoint of recovering the process distributions by solving its forward equation (7.94) with the initial condition $f(x, s^+) = \delta(x - y)$ in order to find first the transition *pdf* and then all the other joint laws. Being however an integro differential equation effectively rules out any possible elementary procedure and hence we will leave aside this point.

Proposition 7.40 *The distributions and the transition probabilities (7.57) of a* ***Ornstein-Uhlenbeck process*** $X(t)$ *satisfy the Fokker-Planck equation*

$$\partial_t f(x, t) = \alpha\partial_x[xf(x, t)] + \frac{D}{2}\,\partial_x^2 f(x, t) \tag{7.95}$$

with $D = 2\alpha\beta^2$. *As a consequence the process is sample continuous.*

Proof Omitted: for the details see Appendix K. We will remark here only that the process sample continuity—announced but not proved in the Proposition 7.29—follows here from the fact that the jump coefficient ℓ of an Ornstein-Uhlenbeck process vanishes and hence the Lindeberg conditions are met. ■

Even in this case the solution procedures of the Eq. (7.95) are less elementary than those of the previous examples and we will neglect them: we will only remark in the end that it would be tedious, but not particularly difficult to check by direct calculation that the transition *pdf* (7.57) is a solution of our equation with the degenerate initial condition $f(x, t) = \delta(x - y)$.

References

1. Sato, K.I.: Lévy Processes and Infinitely Divisible Distributions. Cambridge UP, Cambridge (1999)
2. Applebaum, D.: Lévy Processes and Stochastic Calculus. Cambridge UP, Cambridge (2009)
3. Feller, W.: An Introduction to Probability Theory and Its Applications, vols. I–II. Wiley, New York (1968 and 1971)
4. Gardiner, C.W.: Handbook of Stochastic Methods. Springer, Berlin (1997)
5. Gihman, I.I., Skorohod, A.V.: The Theory of Stochastic Processes. Springer, Berlin (1975)

Chapter 8
An Outline of Stochastic Calculus

8.1 Wienerian White Noise

For simplicity again, in this chapter we will only consider processes with just one component. We already remarked in the Sect. 6.3 that a *white noise* is a singular process whose main properties can be traced back to the non differentiability of some processes. As a first example we have shown indeed that the Poisson impulse process (6.63) and its associated compensated version (6.65) are white noises entailed by the formal derivation respectively of a simple Poisson process $N(t)$ and of its compensated variant $\widetilde{N}(t)$. In the same vein we have shown then in the Example 6.22 that also the formal derivative of the Wiener process $W(t)$—not differentiable according to the Proposition 6.18—meets the conditions (6.69) to be a white noise, and in the Appendix H we also hinted that the role of the fluctuating force $B(t)$ in the Langevin equation (6.78) for the Brownian motion is actually played by such a white noise $\dot{W}(t)$. We can now give a mathematically more cogent justification for this identification in the framework of the Markovian diffusions.

The Langevin equation (6.78) is a particular case of the more general equation

$$\dot{X}(t) = a(X(t), t) + b(X(t), t)\, Z(t) \tag{8.1}$$

where $a(x, t)$ and $b(x, t)$ are given functions and $Z(t)$ is a process with $\boldsymbol{E}\,[Z(t)] = 0$ and uncorrelated with $X(t)$. From a formal integration of (8.1) we find

$$X(t) = X(t_0) + \int_{t_0}^{t} a(X(s), s)\, ds + \int_{t_0}^{t} b(X(s), s) Z(s)\, ds$$

so that, being $X(t)$ assembled as a combination of $Z(s)$ values with $t_0 < s < t$, to secure the non correlation of $X(t)$ and $Z(t)$ we should intuitively require also the non correlation of $Z(s)$ and $Z(t)$ for every pair $s \neq t$. Since moreover $Z(t)$ is presumed

N. Cufaro Petroni, *Probability and Stochastic Processes for Physicists*, UNITEXT for Physics, https://doi.org/10.1007/978-3-030-48408-8_8

to be wildly irregular, we are also led to suppose that its variance—namely here just $\boldsymbol{E}\left[Z^2(t)\right]$—is very large, so that finally, for a suitable constant $D > 0$, it will be quite natural to assume that

$$\boldsymbol{E}\left[Z(t)Z(s)\right] = D\,\delta(t-s)$$

namely that $Z(t)$ is a stationary white noise with vanishing expectation and intensity D. We will suppose in fact that $Z(s)$ and $Z(t)$ are even independent[1] for $s \neq t$. If finally the Eq. (8.1) is intended to describe physical phenomena similar to the Brownian motion, all the involved processes will be obviously supposed to be sample continuous. We will show now that all these hypotheses entail that the white noise $Z(t)$ can only be a Wienerian white noise $\dot{W}(t)$.

Proposition 8.1 *If $Z(t)$ is a stationary white noise of intensity $D > 0$ with $Z(s)$ and $Z(t)$ independent for $s \neq t$, and if the process*

$$W(t) = \int_{t_0}^{t} Z(s)\,ds \tag{8.2}$$

is sample continuous, then $W(t)$ is a Wiener process with diffusion coefficient D.

Proof To prove the result it will be enough to show that the distributions of the process $W(t)$ in (8.2) obey to the Fokker-Planck equation (7.92) of a Wiener process. Let us remark first that the increments of $W(t)$ on non overlapping intervals $t_1 < t_2 \leq t_3 < t_4$ are

$$W(t_2) - W(t_1) = \int_{t_1}^{t_2} Z(s)\,ds \qquad W(t_4) - W(t_3) = \int_{t_3}^{t_4} Z(s)\,ds$$

namely are sums of *rv*'s $Z(s)$ independent by hypothesis, and are therefore themselves independent. According to the Proposition 7.9, $W(t)$ is thus a Markov process, and since it is sample continuous by hypothesis it turns out to be a diffusion and its distributions will satisfy the Fokker–Planck equation (with $\ell = 0$) discussed in the Sect. 7.2.3.2. To find out now what a particular diffusion $W(t)$ is, it will be enough to calculate the equation coefficients (7.67) and (7.69) that in our one-dimensional setting are

$$A(x,t) = \lim_{\epsilon\to 0^+}\lim_{\Delta t\to 0}\int_{|y-x|<\epsilon}\frac{y-x}{\Delta t}\,f_W(y,t+\Delta t\,|\,x,t)\,dy$$

$$B(x,t) = \lim_{\epsilon\to 0^+}\lim_{\Delta t\to 0}\int_{|y-x|<\epsilon}\frac{(y-x)^2}{\Delta t}\,f_W(y,t+\Delta t\,|\,x,t)\,dy$$

To this end remark first that since $W(t)$ is sample continuous the Lindeberg conditions (7.47) require that

[1]This is not a very restrictive hypothesis: since our processes will turn out to be Gaussian, independence and non correlation happen to be quite equivalent.

$$\lim_{\Delta t \to 0} \int_{|y-x|>\epsilon} \frac{1}{\Delta t} f_W(y, t + \Delta t \,|\, x, t)\, dy = 0 \qquad \forall \epsilon > 0$$

namely that, with $\Delta t \to 0$, the support of $f_W(y, t + \Delta t \,|\, x, t)$ will quickly shrink into $[x - \epsilon\,,\, x + \epsilon]$. As a consequence the A and B defining formulas can be simplified by extending the integration interval to $(-\infty, +\infty)$ without changing the final result: we thus have

$$\begin{aligned}
A(x,t) &= \lim_{\Delta t \to 0} \int_{-\infty}^{+\infty} \frac{y-x}{\Delta t} f(y, t + \Delta t \,|\, x, t)\, dy \\
&= \lim_{\Delta t \to 0} \boldsymbol{E}\left[\frac{\Delta W(t)}{\Delta t} \,\middle|\, W(t) = x \right] \\
B(x,t) &= \lim_{\Delta t \to 0} \int_{-\infty}^{+\infty} \frac{(y-x)^2}{\Delta t} f(y, t + \Delta t \,|\, x, t)\, dy \\
&= \lim_{\Delta t \to 0} \boldsymbol{E}\left[\frac{[\Delta W(t)]^2}{\Delta t} \,\middle|\, W(t) = x \right]
\end{aligned}$$

On the other hand from the properties of $Z(t)$ we know that

$$\begin{aligned}
\boldsymbol{E}\left[\Delta W(t) \,|\, W(t) = x\right] &= \boldsymbol{E}\left[\int_t^{t+\Delta t} Z(s)\, ds \,\middle|\, W(t) = x \right] \\
&= \int_t^{t+\Delta t} \boldsymbol{E}\left[Z(s)\right] ds = 0 \\
\boldsymbol{E}\left[[\Delta W(t)]^2 \,|\, W(t) = x\right] &= \boldsymbol{E}\left[\int_t^{t+\Delta t} Z(s)\, ds \int_t^{t+\Delta t} Z(s')\, ds' \,\middle|\, W(t) = x \right] \\
&= \int_t^{t+\Delta t} ds \int_t^{t+\Delta t} ds' \boldsymbol{E}\left[Z(s)Z(s')\right] \\
&= D \int_t^{t+\Delta t} ds \int_t^{t+\Delta t} ds' \delta(s - s') \\
&= D \int_t^{t+\Delta t} ds = D\Delta t
\end{aligned}$$

and hence we finally get

$$A(x,t) = 0 \qquad B(x,t) = D$$

that is the coefficients of the Wiener Fokker-Planck equation (7.92). ■

From the previous proposition it follows thus that a $W(t)$ defined as in (8.2) is a Wiener process, and hence that its formal derivative $Z(t) = \dot{W}(t)$ is a Wienerian white noise. This noise plays the role of a random force in the Langevin equation (8.1) that however is still not well defined exactly because of the singular character of this white noise. To correctly address this problem we will then remark—as already done in the Sect. 6.3—that, while the derivative of a Wiener process $W(t)$ does not exist, we can hope to give a precise meaning to its differential $dW(t)$ first understood as the limit for $\Delta t \to 0$ of the increment $\Delta W(t) = W(t + \Delta t) - W(t)$, and then as a shorthand notation coming from the integral

$$\int_{t_0}^{t} dW(s) = W(t) - W(t_0)$$

If we can manage to do that, we will be able to reformulate the Eq. (8.1) rather in terms of differentials, than in terms of derivatives, in such a way that—by replacing the problematic notation $Z(t)dt = \dot{W}(t)dt$ with $dW(t)$—its new layout will be

$$dX(t) = a(X(t), t)dt + b(X(t), t)dW(t)$$

understood indeed as a shorthand notation for the finite, integral expression

$$X(t) = X(t_0) + \int_{t_0}^{t} a(X(s), s)\, ds + \int_{t_0}^{t} b(X(s), s)\, dW(s)$$

We must say at once however that, while the first integral

$$\int_{t_0}^{t} a(X(s), s)\, ds$$

can be considered as well defined based on the remarks already made in the Sect. 5.4, it is instead still an open problem the meaning to give to the second integral

$$\int_{t_0}^{t} b(X(s), s)\, dW(s) \tag{8.3}$$

where the measure $dW(s)$ should be defined using a Wiener process: a case not considered in our previous discussions. A coherent definition of this new kind of integrals will be the topic of the next section and will be crucial to introduce the stochastic calculus.

8.2 Stochastic Integration

There are several kinds of stochastic integrals that turn out to be well defined under a variety of conditions: in any case, when they exist, they always are *rv*'s. We have already discussed in the Proposition 5.8 a few elementary requirements needed to ensure the *ms*-convergence of the simplest case of stochastic integral (5.8) defined according to a generalized Riemann procedure with the measure dt. This definition can also be easily generalized in a Lebesgue-Stieltjes form as

$$\int_a^b Y(t)\,dx(t)$$

where $Y(t)$ is again a process, while now $x(t)$ is a function that in general must be supposed of *bounded variation*.[2] Under this hypothesis, and a set of rather wide requirements on the process $Y(t)$, it is possible to prove (see [2], p. 62) that the previous integral not only exists in *ms*, but also converges in the sense of Lebesgue-Stieltjes for almost every trajectory of $Y(t)$. Basically the previous integral turns out to be well defined, in a rather traditional sense, trajectory by trajectory. The problem is instead harder when the integrator $x(t)$ becomes a stochastic process $X(t)$, because in this case we can not suppose that its trajectories are of bounded variation, so that the usual procedures are no longer able to coherently ensure the convergence of the integral. The typical case with which we will have to deal in the rest of these lessons is that in which the integrator is precisely the Wiener process $W(t)$: its trajectories in fact—being nowhere differentiable—are not of bounded variation.

8.2.1 Wiener Integral

Take first the integrals

$$\int_a^b y(t)\,dX(t) \tag{8.4}$$

where $y(t)$ is a non random function, while $X(t)$ is a process: we have already hinted that a trajectory by trajectory definition of (8.4) following a Lebesgue-Stieltjes procedure can not be adopted because here, generally speaking, the process trajectories no longer are of bounded variation: we can not presume indeed—as the Wiener pro-

[2] A function $w(x)$ defined on $[a, b]$ is said of ***bounded variation*** if it exists $C > 0$ such that

$$\sum_{k=1}^{n} |w(x_k) - w(x_{k-1})| < C$$

for every finite partition $a = x_0 < x_1 < \cdots < x_n = b$ of $[a, b]$; in this case the quantity

$$\mathcal{V}[w] = \sup_{\mathcal{D}} \sum_{k=1}^{n} |w(x_k) - w(x_{k-1})|$$

where $\mathcal{D}$ is the set of the finite partitions of $[a, b]$, is called the *total variation* of w. It is known that the Lebesgue-Stieltjes integral

$$\int_a^b f(x)\,dw(x)$$

can be coherently defined when $w(x)$ is a function of bounded variation. Remark that every function of bounded variation is (almost everywhere) differentiable: for further details see [1], pp. 328–332.

cess shows—that the trajectories are differentiable, and hence we can not consider them as bounded variation functions (see the Footnote 2 in the present section). The most widespread form of this kind of integrals occurs when the random integrator is a Wiener process

$$\int_a^b y(t)\,dW(t) \tag{8.5}$$

and in this case we will call it ***Wiener integral***. Even in its more general form (8.4), however, this integral can be coherently defined when we are dealing with

- uncorrelated increments processes $X(t)$ (the Wiener process, for example, has independent, and hence uncorrelated, increments)
- Lebesgue square integrable functions $y(t)$

and in this case the following procedure (here only briefly summarized: for details see [2], pp. 426–433) is adopted:

1. we first define it in an elementary way for *step functions* $\varphi(t)$

$$\int_a^b \varphi(t)\,dX(t)$$

2. we then take a sequence of step functions $\varphi_n(t)$ *ms*-convergent to $y(t)$ (it is proven that such a sequence exists and that its particular choice is immaterial)
3. we finally define the integral (8.4) as the *ms*-limit of the *rv*'s sequence

$$\int_a^b \varphi_n(t)\,dX(t)$$

It is possible to show that this definition is perfectly consistent, and that, if $y(t)$ is also continuous, the integral (8.4) can also be calculated following a standard Riemann procedure:

1. take a partition $a = t_0 < t_1 < \cdots < t_n = b$ of the integration interval
2. choose the arbitrary points τ_j in every $[t_j, t_{j+1}]$ and take

$$\delta = \max_j \{t_{j+1} - t_j\}$$

3. calculate finally the integral as the *ms* limit

$$\lim_{n,\delta\to 0}\text{-}ms \sum_{j=0}^{n-1} y(\tau_j)\big[X(t_{j+1}) - X(t_j)\big]$$

When both these integrals do in fact exist, the second, more familiar, Riemann procedure leads to a result which coincides with that defined within the first procedure, and this happens regardless of both the particular partition sequence selected, and the choice of the points τ_j inside every sub-interval $[t_j, t_{j+1}]$. In particular in this way a precise meaning is ascribed to the Wiener integrals (8.5).

8.2.2 Itō Integral

The previous integral (8.4) is a particular case of the more general type

$$\int_a^b Y(t)\, dX(t) \tag{8.6}$$

where both $X(t)$ and $Y(t)$ are now *sp*'s: the integral (8.3) at the end of the previous section is an example of this kind. A consistent definition of (8.6) is not an elementary one (for further details see [3], Sect. 3.2 and [4], Chap. 3) and requires a new procedure pioneered by K. Itō (1944) in the case of Wienerian integrators

$$\int_a^b Y(t)\, dW(t) \tag{8.7}$$

and later extended to a wider class of integrators $X(t)$, slightly narrower anyway than that for the integrals of the Sect. 8.2.1. This new definition requires moreover for the integrand process $Y(t)$ a few general conditions, the most important of which for the Itō integrals like (8.7) is its *non-anticipativity* w.r.t. a Wiener process $W(t)$.

Definition 8.2 Take a Wiener process $W(t)$, and the growing family of σ-algebras $\mathcal{F}_t = \sigma\{W(s),\ s \le t\}$ generated by $W(t)$ (its **natural filtration**): we will say that the process $Y(t)$ is **non-anticipative** w.r.t. $W(t)$ if

- $Y(t)$ is $\mathcal{F}_t$-measurable for every $t > 0$
- $Y(t)$ is independent from $W(s) - W(t)$ for every $s > t > 0$

that is if $Y(t)$ depends on the past (and the present) of $W(t)$, but not on its future.

This concept, which expresses a rather natural requirement of causality, is essential for a rigorous definition of the Itō integral, and subsequently of the Itō stochastic differential equations, in the sense that a number of important results can be deduced only with this assumption. For the time being we will just remark that it is easy to check that, if $Y(t)$ is non-anticipative, then $W(t)$ itself and the following processes

$$\int_{t_0}^t h[W(s)]\, ds \qquad \int_{t_0}^t h[W(s)]\, dW(s) \qquad \int_{t_0}^t Y(s)\, ds \qquad \int_{t_0}^t Y(s)\, dW(s)$$

are all non-anticipative.

In the following we will always suppose, among others, that the integrand $Y(t)$ is non-anticipative w.r.t. $W(t)$, and within these hypotheses we will define the **Itō**

integral according to a procedure similar to that adopted for the Wiener integral in the Sect. 8.2.1:

1. we first define the elementary Itō integral for random *step functions* $\Phi(\omega; t)$ (particular non-anticipative *sp*'s)

$$\int_a^b \Phi(\omega; t)\, dW(\omega; t)$$

2. we take then a sequence $\Phi_n(t)$ of such step functions converging in *ms* to the given non anticipative *sp* $Y(t)$ (we will not prove that such a sequence exists and that its particular choice is immaterial)
3. the Itō (8.7) integral is finally defined as the *ms*-limit of the following sequence of *rv*'s

$$\int_a^b \Phi_n(t)\, dW(t)$$

Even in this case, of course, it is possible to prove that the limit is independent from the particular sequence $\Phi_n(t)$ chosen, so that the definition is perfectly consistent, but this new procedure, despite an apparent analogy with that defining the Wiener integral, introduces two relevant changes:

- like the Wiener integral (8.5), by adopting an ***appropriate Riemann procedure*** the Itō integral can also be calculated as

$$\lim_{n,\delta\to 0}\text{-}ms \sum_{j=0}^{n-1} Y(t_j)\big[W(t_{j+1}) - W(t_j)\big] \tag{8.8}$$

  but now the values $Y(t_j)$ of the integrand *must* always be taken in the left endpoint of the interval $[t_j, t_{j+1}]$, and not in an arbitrary τ_j within it: it is indeed possible to show (1) that the value of the Riemann limit (8.8) *depends* on this choice, and (2) that only with the choice $Y(t_j)$ it is possible to recover the correct value of the Itō integral previously defined with the Itō procedure; we will show later an explicit example of this behavior
- the definition of the Itō integral does not come into being without an additional cost: in particular it entails a ***new stochastic calculus*** with rules that deviate from those of the ordinary calculus; a whiff of this important innovation—an innovation to which we must learn to adapt to take advantage of it—can be found in the Appendix H displaying the possible mistakes induced by a careless use of the usual calculus: we will devote a sizable part of the subsequent sections to a detailed review of these new rules

A few remarks about possible alternative definitions of stochastic integrals, like the *Stratonovich integral* that famously would preserve the usual rules of calculus, can be finally found in the Appendix L along with the motivations for not adopting them here.

8.3 Itō Stochastic Calculus

All along the following sections we will adopt a heuristic standpoint, and we will calculate the Itō integrals (8.7) as the *ms*-limit of the Riemann sums (8.8). Moreover the Wiener process $W(t)$ with diffusion coefficient D will be supposed to satisfy arbitrary initial conditions $W(t_0) = w_0$, $\boldsymbol{P}$-a.s. when t_0 is the left endpoint of the integration interval. It will be expedient then to slightly adjust the results of the Propositions 6.16, 6.17 and 6.18: if $W_0(t) \sim \mathfrak{N}(0, Dt)$ denotes the process with conditions $W_0(0) = 0$, $\boldsymbol{P}$-a.s., we will have $W(t) = W_0(t - t_0) + w_0$ defined for $t \geq t_0$ so that $W(t) \sim \mathfrak{N}(w_0, D(t - t_0))$, and

$$\boldsymbol{E}\left[W(t)\right] = w_0 \qquad \boldsymbol{V}\left[W(t)\right] = D(t - t_0) \tag{8.9}$$

$$\boldsymbol{E}\left[W(s)W(t)\right] = D\min\{s - t_0, t - t_0\} + w_0^2 \tag{8.10}$$

For short, moreover, for every Riemann partition $t_1 < \cdots < t_n$ we will adopt the synthetic notations $(j = 1, \ldots, n)$

$$W_j = W(t_j) \qquad \Delta W_j = W_j - W_{j-1} \qquad \Delta t_j = t_j - t_{j-1}$$

8.3.1 Elementary Integration Rules

Lemma 8.3 *If $W(t)$ is a Wiener process with $W(t_0) = w_0$, $\boldsymbol{P}$-a.s., then*

$$W_j \sim \mathfrak{N}(w_0, D(t_j - t_0)) \qquad \Delta W_j \sim \mathfrak{N}(0, D\Delta t_j) \tag{8.11}$$

$$\boldsymbol{E}\left[(\Delta W_j)^4\right] = 3D^2(\Delta t_j)^2 \tag{8.12}$$

Proof The first relation in (8.11) follows from fact that $W(t) \sim \mathfrak{N}(w_0, D(t - t_0))$. As for the second relation in (8.11), being the increments $\Delta W(t)$ Gaussian according to the Proposition 6.16, and keeping into account (8.9) and (8.10), it will be enough to remark that

$$\begin{aligned}
\boldsymbol{E}\left[\Delta W_j\right] &= \boldsymbol{E}\left[W_j - W_{j-1}\right] = w_0 - w_0 = 0 \\
\boldsymbol{V}\left[\Delta W_j\right] &= \boldsymbol{E}\left[(\Delta W_j)^2\right] = \boldsymbol{E}\left[W_j^2 + W_{j-1}^2 - 2W_jW_{j-1}\right] \\
&= w_0^2 + D(t_j - t_0) + w_0^2 + D(t_{j-1} - t_0) - 2[w_0^2 + D(t_{j-1} - t_0)] \\
&= D(t_j - t_{j-1}) = D\Delta t_j
\end{aligned}$$

As for (8.12) first remark that if $X \sim \mathfrak{N}(0, \sigma^2)$, an integration by parts leads to

$$E\left[X^4\right] = \int_{-\infty}^{+\infty} x^4 \frac{e^{-x^2/2\sigma^2}}{\sigma\sqrt{2\pi}} dx = -\sigma^2 \int_{-\infty}^{+\infty} x^3 \frac{d}{dx}\left(\frac{e^{-x^2/2\sigma^2}}{\sigma\sqrt{2\pi}}\right) dx$$
$$= 3\sigma^2 \int_{-\infty}^{+\infty} x^2 \frac{e^{-x^2/2\sigma^2}}{\sigma\sqrt{2\pi}} dx = 3\sigma^4 = 3E\left[X^2\right]^2$$

and then that the result follows from (8.11), namely from $\Delta W_j \sim \mathfrak{N}(0, D\Delta t_j)$. ∎

Proposition 8.4 *If $W(t)$ is a Wiener process with $W(t_0) = w_0$, $\boldsymbol{P}$-a.s., then*

$$\int_{t_0}^{t} W(s)\, dW(s) = \frac{1}{2}\left[W^2(t) - W^2(t_0) - D(t - t_0)\right] \tag{8.13}$$
$$E\left[\int_{t_0}^{t} W(s)\, dW(s)\right] = 0 \tag{8.14}$$

Proof Remark first of all that the term $\frac{1}{2}D(t - t_0)$ in (8.13) is totally alien to the usual formula of the integral calculus that is instead confined to the first two terms: this is a first example of the quantitative changes introduced by the Itō stochastic calculus w.r.t. the ordinary calculus.

To prove (8.13) let us begin by remarking that in the present instance the Riemann sums (8.8) take the particular form

$$S_n = \sum_{j=1}^{n} W_{j-1}(W_j - W_{j-1}) = \sum_{j=1}^{n} W_{j-1}\Delta W_j$$
$$= \frac{1}{2}\sum_{j=1}^{n}\left[(W_{j-1} + \Delta W_j)^2 - W_{j-1}^2 - (\Delta W_j)^2\right]$$
$$= \frac{1}{2}\sum_{j=1}^{n}\left[W_j^2 - W_{j-1}^2 - (\Delta W_j)^2\right] = \frac{1}{2}\left[W^2(t) - W^2(t_0)\right] - \frac{1}{2}\sum_{j=1}^{n}(\Delta W_j)^2$$

so that the result will be secured if we will be able to prove that

$$\lim_n\text{-}ms \sum_{j=1}^{n}(\Delta W_j)^2 = D(t - t_0) \tag{8.15}$$

namely, according to the Theorem 4.6, that

$$\lim_n E\left[\sum_{j=1}^{n}(\Delta W_j)^2\right] = D(t - t_0) \qquad \lim_n V\left[\sum_{j=1}^{n}(\Delta W_j)^2\right] = 0 \tag{8.16}$$

The first result in (8.16) follows from the Lemma 8.3 because for every n it is

$$\boldsymbol{E}\left[\sum_{j=1}^{n}(\Delta W_j)^2\right] = \sum_{j=1}^{n}\boldsymbol{E}\left[(\Delta W_j)^2\right] = \sum_{j=1}^{n}D(t_j - t_{j-1}) = D(t - t_0)$$

and hence also its limit for $n \to \infty$ has the same value. As for the second limit in (8.16) remark first that from the previous result we have

$$\begin{aligned}\boldsymbol{V}\left[\sum_{j=1}^{n}(\Delta W_j)^2\right] &= \boldsymbol{E}\left[\left(\sum_{j=1}^{n}(\Delta W_j)^2 - D(t - t_0)\right)^2\right] \\ &= \boldsymbol{E}\left[\sum_{j=1}^{n}(\Delta W_j)^4 + 2\sum_{j<k}(\Delta W_j)^2(\Delta W_k)^2\right. \\ &\qquad \left. -2D(t - t_0)\sum_{j=1}^{n}(\Delta W_j)^2 + D^2(t - t_0)^2\right]\end{aligned}$$

and then that the expectations can be calculated again by keeping into account the Lemma 8.3—in particular the formula (8.12)—and the increments independence in a Wiener process that for $j < k$ entails

$$\boldsymbol{E}\left[(\Delta W_j)^2(\Delta W_k)^2\right] = \boldsymbol{E}\left[(\Delta W_j)^2\right]\boldsymbol{E}\left[(\Delta W_k)^2\right] = D^2(t_j - t_{j-1})(t_k - t_{k-1})$$

Rearranging now all the terms of the previous expression, and recalling that in the Riemann procedure

$$\sum_{j=1}^{n}(t_j - t_{j-1}) = t - t_0 \qquad \delta = \max_j\{t_j - t_{j-1}\} \xrightarrow{n} 0$$

overall we will find

$$\begin{aligned}\boldsymbol{V}\left[\sum_{j=1}^{n}(\Delta W_j)^2\right] &= 3D^2\sum_{j=1}^{n}(t_j - t_{j-1})^2 + 2D^2\sum_{j<k}(t_j - t_{j-1})(t_k - t_{k-1}) \\ &\qquad -2D^2(t - t_0)\sum_{j=1}^{n}(t_j - t_{j-1}) + D^2(t - t_0)^2 \\ &= 2D^2\sum_{j=1}^{n}(t_j - t_{j-1})^2 + D^2\sum_{j,k=1}^{n}(t_j - t_{j-1})(t_k - t_{k-1}) - D^2(t - t_0)^2 \\ &= 2D^2\sum_{j=1}^{n}(t_j - t_{j-1})^2 \le 2D^2\max_j\{t_j - t_{j-1}\}\sum_{j=1}^{n}(t_j - t_{j-1}) \\ &= 2\delta D^2(t - t_0) \xrightarrow{n} 0\end{aligned}$$

The convergence (8.15) then holds, and the result (8.13) is proved. To check finally (8.14) we just remark that from (8.13) it is

$$\boldsymbol{E}\left[\int_{t_0}^{t} W(s)\,dW(s)\right] = \frac{1}{2}\,\boldsymbol{E}\left[W^2(t) - w_0^2 - D(t-t_0)\right] = 0$$

where we took advantage of the fact that $W(t) \sim \mathfrak{N}(w_0, D(t-t_0))$. ■

Example 8.5 In the Sect. 8.2.2 we stated without proof that the right value of an Itō integral like (8.13) can also be recovered as a *ms*-limit of the Riemann sums (8.8) where however the integrand must always be calculated in the left endpoints of the partition intervals. Without going into details, we can now show that the result of the Riemann procedure to calculate (8.13) would have been different if we had not taken the integrand in the left endpoints, as for instance in

$$S_n = \sum_{j=1}^{n} W(\tau_j)\left[W(t_j) - W(t_{j-1})\right]$$

where now τ_j are arbitrary points in $[t_{j-1}, t_j]$. To prove without unnecessary complications that the *ms*-limit of the sequence S_n does in fact depend on the choice of the τ_j it will be enough indeed to point out this dependence only for the limit of their expectations $\boldsymbol{E}[S_n]$, because if the limit of the expectations is contingent on the choice of τ_j, then also the *ms*-limit of S_n must depend on them. We have in fact from (8.10)

$$\begin{aligned}
\boldsymbol{E}[S_n] &= \boldsymbol{E}\left[\sum_{j=1}^{n} W(\tau_j)\left[W(t_j) - W(t_{j-1})\right]\right] \\
&= \sum_{j=1}^{n}\left(\boldsymbol{E}\left[W(\tau_j)W(t_j)\right] - \boldsymbol{E}\left[W(\tau_j)W(t_{j-1})\right]\right) \\
&= \sum_{j=1}^{n}\left[w_0^2 + D(\tau_j - t_0) - w_0^2 - D(t_{j-1} - t_0)\right] \\
&= \sum_{j=1}^{n}\left[D(\tau_j - t_0) - D(t_{j-1} - t_0)\right] = D\sum_{j=1}^{n}(\tau_j - t_{j-1})
\end{aligned}$$

Take now a parameter $\alpha \in [0, 1]$ identifying the position of τ_j within the jth interval according to

$$\tau_j = \alpha t_j + (1-\alpha)t_{j-1}$$

then for every n we will have

$$E[S_n] = \alpha D \sum_{j=1}^{n} (t_j - t_{j-1}) = \alpha D(t - t_0)$$

so that—taking for granted that we are entitled to exchange the Riemann *ms*-limit with the expectation—we find

$$E\left[\lim_n \text{-}ms\, S_n\right] = \lim_n E[S_n] = \alpha D(t - t_0)$$

a result that apparently depends on α, namely on the location of τ_j within the interval $[t_{j-1}, t_j]$: the right result for the Itō integral being in any case (8.14)—we will abstain however from giving here an independent proof of this statement—this value turns out to be recovered only with $\alpha = 0$, namely when τ_j is the left endpoint of the interval $[t_{j-1}, t_j]$.

8.3.2 Expectation and Covariance

Proposition 8.6 *If $G(t)$ and $H(t)$ non-anticipative processes w.r.t. a wiener process $W(t)$, then*

$$E\left[\int_{t_0}^{t} G(s)\, dW(s)\right] = 0 \tag{8.17}$$

$$E\left[\int_{t_0}^{t} G(s)\, dW(s) \int_{t_0}^{t} H(s')\, dW(s')\right] = D \int_{t_0}^{t} E[G(s)H(s)]\, ds \tag{8.18}$$

Proof The formula (8.17) generalizes (8.14): from both the non-anticipativity of $G(t)$ and the Lemma 8.3 we indeed have for the Riemann sums

$$E\left[\sum_{j=1}^{n} G_{j-1} \Delta W_j\right] = \sum_{j=1}^{n} E\left[G_{j-1}\right] E\left[\Delta W_j\right] = 0$$

and taking as usual for granted that we are entitled to exchange the Riemann *ms*-limit with the expectations, the result easily follows. As for the covariance formula (8.18), from the non-anticipativity, the Lemma 8.3 and the increment independence we have

$$E\left[\sum_{j=1}^{n} G_{j-1} \Delta W_j \sum_{k=1}^{n} H_{k-1} \Delta W_k\right]$$

$$= E\left[\sum_{j=1}^{n} G_{j-1} H_{j-1} (\Delta W_j)^2\right] + E\left[\sum_{k>j} (G_{j-1} H_{k-1} + G_{k-1} H_{j-1}) \Delta W_j \Delta W_k\right]$$

$$
\begin{aligned}
&= \sum_{j=1}^{n} \boldsymbol{E}\left[G_{j-1}H_{j-1}\right] \boldsymbol{E}\left[(\Delta W_j)^2\right] \\
&\qquad\qquad + \sum_{k>j} \boldsymbol{E}\left[(G_{j-1}H_{k-1} + G_{k-1}H_{j-1})\Delta W_j\right] \boldsymbol{E}\left[\Delta W_k\right] \\
&= \sum_{j=1}^{n} \boldsymbol{E}\left[G_{j-1}H_{j-1}\right] D\Delta t_j
\end{aligned}
$$

and the result follows again by exchanging the *ms*-limit with the expectation. ■

We can now look at the remarks on the Wiener white noise of the Proposition 8.1 from a new, reversed. standpoint

Corollary 8.7 *If $W(t)$ with $W(t_0) = w_0$ is a Wiener process with diffusion coefficient D, a process $Z(t)$ such that $\boldsymbol{E}[Z(t)] = 0$, and $dW(t) = Z(t)\,dt$, can only be a stationary (Wienerian) white noise of intensity D.*

Proof This is an immediate consequence of the (8.18): take two arbitrary, non-anticipative processes $G(t)$ and $H(t)$ independent from $Z(t)$, then—freely exchanging expectations and integrals into (8.18)—from our hypotheses it follows that

$$
\begin{aligned}
D\int_{t_0}^{t} \boldsymbol{E}[G(s)H(s)]\,ds &= \boldsymbol{E}\left[\int_{t_0}^{t} G(s)\,dW(s) \int_{t_0}^{t} H(s')\,dW(s')\right] \\
&= \boldsymbol{E}\left[\int_{t_0}^{t} ds \int_{t_0}^{t} ds'\, G(s)H(s')Z(s)Z(s')\right] \\
&= \int_{t_0}^{t} ds \int_{t_0}^{t} ds' \boldsymbol{E}\left[G(s)H(s')Z(s)Z(s')\right] \\
&= \int_{t_0}^{t} ds \int_{t_0}^{t} ds' \boldsymbol{E}\left[G(s)H(s')\right] \boldsymbol{E}\left[Z(s)Z(s')\right]
\end{aligned}
$$

that can be true only if

$$
\boldsymbol{E}\left[Z(s)Z(s')\right] = D\delta(s - s')
$$

namely if $Z(t)$ is a stationary white noise of intensity D. Since on the other hand we also ask $dW(t) = Z(t)dt$, the said white noise can only be Wienerian. ■

8.3.3 Stochastic Infinitesimals

Proposition 8.8 *Take a non-anticipative process $G(t)$, then with $k = 0, 1, \ldots$ it is*

$$\int_{t_0}^{t} G(s)\,[dW(s)]^{2+k} = \operatorname*{lim\text{-}ms}_{n} \sum_{j=1}^{n} G_{j-1}(\Delta W_j)^{2+k} = \delta_{k0}\,D \int_{t_0}^{t} G(s)\,ds$$

$$\int_{t_0}^{t} G(s)\,ds\,[dW(s)]^{1+k} = \operatorname*{lim\text{-}ms}_{n} \sum_{j=1}^{n} G_{j-1}\left(\Delta W_j\right)^{1+k}\Delta t_j = 0$$

where $\delta_{k\ell}$ is the Kronecker symbol; from now on we will also adopt the shorthand notation

$$[dW(t)]^{2+k} = \delta_{k0}\,D\,dt \qquad [dW(t)]^{1+k}\,dt = 0 \qquad k = 0, 1, \dots \tag{8.19}$$

Proof These results give a precise meaning to our statements of the Sect. 6.3 where we had surmised that $dW(t)$ behaves indeed as an infinitesimal of the order $\sqrt{dt}$. More precisely the present proposition entitle us to neglect in the calculations all the terms like $dW(t)\,dt$, $[dW(t)]^3$, ... because they are infinitesimals of order higher than dt, but it also urges us to keep the terms like $[dW(t)]^2 = D\,dt$ that—against their semblance—are in fact of the order dt.

To avoid redundancy we will confine ourselves to prove only the non zero formula $[dW(t)]^2 = D\,dt$, namely that

$$\int_{t_0}^{t} G(s)\left[dW(s)\right]^2 = D\int_{t_0}^{t} G(s)\,ds$$

neglecting instead to check—in a similar way—all the other vanishing results. The Riemann procedure requires then to verify that

$$\operatorname*{lim\text{-}ms}_{n} \sum_{j=1}^{n} G_{j-1}(\Delta W_j)^2 = D \operatorname*{lim\text{-}ms}_{n} \sum_{j=1}^{n} G_{j-1}\Delta t_j$$

namely, in an equivalent setting, that

$$\operatorname*{lim\text{-}ms}_{n} \sum_{j=1}^{n} \left[G_{j-1}(\Delta W_j)^2 - G_{j-1}D\Delta t_j\right] = \lim_{n} \mathcal{E}_n = 0 \tag{8.20}$$

where for short we have defined

$$\mathcal{E}_n = \boldsymbol{E}\left[\left|\sum_{j=1}^{n} G_{j-1}\big[(\Delta W_j)^2 - D\Delta t_j\big]\right|^2\right]$$

$$= \boldsymbol{E}\left[\sum_{j=1}^{n} G_{j-1}^2\big[(\Delta W_j)^2 - D\Delta t_j\big]^2 \right.$$
$$\left. +2\sum_{j<k} G_{j-1}G_{k-1}\big[(\Delta W_j)^2 - D\Delta t_j\big]\big[(\Delta W_k)^2 - D\Delta t_k\big]\right]$$

Because of the non-anticipativity of $G(t)$, the terms G_{j-1}^2 are independent from $(\Delta W_j)^2 - D\Delta t_j$, while the $G_{j-1}G_{k-1}\big[(\Delta W_j)^2 - D\Delta t_j\big]$ turn out to be independent from $(\Delta W_k)^2 - D\Delta t_k$; as a consequence—taking also advantage of the Lemma 8.3 – the second term of $\mathcal{E}_n$ vanishes

$$\begin{aligned}\boldsymbol{E}\left[G_{j-1}G_{k-1}\big[(\Delta W_j)^2 - D\Delta t_j\big]\big[(\Delta W_k)^2 - D\Delta t_k\big]\right] \\ = \boldsymbol{E}\left[G_{j-1}G_{k-1}\big[(\Delta W_j)^2 - D\Delta t_j\big]\right]\boldsymbol{E}\left[(\Delta W_k)^2 - D\Delta t_k\right] = 0\end{aligned}$$

while for the first we have

$$\begin{aligned}\boldsymbol{E}\left[\big[(\Delta W_j)^2 - D\Delta t_j\big]^2\right] &= \boldsymbol{E}\left[(\Delta W_j)^4\right] + (D\Delta t_j)^2 - 2D\Delta t_j\boldsymbol{E}\left[(\Delta W_j)^2\right] \\ &= 3(D\Delta t_j)^2 + (D\Delta t_j)^2 - 2(D\Delta t_j)^2 = 2(D\Delta t_j)^2\end{aligned}$$

and hence overall we find

$$\mathcal{E}_n = 2D^2\sum_{j=1}^{n}(\Delta t_j)^2\,\boldsymbol{E}\left[G_{j-1}^2\right] \le 2D^2\max_j\{\Delta t_j\}\sum_{j=1}^{n}\Delta t_j\,\boldsymbol{E}\left[G_{j-1}^2\right]$$

The result (8.20) is then secured if we plausibly require that

$$\lim_{n,\delta\to 0}\sum_{j=1}^{n}\Delta t_j\,\boldsymbol{E}\left[G_{j-1}^2\right] = \int_{t_0}^{t}\boldsymbol{E}\left[G^2(s)\right]ds < +\infty$$

because in the Riemann limit it is $\max_j\{\Delta t_j\} = \delta \to 0$. ■

Proposition 8.9 *Take a Wiener process* $W(t)$ *with* $W(t_0) = w_0$, *then for* $n = 1, 2, \ldots$ *we have*

$$\int_{t_0}^{t} W^n(s)\,dW(s) = \frac{W^{n+1}(t) - W^{n+1}(t_0)}{n+1} - \frac{nD}{2}\int_{t_0}^{t} W^{n-1}(s)\,ds \qquad (8.21)$$

Proof The result (8.21) generalizes (8.13) and can be easily deduced by taking advantage of the shorthand notations about the order of infinitesimals in the Proposition 8.8: we have indeed

$$
\begin{aligned}
dW^{n+1}(t) &= W^{n+1}(t+dt) - W^{n+1}(t) = \big[W(t) + dW(t)\big]^{n+1} - W^{n+1}(t) \\
&= \sum_{k=0}^{n+1} \binom{n+1}{k} W^{n+1-k}(t)[dW(t)]^k - W^{n+1}(t) \\
&= \sum_{k=1}^{n+1} \binom{n+1}{k} W^{n+1-k}(t)[dW(t)]^k \\
&= \binom{n+1}{1} W^n(t)dW(t) + \binom{n+1}{2} W^{n-1}(t)\big[dW(t)\big]^2 \\
&= (n+1)W^n(t)dW(t) + \frac{(n+1)n}{2} W^{n-1}(t)Ddt
\end{aligned}
$$

and therefore

$$
\begin{aligned}
W^{n+1}(t) - W^{n+1}(t_0) &= \int_{t_0}^{t} dW^{n+1}(s) \\
&= (n+1)\int_{t_0}^{t} W^n(s)\,dW(s) + \frac{(n+1)n}{2} D \int_{t_0}^{t} W^{n-1}(s)\,ds
\end{aligned}
$$

so that the formula (8.21) results immediately. ■

It is apparent then from the previous proposition that here too the usual results of the ordinary calculus are complemented with an additional term explicitly depending on the existence of a non vanishing diffusion coefficient D.

8.3.4 Differentiation Rules

Proposition 8.10 *If $g(x,t)$ is at least twice differentiable in x and once in t, and if $W(t)$ is a Wiener process, then within the notations*

$$
g_x = \partial_x g \qquad g_{xx} = \partial_x^2 g \qquad g_t = \partial_t g
$$

the following differentiation rule holds

$$
dg\big(W(t),t\big) = \left[g_t\big(W(t),t\big) + \frac{D}{2}\, g_{xx}\big(W(t),t\big)\right] dt + g_x\big(W(t),t\big)\, dW(t) \tag{8.22}
$$

Proof Taking into account the Proposition 8.8 we have

$$
\begin{aligned}
dg\big(W(t),t\big) &= g\big(W(t+dt),t+dt\big) - g\big(W(t),t\big) \\
&= \Big[g\big(W(t+dt),t+dt\big) - g\big(W(t),t+dt\big)\Big] \\
&\qquad\qquad + \Big[g\big(W(t),t+dt\big) - g\big(W(t),t\big)\Big] \\
&= \Big[g\big(W(t),t+dt\big) + g_x\big(W(t),t+dt\big)dW(t) \\
&\qquad + \frac{1}{2}\,g_{xx}\big(W(t),t+dt\big)\big[dW(t)\big]^2 + \cdots - g\big(W(t),t+dt\big)\Big] \\
&\quad + \Big[g\big(W(t),t\big) + g_t\big(W(t),t\big)dt \\
&\qquad + \frac{1}{2}\,g_{tt}\big(W(t),t\big)(dt)^2 + \cdots - g\big(W(t),t\big)\Big] \\
&= \big[g_x\big(W(t),t\big) + g_{xt}\big(W(t),t\big)dt + \cdots\big]dW(t) \\
&\quad + \frac{1}{2}\Big[g_{xx}\big(W(t),t\big) + g_{xxt}\big(W(t),t\big)dt + \cdots\Big]\big[dW(t)\big]^2 + \cdots \\
&\quad + g_t\big(W(t),t\big)dt + \frac{1}{2}g_{tt}\big(W(t),t\big)(dt)^2 + \cdots \\
&= g_x\big(W(t),t\big)dW(t) + \frac{D}{2}\,g_{xx}\big(W(t),t\big)dt + g_t\big(W(t),t\big)dt
\end{aligned}
$$

namely the stated result (8.22). ■

Example 8.11 In a nutshell the stochastic differentiation requires that we consider $\big[dW(t)\big]^2$ as an infinitesimal of the same order of dt, and not—as one could presume from its external semblance—of higher order. In particular this entails the existence of new terms that would not be otherwise acceptable. For instance, in a geometric Wiener process (6.58) $X(t) = e^{W(t)}$, within our notation it is $g(x,t) = e^x$ and hence

$$
dX(t) = d\left(e^{W(t)}\right) = e^{W(t)}\,dW(t) + \frac{D}{2}\,e^{W(t)}\,dt
$$

In the same way, for $X(t) = W^2(t)$, namely if $g(x,t) = x^2$, we get

$$
dX(t) = d\left(W^2(t)\right) = 2W(t)\,dW(t) + D\,dt
$$

From these examples we understand first that the unconventional additional terms in dt apparently follow from the second order terms in $dW(t)$, and second that they are branded by the presence of the diffusion coefficient D: when this possibly vanishes the process degenerates into deterministic trajectories, and we recover the usual differentiation rules. Unsurprisingly these remarks—suitably tailored—can be extended to all the other formulas met hitherto in the stochastic calculus as for example (8.13), (8.18), (8.19), (8.21) and (8.22).

The new differentiation rules also prompt a generalization of the *integration by parts* formulas: in the usual calculus we know for instance that

$$
d\big[x(t)h(x(t),t)\big] = x(t)\,dh(x(t),t) + h(x(t),t)\,dx(t)
$$

from which the following formula stems

$$\int_a^b h(x(t),t)\,dx(t) = \big[x(t)h(x(t),t)\big]_a^b - \int_a^b x(t)\,dh(x(t),t)$$

This expression is reduced to the most familiar one when $h(x,t) = h(t)$ does not depend on x: in this case we have indeed

$$d\big[x(t)h(t)\big] = x(t)\,dh(t) + h(t)\,dx(t) = \big[x(t)\dot{h}(t) + \dot{x}(t)h(t)\big]\,dt$$

namely the well known formula

$$\int_a^b h(t)\dot{x}(t)\,dt = \big[h(t)x(t)\big]_a^b - \int_a^b \dot{h}(t)x(t)\,dt$$

The stochastic calculus requires a modification of these results, but the differences w.r.t. the usual formulas are perceptible only when $h(x,t)$ also depend on x.

Proposition 8.12 Integration by parts: *If $h(x,t)$ is at least twice differentiable in x and once in t, and if $W(t)$ is a Wiener process with $W(t_0) = w_0$, the integration by parts rule is*

$$\begin{aligned}\int_{t_0}^t h\big(W(s),s\big)\,dW(s) = \Big[W(s)h\big(W(s),s\big)\Big]_{t_0}^t - \int_{t_0}^t W(s)\,dh\big(W(s),s\big)\\ - D\int_{t_0}^t h_x\big(W(s),s\big)\,ds \qquad (8.23)\end{aligned}$$

Proof From the differentiation rule (8.22) with $g(x,t) = xh(x,t)$ we get

$$g_t = xh_t \qquad g_x = h + xh_x \qquad g_{xx} = 2h_x + xh_{xx}$$

and therefore

$$\begin{aligned}d\big[W(t)h\big(W(t),t\big)\big] &= dg\big(W(t),t\big)\\ &= \Big[W(t)h_t\big(W(t),t\big) + \frac{D}{2}\Big(2h_x\big(W(t),t\big) + W(t)h_{xx}\big(W(t),t\big)\Big)\Big]dt\\ &\quad + \big[h\big(W(t),t\big) + W(t)h_x\big(W(t),t\big)\big]dW(t)\\ &= W(t)\Big[\Big(h_t\big(W(t),t\big) + \frac{D}{2}h_{xx}\big(W(t),t\big)\Big)dt + h_x\big(W(t),t\big)dW(t)\Big]\\ &\quad + h\big(W(t),t\big)dW(t) + Dh_x\big(W(t),t\big)dt\\ &= W(t)dh\big(W(t),t\big) + h\big(W(t),t\big)dW(t) + Dh_x\big(W(t),t\big)dt\end{aligned}$$

and the formula follows by integration. ∎

8.4 Stochastic Differential Equations (*SDE*)

8.4.1 Stochastic Differentials and Itō Formula

Definition 8.13 We say that a process $X(t)$ admits in $[0, T]$ the **stochastic differential**

$$dX(t) = A(t)\,dt + B(t)\,dW(t) \tag{8.24}$$

when for every t_0, t with $0 \le t_0 < t \le T$ it can be represented as

$$X(t) = X(t_0) + \int_{t_0}^{t} A(s)\,ds + \int_{t_0}^{t} B(s)\,dW(s)$$

where $W(t)$ is a Wiener process with $W(t_0) = w_0$, and the processes $A(t),\ B(t)$ are such that

$$\boldsymbol{P}\left\{\int_0^T |A(t)|\,dt < +\infty\right\} = 1 \qquad \boldsymbol{P}\left\{\int_0^T |B(t)|^2\,dt < +\infty\right\} = 1$$

Proposition 8.14 Itō formula: *If $X(t)$ admits the stochastic differential (8.24), and if $g(x,t)$ is a function at least twice differentiable in x and once in t, then also $g\big(X(t),t\big)$ admits the following stochastic differential*

$$\begin{aligned} d\,g\big(X(t),t\big) &= \left[g_t\big(X(t),t\big) + \frac{D}{2}B^2(t)g_{xx}\big(X(t),t\big)\right]dt + g_x\big(X(t),t\big)\,dX(t) \qquad (8.25)\\ &= \left[g_t\big(X(t),t\big) + A(t)g_x\big(X(t),t\big) + \frac{D}{2}B^2(t)g_{xx}\big(X(t),t\big)\right]dt + B(t)g_x\big(X(t),t\big)\,dW(t) \end{aligned}$$

Proof The Itō formula (8.25) generalizes (8.22) that is recovered for $A(t) = 0$ and $B(t) = 1$, namely when from (8.24) it is $X(t) = W(t)$. To prove (8.25) remark first that from (8.24) and (8.19) we have

$$\big[dX(t)\big]^2 = \big[A(t)dt\big]^2 + \big[B(t)dW(t)\big]^2 + 2A(t)B(t)dW(t)dt = B^2(t)Ddt$$

and then that, retracing the proof of (8.22) with $X(t)$ instead of $W(t)$, it is

$$\begin{aligned} dg\big(X(t),t\big) &= \big[g_x\big(X(t),t\big) + g_{xt}\big(X(t),t\big)dt + \cdots\big]dX(t)\\ &\qquad + \frac{1}{2}\Big[g_{xx}\big(X(t),t\big) + g_{xxt}\big(X(t),t\big)dt + \cdots\Big]\big[dX(t)\big]^2 + \cdots\\ &\qquad + g_t\big(X(t),t\big)dt + \frac{1}{2}\,g_{tt}\big(X(t),t\big)(dt)^2 + \cdots\\ &= g_x\big(X(t),t\big)\big[A(t)\,dt + B(t)\,dW(t)\big]\\ &\qquad + \frac{D}{2}B^2(t)g_{xx}\big(X(t),t\big)dt + g_t\big(X(t),t\big)dt \end{aligned}$$

so that the Itō formula (8.25) immediately follows. ■

8.4.2 The SDE's and Their Solutions

Definition 8.15 We call **stochastic differential equation** (*SDE*) the equation

$$dX(t) = a\big(X(t), t\big)\,dt + b\big(X(t), t\big)\,dW(t) \qquad 0 \le t_0 < t \le T \qquad (8.26)$$
$$X(t_0) = X_0 \qquad \boldsymbol{P}\text{-a.s.}$$

where $W(t)$ is a Wiener process with $W(t_0) = w_0$, and X_0 a *rv* independent from $W(t)$. We also say that a process $X(t)$ is a **solution** if it admits (8.26) as stochastic differential, that is if

$$X(t) = X_0 + \int_{t_0}^{t} a\big(X(s), s\big)\,ds + \int_{t_0}^{t} b\big(X(s), s\big)\,dW(s) \qquad (8.27)$$

This solution is said to be **unique** if, for every pairs $X_1(t),\ X_2(t)$ of solutions it is

$$\boldsymbol{P}\left\{\sup_{t_0 \le t \le T} |X_1(t) - X_2(t)| > 0\right\} = 0$$

The solutions of (8.26) can be contrived by following several approximation procedures:

1. take the following *sequence of approximating processes*

$$X_0(t) = X_0$$
$$X_n(t) = X_0 + \int_{t_0}^{t} a\big(X_{n-1}(s), s\big)\,ds + \int_{t_0}^{t} b\big(X_{n-1}(s), s\big)\,dW(s)$$

   and investigate its (distribution) limit process for $n \to \infty$; this is the recursive procedure usually adopted to prove the theorems of existence and unicity;
2. produce the *trajectories of the solution process* with the recursive method generally used to generate simulations: take n arbitrary instants (usually equidistant)

$$t_0 < t_1 < \cdots < t_n = t \le T$$

   build the samples starting with an initial value x_0 according to the following procedure

$$x_{j+1} = x_j + a(x_j, t_j)\Delta t_j + b(x_j, t_j)\Delta w_j \qquad j = 0, 1, \ldots, n-1$$

where

$$x_j = x(t_j) \qquad \Delta t_j = t_{j+1} - t_j \qquad \Delta w_j = w(t_{j+1}) - w(t_j)$$

and $w(t)$ is a sample of the Wiener process $W(t)$; the values Δw_j of ΔW_j are drawn independently from the x_j. For every value x_0 and for every Wiener sample $w(t)$ we get a possible discretized trajectory. Go then to the limit $n \to \infty$: the solution *exists* if such a limit exists for almost every sample $w(t)$ of the Wiener process; this solution is moreover *unique* if for almost every sample $w(t)$ of the Wiener process the limit trajectory is unique.

Theorem 8.16 Theorem of existence and uniqueness: *The solution $X(t)$ of the SDE (8.26) exists and is unique if the Lipschitz conditions are met, that is if there exist two numbers k_1 and k_2 such that*

$$|a(x,t) - a(y,t)| + |b(x,t) - b(y,t)| \leq k_1 |x - y|$$
$$|a(x,t)|^2 + |b(x,t)|^2 \leq k_2 (1 + |x|^2)$$

for every x, y and $t \in [0, T]$. This solution is ***sample continuous (diffusion)*** *and non anticipative w.r.t. $W(t)$.*

Proof Omitted: see [3], p. 289 and [4], p. 66. The proof essentially consists in checking that the sequence of processes $X_n(t)$ generated wit the procedure 1 converges $\boldsymbol{P}$-a.s. and uniformly in $[0, T]$. Since however it may happen that the functions $a(x,t)$ and $b(x,t)$ do not conform to the Lipschitz conditions, it is also usual to define the so-called ***weak solutions*** instead of the ***strong solutions*** of the Definition 8.15: for more details we will only refer to the literature cited for this proof. ■

Corollary 8.17 Change of variable: *If $X(t)$ is a solution of the SDE (8.26) and $g(x,t)$ is a function at least twice differentiable in x and once in t, then for the process $Y(t) = g\big(X(t), t\big)$ we find*

$$dg\big(X(t),t\big) = \left[g_t\big(X(t),t\big) + a\big(X(t),t\big)g_x\big(X(t),t\big) + \frac{D}{2}b^2\big(X(t),t\big)g_{xx}\big(X(t),t\big)\right]dt + b\big(X(t),t\big)g_x\big(X(t),t\big)\,dW(t) \qquad (8.28)$$

that can always be put in the form of a new SDE for $Y(t)$ whenever a function $h(y,t)$ can be found to implement the inverse transformation $X(t) = h\big(Y(t), t\big)$.

Proof Just take advantage of the Itō formula (8.25). ■

It should be remembered moreover that in addition to the Itō *SDE*'s, the so-called ***random differential equations*** can also be defined, the characteristic of which is that they deal with processes that in general are not Markovian, but are derivable (in *ms*) and satisfy equations whose coefficients are other given processes: see in this respect the remarks about the trade off between regularity and Markovianity of the process paths in the Sect. 7.1.7, and the discussion about the lack of Markovianity in the position process of a Brownian motion in the Proposition 9.1. We do not have here the space to dwell also on these equations which are widely used in the applications (for instance in electronics and signal processing: see [5], Sect. 10.2) and we will refer instead to the existing literature [6] for further details.

8.4.3 SDE's and Fokker-Planck Equations

In the following sections we will suppose to take the expectations by keeping into account *all the initial conditions* required on the involved processes ($X(t)$, $W(t)$ and even others, if need be), namely by means of the corresponding conditional distributions.

Proposition 8.18 *Every solution of the SDE (8.26) is a Markov process.*

Proof We will confine the discussion to an intuitive justification. Take the sample trajectories of $X(t)$ according to the procedure 2, and $X(s) = y$ for $s > t_0$: the evolution of $X(t)$ for $t > s$ is apparently contingent only on the sample $w(t)$ of $W(t)$ for $t > s$. Since on the other hand $X(t)$ is non anticipative, the *rv*'s $X(t')$ with $t' < s$, and $W(t)$ with $t > s$ are independent: as a consequence, when y is known, the values of $X(t)$ with $t > s$, and those with $t' < s$ will be independent, so that $X(t)$ will turn out to be a Markov process. ■

Proposition 8.19 *Take an ac solution $X(t)$ of the SDE (8.26) with $X(t_0) = X_0$, $\boldsymbol{P}$-a.s., then its pdf will be a solution of the Fokker-Planck equation*

$$\partial_t f(x,t) = -\partial_x [A(x,t) f(x,t)] + \frac{1}{2}\partial_x^2 [B(x,t) f(x,t)] \qquad f(x,t_0) = f_0(x) \tag{8.29}$$

where f_0 is the pdf of X_0, and

$$A(x,t) = a(x,t) \qquad B(x,t) = D\,b^2(x,t) \tag{8.30}$$

In particular the transition pdf $f(x,t\,|\,x_0,t_0)$ results from the degenerate initial condition $f(x,t_0) = \delta(x - x_0)$, that is $X(t_0) = x_0$, $\boldsymbol{P}$-a.s.

Proof We already know from the Theorem 8.16 and the Proposition 8.18 that a solution of the *SDE* (8.26) is a sample continuous Markov process, and hence its transition *pdf* is a solution of a Fokker-Planck equation (7.81). Take then $X(t_0) = x_0$, $\boldsymbol{P}$-a.s., and a function $h(x)$ twice differentiable in x: from the change of variable formula (8.28) we find

$$dh\big(X(t)\big) = \left[a\big(X(t),t\big)h'\big(X(t)\big) + \frac{D}{2}\,b^2\big(X(t),t\big)h''\big(X(t)\big)\right]dt + b\big(X(t),t\big)h'\big(X(t)\big)\,dW(t)$$

Since moreover $X(t)$ is non anticipative, we have

$$E\left[b\big(X(t),t\big)h'\big(X(t)\big)\,dW(t)\right] = E\left[b\big(X(t),t\big)h'\big(X(t)\big)\right]E\left[dW(t)\right] = 0$$

and hence integrating by parts

$$\begin{aligned} E\left[dh\big(X(t)\big)\right] &= E\left[a\big(X(t),t\big)h'\big(X(t)\big) + \frac{D}{2}\,b^2\big(X(t),t\big)h"\big(X(t)\big)\right]dt \\ &= \int_{-\infty}^{+\infty}\left[a(x,t)h'(x) + \frac{D}{2}\,b^2(x,t)h"(x)\right]f(x,t\,|\,x_0,t_0)\,dx\,dt \\ &= \int_{-\infty}^{+\infty}\left[-\,\partial_x\left[a(x,t)f(x,t\,|\,x_0,t_0)\right]\right. \\ &\qquad\left.+\frac{D}{2}\,\partial_x^2\left[b^2(x,t)f(x,t\,|\,x_0,t_0)\right]\right]h(x)\,dx\,dt \end{aligned}$$

On the other hand it is also

$$\begin{aligned} E\left[dh\big(X(t)\big)\right] &= dE\left[h\big(X(t)\big)\right] = \frac{d}{dt}E\left[h\big(X(t)\big)\right]dt \\ &= \int_{-\infty}^{+\infty} h(x)\partial_t f(x,t\,|\,x_0,t_0)\,dx\,dt \end{aligned}$$

and comparing the two expressions the result for the transition *pdf* follows from the arbitrariness of $h(x)$. The equation for general, non degenerate initial conditions easily results finally from that for the transition *pdf*. ■

Taking into account the role played by the coefficients A and B in the *forward equations* (see Sect. 7.2.3), the Proposition 8.19 imply in fact that also the coefficients a and b of the (8.26) are to be understood respectively as a *drift velocity* and a *diffusion term*.

8.5 Notable *SDE*'s

We already know that the law of a Markov process $X(t)$ can be completely specified by its *pdf*'s $f(x,t)$ and $f(x,t\,;\,y,s)$: if moreover $X(t)$ is a *Gaussian process* (see Sect. 7.1.10) these *pdf*'s are in their turn totally determined by $E[X(t)]$ and $\boldsymbol{cov}[X(t), X(s)]$: we have indeed that

$$f(x,t) = \mathfrak{N}\big(E[X(t)]\,,\,V[X(t)]\big) \qquad f(x,t;\,y,s) = \mathfrak{N}\big(\boldsymbol{b},\,\mathbb{A}\big)$$

where

$$\boldsymbol{b} = \begin{pmatrix} \boldsymbol{E}[X(t)] \\ \boldsymbol{E}[X(s)] \end{pmatrix} \qquad \mathbb{A} = \begin{pmatrix} \boldsymbol{V}[X(t)] & \boldsymbol{cov}[X(s), X(t)] \\ \boldsymbol{cov}[X(t), X(s)] & \boldsymbol{V}[X(s)] \end{pmatrix}$$

These remarks will be instrumental in the following to calculate the distributions of a few notable *SDE*'s solutions. Remember finally that we will usually take initial conditions in an arbitrary $t_0 \geq 0$, and in particular the degenerate condition $W(s) = y$ to select the transition *pdf* $f(x, t|y, s)$.

8.5.1 SDE's with Constant Coefficients

The simplest *SDE* has constant coefficients $a(x, t) = a,\ b(x, t) = b$, namely

$$dX(t) = a\,dt + b\,dW(t) \qquad X(t_0) = X_0 \tag{8.31}$$

and its solution simply is

$$X(t) = X_0 + a(t - t_0) + b\big[W(t) - w_0\big]$$

The corresponding Fokker-Planck equation according to the Proposition 8.19 is

$$\partial_t f(x, t) = -a\,\partial_x f(x, t) + \frac{Db^2}{2}\partial_x^2 f(x, t) \qquad f(x, t_0) = f_0(x)$$

where f_0 is the *pdf* of X_0. The solution $X(t)$ turns out to be Gaussian if the initial condition X_0 is Gaussian (it is indeed a linear combination of Gaussian *rv*'s), in particular if $X_0 = x_0$, $\boldsymbol{P}$-a.s. It is apparent then that in this case the solution of (8.31) is nothing but a Wiener process slightly modified with a constant drift a and a rescaling b of the diffusion coefficient D, and hence also the transition *pdf* $f(x, t|x_0, t_0)$ is $\mathfrak{N}(x_0 + a(t - t_0)\,,\ Db^2(t - t_0))$. Of course if in particular $a = 0$ and $b = 1$, $X(t)$ exactly coincides with a Wiener process complying with the Fokker-Planck equation (7.92). Remark that another arbitrary initial condition X_0 would instead produce a process $X(t)$ with the same Wienerian transition *pdf*, but with different, non Gaussian joint laws.

8.5.2 SDE's with Time Dependent Coefficients

With time dependent coefficients $a(t),\ b(t)$ the *SDE* (8.26) becomes

$$dX(t) = a(t)\,dt + b(t)\,dW(t) \qquad X(t_0) = X_0, \quad \boldsymbol{P}\text{-a.s.} \tag{8.32}$$

and its formal explicit solution is

$$X(t) = X_0 + \int_{t_0}^{t} a(t')\, dt' + \int_{t_0}^{t} b(t')\, dW(t') \tag{8.33}$$

Even in this case—being a Wiener integral apparently Gaussian—the solution is Gaussian if X_0 is Gaussian too, and in particular if $X_0 = x_0$. The corresponding Fokker–Planck equation moreover is

$$\partial_t f(x,t) = -a(t)\partial_x f(x,t) + \frac{D}{2} b(t)^2 \partial_x^2 f(x,t) \qquad f(x,t_0) = f_0(x)$$

To find the process distribution it will then be enough to have the transition *pdf* that is selected by the degenerate initial condition $X(t_0) = x_0$, $\boldsymbol{P}$-a.s.: all the other solutions will then follow from the Chapman-Kolmogorov equation (7.16) with arbitrary initial conditions $f_0(x)$.

Proposition 8.20 *The solution $X(t)$ of the SDE*

$$dX(t) = a(t)\, dt + b(t)\, dW(t) \qquad X(t_0) = x_0, \quad \boldsymbol{P}\text{-a.s.} \tag{8.34}$$

is a Gaussian process with

$$m(t) = \boldsymbol{E}[X(t)] = x_0 + \int_{t_0}^{t} a(t')\, dt' \qquad \boldsymbol{cov}[X(s), X(t)] = D \int_{t_0}^{s \wedge t} b^2(t')\, dt' \tag{8.35}$$

where $s \wedge t = \min\{s,t\}$, and hence its transition pdf $f(x,t|x_0,t_0)$ is $\mathfrak{N}(m(t)$ $\sigma^2(t))$ with $\sigma^2(t) = \boldsymbol{V}[X(t)] = \boldsymbol{cov}[X(t), X(t)]$ deduced from (8.35).

Proof To prove that $X(t)$ of (8.33) is a Gaussian process we can take advantage of the point 2 in the Proposition 4.20 by showing that every linear combination of the *rv*'s $X(t_1), \dots, X(t_n)$ is Gaussian too: we will neglect however to check that explicitly. Being $X(t)$ a Gaussian process, to get its law it will then be enough to find its expectation and covariance: from (8.33) with $X_0 = x_0$ the expectation is

$$m(t) = \boldsymbol{E}[X(t)] = \boldsymbol{E}[X_0] + \int_{t_0}^{t} a(t')\, dt' + \int_{t_0}^{t} b(t')\, \boldsymbol{E}\left[dW(t')\right] = x_0 + \int_{t_0}^{t} a(t')\, dt'$$

while the covariance, taking $t_0 < s < t$, follows from the previous results and is

$$\begin{aligned}\boldsymbol{cov}[X(s), X(t)] &= \boldsymbol{E}\left[\big(X(t) - \boldsymbol{E}[X(t)]\big)\big(X(s) - \boldsymbol{E}[X(s)]\big)\right] \\ &= \boldsymbol{E}\left[\int_{t_0}^{t} b(t')\, dW(t') \int_{t_0}^{s} b(s')\, dW(s')\right]\end{aligned}$$

Since moreover the increments of $W(t)$ on non overlapping intervals are independent, from (8.18) we have

$$\begin{aligned}\boldsymbol{cov}\,[X(s),X(t)] &= \boldsymbol{E}\left[\int_{t_0}^{s} b(t')\,dW(t')\int_{t_0}^{s} b(s')\,dW(s')\right]\\ &\qquad + \boldsymbol{E}\left[\int_{s}^{t} b(t')\,dW(t')\int_{t_0}^{s} b(s')\,dW(s')\right]\\ &= D\int_{t_0}^{s} b^2(t')\,dt'\end{aligned}$$

that is (8.35) for arbitrary s and t. This also entails in particular that

$$\sigma^2(t) = \boldsymbol{V}\,[X(t)] = \boldsymbol{cov}\,[X(t),X(t)] = D\int_{t_0}^{t} b^2(t')\,dt'$$

so that in general $X(t) \sim \mathfrak{N}\,(m(t)\,,\,\sigma^2(t))$ and its *pdf* also apparently coincides with the transition *pdf* $f(x,t|x_0,t_0)$. ■

8.5.3 SDE's with No Drift and x-Linear Diffusion

Take now an x-linear diffusion coefficient $b(x,t) = cx$ with $c > 0$, and for simplicity a vanishing drift $a(x,t) = 0$: our *SDE* then becomes

$$dX(t) = cX(t)\,dW(t) \qquad\qquad X(t_0) = X_0 > 0, \quad \boldsymbol{P}\text{-a.s.} \tag{8.36}$$

while the corresponding Fokker–Planck equation, with $A(x,t) = a(x,t) = 0$ and $B(x,t) = Db^2(x,t) = Dc^2x^2$, is now

$$\partial_t f(x,t) = \frac{Dc^2}{2}\,\partial_x^2\big[x^2 f(x,t)\big] \qquad\qquad f(x,t_0) = f_0(x)$$

To solve (8.36) it is expedient to change the variable according to the transformation $g(x) = \ln x$

$$Y(t) = g(X(t)) = \ln X(t) \qquad\qquad Y(t_0) = Y_0 = \ln X_0$$

The new *SDE* for $Y(t)$ can now be found from (8.28): since it is

$$g(x,t) = \ln x \qquad g_x(x,t) = \frac{1}{x} \qquad g_{xx}(x,t) = -\frac{1}{x^2} \qquad g_t(x,t) = 0$$

from (8.28) immediately follows that

$$dY(t) = -\frac{Dc^2}{2}\,dt + c\,dW(t) \qquad\qquad Y(t_0) = Y_0, \quad \boldsymbol{P}\text{-a.s.} \tag{8.37}$$

Remark that the first term in the r.h.s. of this equation would not be there by adopting the usual differentiation rules: this additional constant drift term, which would be signally absent in the non stochastic calculus, is indeed a byproduct of the Itō formula. The *SDE* (8.37) has now constant coefficients as in the Eq. (8.31) discussed in the Sect. 8.5.1, and hence its solution simply is

$$Y(t) = Y_0 - \frac{Dc^2}{2}(t - t_0) + c\big[W(t) - w_0\big] \tag{8.38}$$

namely a modified Wiener process plus an independent initial *rv*, so that going back to the original variables with $h(x) = e^x$ we finally find

$$X(t) = h(Y(t)) = e^{Y(t)} = X_0\, e^{-Dc^2(t-t_0)/2} e^{c\,[W(t)-w_0]} \tag{8.39}$$

Proposition 8.21 *The process $Y(t)$ (8.38) solution of the SDE (8.37) with Gaussian initial condition $Y_0 \sim \mathfrak{N}(y_0, \sigma_0^2)$ is Gaussian with distribution at time t*

$$Y(t) \sim \mathfrak{N}\left(y_0 - \frac{Dc^2}{2}(t - t_0)\,,\ \sigma_0^2 + Dc^2(t - t_0)\right) \tag{8.40}$$

and with autocovariance

$$\boldsymbol{cov}\,[Y(s), Y(t)] = \sigma_0^2 + Dc^2 \min\{t - t_0\,,\ s - t_0\} \tag{8.41}$$

Proof The process $Y(t)$ is apparently Gaussian if Y_0 is Gaussian because (8.38) always turns out to be a linear combination of Gaussian *rv*'s. Remark that on the other hand $X(t)$ in (8.39) is still Markovian, but it is not Gaussian, as we will see later. Nevertheless we will be able to find the transition *pdf* of $X(t)$, and thus all its other distributions from the Chapman-Kolmogorov equations. The law of $Y(t)$ (8.38) is thus completely determined by $\boldsymbol{E}\,[Y(t)]$ and $\boldsymbol{cov}\,[Y(s), Y(t)]$: from (8.38) we first have

$$\boldsymbol{E}\,[Y(t)] = y_0 - \frac{Dc^2}{2}(t - t_0) \tag{8.42}$$

For the autocovariance it is expedient to define $\widetilde{Y}_0 = Y_0 - y_0$ and the centered processes $\widetilde{W}(t) = W(t) - w_0$ and

$$\widetilde{Y}(t) = Y(t) - \boldsymbol{E}\,[Y(t)] = Y_0 - y_0 + c[W(t) - w_0] = \widetilde{Y}_0 + c\widetilde{W}(t)$$

From (8.9) and (8.10) we then have

$$\boldsymbol{E}\left[\widetilde{W}(t)\right] = 0 \qquad \boldsymbol{E}\left[\widetilde{W}(s)\widetilde{W}(t)\right] = D\min\{s - t_0\,,\ t - t_0\}$$

so that—keeping also into account the independence of Y_0 and $W(t)$—we eventually find the required form (8.41) for the autocovariance:

$$\begin{aligned}\boldsymbol{cov}\left[Y(s), Y(t)\right] &= \boldsymbol{E}\left[\widetilde{Y}(s)\widetilde{Y}(t)\right] = \boldsymbol{E}\left[(\widetilde{Y}_0 + c\widetilde{W}(s))(\widetilde{Y}_0 + c\widetilde{W}(t))\right] \\ &= \boldsymbol{E}\left[\widetilde{Y}_0^2\right] + c^2\boldsymbol{E}\left[\widetilde{W}(s)\widetilde{W}(t)\right] \\ &= \sigma_0^2 + Dc^2 \min\{t - t_0\,,\ s - t_0\}\end{aligned}$$

This also entails that $\boldsymbol{V}\left[Y(t)\right] = \sigma_0^2 + Dc^2(t - t_0)$ and hence the form (8.40) for the distribution of $Y(t)$. ■

Proposition 8.22 *The distribution of $X(t)$ solution of the SDE (8.36) with log-normal initial conditions $X(t_0) = X_0 = e^{Y_0} \sim \mathfrak{ln}\mathfrak{N}(y_0, \sigma_0^2)$ is the log-normal*

$$X(t) \sim \mathfrak{ln}\mathfrak{N}\left(y_0 - \frac{Dc^2}{2}(t - t_0)\,,\ \sigma_0^2 + Dc^2(t - t_0)\right) \tag{8.43}$$

and, with $x_0 = e^{y_0}$, we also have

$$\boldsymbol{E}\left[X(t)\right] = x_0\, e^{\sigma_0^2/2} \qquad \boldsymbol{V}\left[X(t)\right] = x_0^2\, e^{\sigma_0^2}\left(e^{\sigma_0^2 + Dc^2(t-t_0)} - 1\right) \tag{8.44}$$

$$\boldsymbol{cov}\left[X(s), X(t)\right] = x_0^2\, e^{2\sigma_0^2}\left(e^{Dc^2(t-t_0)\wedge(s-t_0)} - 1\right) \tag{8.45}$$

The log-normal transition pdf is easily recovered from (8.43) by taking $\sigma_0 = 0$, that is by choosing a degenerate initial condition.

Proof The process $X(t) = e^{Y(t)}$ is the exponential of the Gaussian process $Y(t)$ discussed in the Proposition 8.21 with distribution (8.40), and hence its distribution at the time t is the log-normal (8.43). From (3.65) it is then straightforward to calculate the expectation and the variance listed in (8.44). As for the autocovariance we remark first that $X(s)X(t) = e^{Y(s)+Y(t)}$, and that from the Proposition 4.20 it follows that the *rv*'s $Y(s) + Y(t)$ always are Gaussian. From (8.42) we have moreover

$$\boldsymbol{E}\left[Y(s) + Y(t)\right] = 2y_0 - \frac{Dc^2}{2}(s + t - 2t_0)$$

while from the Proposition 3.29 and from (8.41) we have

$$\begin{aligned}\boldsymbol{V}\left[Y(s) + Y(t)\right] &= \boldsymbol{V}\left[Y(s)\right] + \boldsymbol{V}\left[Y(t)\right] + 2\boldsymbol{cov}\left[Y(s), Y(t)\right] \\ &= 4\sigma_0^2 + Dc^2\left[(s + t - 2t_0) + 2\min\{t - t_0\,,\ s - t_0\}\right]\end{aligned}$$

By summarizing we have then that

$$X(s)X(t) \sim \ln\mathfrak{N}\left(2y_0 - \frac{Dc^2}{2}(s+t-2t_0)\,,\right.$$
$$\left.4\sigma_0^2 + Dc^2\left[(s+t-2t_0) + 2(t-t_0)\wedge(s-t_0)\right]\right)$$

so that from (3.65) it is

$$\boldsymbol{E}\left[X(s)X(t)\right] = e^{2y_0+2\sigma_0^2+Dc^2(t-t_0)\wedge(s-t_0)} = x_0^2\, e^{2\sigma_0^2}\, e^{Dc^2(t-t_0)\wedge(s-t_0)}$$

and finally from (8.44) we can deduce the autocovariance (8.45). All these results also entail that the transition *pdf* $f(x,t|x_0,t_0)$ of the process $X(t)$ is the log-normal

$$\ln\mathfrak{N}\left(\ln x_0 - \frac{Dc^2}{2}(t-t_0)\,,\ Dc^2(t-t_0)\right)$$

that is recovered for $\sigma_0 = 0$, namely with the degenerate initial condition $X(t_0) = x_0 = e^{y_0}$: taking then advantage of the Chapman-Kolmogorov equations and of the chain rule we are therefore in a position to find also the complete law of the process. Remark that now, since $X(t)$ is no longer a Gaussian process, the said global law of the process could not be deduced only from the knowledge of $\boldsymbol{E}[X(t)]$ and $\boldsymbol{cov}[X(s), X(t)]$, so that our explicit form of the transition *pdf* plays a crucial role in the characterization of the process $X(t)$. ■

8.5.4 SDE's with x-Linear Drift and Constant Diffusion

Take now $a(x,t) = -\alpha x$ with $\alpha > 0$, and $b(x,t) = 1$: our *SDE* will then be

$$dX(t) = -\alpha X(t)dt + dW(t) \qquad X(t_0) = X_0 \quad \boldsymbol{P}\text{-a.s.} \tag{8.46}$$

With the usual coefficient transformations

$$A(x,t) = a(x,t) = -\alpha x \qquad B(x,t) = Db^2(x,t) = D$$

we then find the Fokker-Planck equation of an Ornstein-Uhlenbeck process

$$\partial_t f(x,t) = \alpha\partial_x\left[xf(x,t)\right] + \frac{D}{2}\partial_x^2 f(x,t) \qquad f(x,t_0) = f_0(x) \tag{8.47}$$

that has been put forward in the Proposition 7.40: we already know its solutions, but we will deduce them again here as an application of the stochastic calculus. To find the solution of (8.46) consider the transformed process

$$Y(t) = X(t)\,e^{\alpha(t-t_0)} \qquad Y(t_0) = X_0$$

whose *SDE* follows from (8.28) with $g(x,t) = xe^{\alpha(t-t_0)}$: since it is

$$g_x(x,t) = e^{\alpha(t-t_0)} \qquad g_{xx}(x,t) = 0 \qquad g_t(x,t) = \alpha x e^{\alpha(t-t_0)}$$

we find that $Y(t)$ is a solution of the *SDE*

$$dY(t) = e^{\alpha(t-t_0)}\, dW(t) \tag{8.48}$$

with time dependent coefficients like (8.32) so that

$$Y(t) = X_0 + \int_{t_0}^{t} e^{\alpha(s-t_0)} dW(s)$$

Recalling then that $X(t) = e^{-\alpha(t-t_0)}Y(t)$, the solution of (8.46) will be

$$X(t) = X_0\, e^{-\alpha(t-t_0)} + \int_{t_0}^{t} e^{-\alpha(t-s)} dW(s) \tag{8.49}$$

which is Gaussian if X_0 is Gaussian, in particular when $X_0 = x_0$, $\boldsymbol{P}$-a.s.

Proposition 8.23 *The process $X(t)$ solution of the SDE (8.46) with Gaussian initial conditions $X_0 \sim \mathfrak{N}(x_0, \sigma_0^2)$ is a Gaussian Ornstein-Uhlenbeck process with*

$$X(t) \sim \mathfrak{N}\left(x_0\, e^{-\alpha(t-t_0)}\,,\, \sigma_0^2\, e^{-2\alpha(t-t_0)} + \beta^2\left(1 - e^{-2\alpha(t-t_0)}\right)\right) \tag{8.50}$$

$$\boldsymbol{cov}\,[X(s), X(t)] = \left(\sigma_0^2 - \beta^2\right) e^{-\alpha(s+t-2t_0)} + \beta^2 e^{-\alpha|t-s|} \tag{8.51}$$

where $\beta^2 = D/2\alpha$. The Gaussian transition pdf is easily recovered from (8.50) by taking $\sigma_0 = 0$, that is by choosing a degenerate initial condition.

Proof Since $X(t)$ with our initial conditions is a Gaussian process, it will be enough to find its expectation and its autocovariance. From (8.17) we first have

$$\boldsymbol{E}\,[X(t)] = \boldsymbol{E}\,[X_0]\; e^{-\alpha(t-t_0)} = x_0\, e^{-\alpha(t-t_0)} \tag{8.52}$$

Then for the autocovariance, from (8.17), (8.18) and the independence of the Wiener integrals on non overlapping intervals, we have

$$\begin{aligned}
\boldsymbol{cov}\,[X(s), X(t)] &= \boldsymbol{E}\,[\,(X(t) - \boldsymbol{E}\,[X(t)])\;(X(s) - \boldsymbol{E}\,[X(s)])\,] \\
&= \boldsymbol{E}\left[\left((X_0 - x_0)e^{-\alpha(t-t_0)} + \int_{t_0}^{t} e^{-\alpha(t-t')} dW(t')\right)\cdot\right. \\
&\qquad\qquad \left.\left((X_0 - x_0)e^{-\alpha(s-t_0)} + \int_{t_0}^{s} e^{-\alpha(s-s')} dW(s')\right)\right] \\
&= \boldsymbol{V}\,[X_0]\; e^{-\alpha(s+t-2t_0)} + \boldsymbol{E}\left[\int_{t_0}^{t} e^{-\alpha(t-t')} dW(t') \int_{t_0}^{s} e^{-\alpha(s-s')} dW(s')\right]
\end{aligned}$$

$$= \sigma_0^2 \, e^{-\alpha(s+t-2t_0)} + D \int_{t_0}^{s\wedge t} e^{-\alpha(t+s-2t')} dt'$$

$$= \sigma_0^2 \, e^{-\alpha(s+t-2t_0)} + De^{-\alpha(t+s)} \frac{e^{2\alpha(s\wedge t)} - e^{2\alpha t_0}}{2\alpha} = \left(\sigma_0^2 - \beta^2\right) e^{-\alpha(s+t-2t_0)} + \beta^2 e^{-\alpha|t-s|}$$

since it is easy to check that $s + t - 2(s \wedge t) = |t - s|$. The process $X(t)$ is thus completely specified, and in particular its variance is

$$\boldsymbol{V}\,[X(t)] = \boldsymbol{cov}\,[X(t), X(t)] = \sigma_0^2 \, e^{-2\alpha(t-t_0)} + \beta^2 \left(1 - e^{-2\alpha(t-t_0)}\right) \tag{8.53}$$

From these results we can also deduce the transition *pdf* choosing the initial condition $X_0 = x_0$, $\boldsymbol{P}$-a.s., that is $\sigma_0^2 = 0$: in this case from (8.50) we easily find $X(t) \sim \mathfrak{N}\left(x_0 e^{-\alpha(t-t_0)}, \beta^2(1 - e^{-2\alpha(t-t_0)})\right)$ in agreement with the aforementioned transition *pdf* (7.57) of the Ornstein-Uhlenbeck process. Remark the relative easy of this derivation from the *SDE* (8.46) w.r.t. the less elementary procedures needed to solve the corresponding Fokker-Planck equation 7.40. ■

References

1. Kolmogorov, A.N., Fomin, S.V.: Introductory Real Analysis. Dover, New York (1975)
2. Doob, J.L.: Stochastic Processes. Wiley, New York (1953)
3. Karatzas, I., Shreve, S.E.: Brownian Motion and Stochastic Calculus. Springer, Berlin (1991)
4. Øksendal, B.: Stochastic Differential Equations. Springer, Berlin (2005)
5. Papoulis, A.: Probability, Random Variables and Stochastic Processes. McGraw Hill, Boston (2002)
6. Neckel, T., Rupp, F.: Random Differential Equations in Scientific Computing. Versita, London (2013)

Part III
Physical Modeling

Chapter 9
Dynamical Theory of Brownian Motion

In 1930 L.S. Ornstein and G.F. Uhlenbeck [1] addressed again the problem of elaborating a suitable model for the Brownian motion, and they refined in more detail the Langevin dynamical equation to investigate the phenomenon at time scales shorter than those considered by Einstein [2] and Smoluchowski [3] in 1905-6. We will now give an account of the Ornstein-Uhlenbeck theory adapted to our notations, and we will look into the conditions under which the Einstein-Smoluchowski theory continues to be a good approximation.

9.1 Free Brownian Particle

In the Ornstein-Uhlenbeck theory the position of the Brownian particle is a process $X(t)$ that is supposed to be differentiable, so that the velocity $V(t) = \dot{X}(t)$ always exists. Resuming then the discussion of the Sect. 6.4.2 we will be able to write down the Newton equation of a free, spherical Browniana particle, with mass m and diameter a, as the following system of differential equations

$$\dot{X}(t) = V(t) \tag{9.1}$$

$$m\dot{V}(t) = -6\pi\eta a V(t) + B(t) \tag{9.2}$$

where, as was argued in the Sect. 8.1, $B(t)$ is a Wiener white noise, while η is the environment viscosity. The Eq. (9.2) indicates in particular that there are two kind of forces acting on the particle: a viscous resistance proportional to the velocity $V(t)$, and a random force embodied by a white noise. Given the singular character of $B(t)$ we know however that our system is better presented in terms of *SDE*'s, namely as

$$dX(t) = V(t)\,dt \tag{9.3}$$

$$dV(t) = -\alpha V(t)\,dt + dW(t) \tag{9.4}$$

N. Cufaro Petroni, *Probability and Stochastic Processes for Physicists*, UNITEXT for Physics, https://doi.org/10.1007/978-3-030-48408-8_9

where we have defined

$$\alpha = \frac{6\pi\eta a}{m} \tag{9.5}$$

while $W(t)$ is now a Wiener noise with a suitable diffusion coefficient D affecting the velocity Eq. (9.4). Remark that the Eqs. (9.3) and (9.4) are uncoupled because $X(t)$ only appears in the first one: this will enable us to deal with them one by one, first solving (9.4) for $V(y)$, and then using it in (9.3).

Proposition 9.1 *Take $t_0 = 0$ and degenerate initial conditions $X(0) = x_0$ and $V(0) = v_0$: then the velocity $V(t)$ of a free Brownian motion is a Gaussian Ornstein-Uhlenbeck process with*

$$\boldsymbol{E}\left[V(t)\right] = v_0\, e^{-\alpha t} \tag{9.6}$$

$$\boldsymbol{cov}\left[V(s), V(t)\right] = \beta^2\left(e^{-\alpha|s-t|} - e^{-\alpha(s+t)}\right) \qquad \beta^2 = \frac{D}{2\alpha} \tag{9.7}$$

If moreover k is the Boltzmann constant and T the absolute temperature, we find

$$\beta^2 = \frac{kT}{m} \tag{9.8}$$

The position $X(t)$ instead is not Markovian, but is a Gaussian process with

$$\boldsymbol{E}\left[X(t)\right] = x_0 + \frac{v_0}{\alpha}\left(1 - e^{-\alpha t}\right) \tag{9.9}$$

$$\boldsymbol{cov}\left[X(s), X(t)\right] = \frac{\beta^2}{\alpha^2}\Big[2\alpha(s \wedge t) - 2 + 2e^{-\alpha s} + 2e^{-\alpha t} - e^{-\alpha|s-t|} - e^{-\alpha(s+t)}\Big] \tag{9.10}$$

Proof A simple change in the notation makes clear that the *SDE* (9.4) coincides with the *SDE* (8.46) discussed in the Sect. 8.5.4 so that, with our initial conditions, the solutions of our system are

$$X(t) = x_0 + \int_0^t V(s)\, ds \tag{9.11}$$

$$V(t) = v_0\, e^{-\alpha t} + \int_0^t e^{-\alpha(t-s)}\, dW(s) \tag{9.12}$$

That $V(t)$ in (9.12) is then an Ornstein-Uhlenbeck process with expectation (9.6) and autocovariance (9.7) has already been shown in the Sect. 8.5.4, while the other general features of such a process have been presented in the Sect. 7.1.9 and in the Proposition 7.40.

As for the relation (9.8) we should remember that, according to the Proposition 7.27, when $t \to +\infty$ the velocity distribution converges to the stationary law $\mathfrak{N}(0, \beta^2)$. As a consequence, at the thermodynamical equilibrium, we can resort to the equipartition of energy

$$\frac{1}{2}kT = \frac{1}{2}m\beta^2$$

that immediately entails (9.8).

From (9.11) it follows that the position process $X(t)$ is Gaussian and hence, according to the Sect. 7.1.10, its distribution can be worked out from its expectation and autocovariance. For the expectation from (9.11) it is

$$\boldsymbol{E}\left[X(t)\right] = x_0 + \int_0^t \boldsymbol{E}\left[V(s)\right] ds = x_0 + v_0 \int_0^t e^{-\alpha s} ds = x_0 + \frac{v_0}{\alpha}\left(1 - e^{-\alpha t}\right)$$

namely (9.9). Then for the autocovariance we first prove that

$$\boldsymbol{cov}\left[X(s), X(t)\right] = \int_0^s \int_0^t \boldsymbol{cov}\left[V(s'), V(t')\right] ds'dt' \tag{9.13}$$

Using indeed for convenience the centered processes

$$\widetilde{V}(t) = V(t) - \boldsymbol{E}\left[V(t)\right] = V(t) - v_0 e^{-\alpha t} \tag{9.14}$$

$$\widetilde{X}(t) = X(t) - \boldsymbol{E}\left[X(t)\right] = \int_0^t V(s)\, ds - \frac{v_0}{\alpha}\left(1 - e^{-\alpha t}\right) = \int_0^t \widetilde{V}(s)\, ds \tag{9.15}$$

we easily find that (9.13) holds:

$$\begin{aligned}\boldsymbol{cov}\left[X(s), X(t)\right] &= \boldsymbol{E}\left[\widetilde{X}(s)\widetilde{X}(t)\right] = \int_0^s \int_0^t \boldsymbol{E}\left[\widetilde{V}(s')\widetilde{V}(t')\right] ds'dt' \\ &= \int_0^s \int_0^t \boldsymbol{cov}\left[V(s'), V(t')\right] ds'dt'\end{aligned}$$

From (9.13) and (9.7) we thus have

$$\boldsymbol{cov}\left[X(s), X(t)\right] = \beta^2 \int_0^s \int_0^t \left(e^{-\alpha|s'-t'|} - e^{-\alpha(s'+t')}\right) ds'dt'$$

and (9.10) follows from a tiresome but elementary integration.

To prove finally that $X(t)$ is not Markovian, we will explicitly calculate its transition *pdf* and we will show that it does not comply with the Chapman-Kolmogorov conditions. To find first the two-times joint, Gaussian *pdf* of $X(t)$ let us call for short $b(t)$, $a^2(t)$ and $r(s,t)$ respectively the expectation, the variance and the correlation coefficient as they are deduced from (9.9) and (9.10): we see then the the one-time *pdf* $f(x,t)$ is $\mathfrak{N}(b(t), a^2(t))$, while the two-times *pdf* $f(x,t;y,s)$ is $\mathfrak{N}(\boldsymbol{b}, \mathbb{A})$ with

$$\boldsymbol{b} = \begin{pmatrix} b(s) \\ b(t) \end{pmatrix} \qquad \mathbb{A} = \begin{pmatrix} a^2(s) & a(s)a(t)r(s,t) \\ a(s)a(t)r(s,t) & a^2(t) \end{pmatrix}$$

The transition *pdf* $f(x, t \mid y, s)$ with $s < t$ follows now from the Proposition 3.40 and is $\mathfrak{N}(A(s,t)y + B(s,t)\,,\, C^2(s,t))$ where we have defined

$$A(s,t) = r(s,t)\frac{a(t)}{a(s)} \qquad\qquad B(s,t) = b(t) - r(s,t)\frac{a(t)}{a(s)}b(s)$$
$$C^2(s,t) = a^2(t)[1 - r^2(s,t)]$$

A tedious, direct calculation—whose details we will neglect here – would show that to meet the Chapman-Kolmogorov condition (7.17) we should have

$$r(s,u)r(u,t) = r(s,t) \qquad\qquad s < u < t \tag{9.16}$$

while from (9.10) it is

$$r(s,t) = \frac{2\alpha s - 2 + 2e^{-\alpha s} + 2e^{-\alpha t} - e^{-\alpha(t-s)} - e^{-\alpha(t+s)}}{\sqrt{2\alpha s - 3 + 4e^{-\alpha s} - e^{-2\alpha s}}\,\sqrt{2\alpha t - 3 + 4e^{-\alpha t} - e^{-2\alpha t}}} \qquad s < t$$

and it is possible to check that (9.16)—and hence the Chapman-Kolmogorov condition—does not hold: we can conclude then that in the Ornstein-Uhlenbeck theory the position process $X(t)$ is not Markovian. This result on the one hand is hardly surprising because (9.1) is in fact a *random differential equation* instead of a *SDE* (see Sect. 8.4.2 and [4]), and on the other it is also in apparent disagreement with the Einstein-Smoluchowski theory that, as elucidated in the Sect. 6.4.1, consider the Brownian position as a Wiener process, namely as a Markov process. ■

9.2 Ornstein-Uhlenbeck Versus Einstein-Smoluchowski

To better compare the Einstein-Smoluchowski theory of the Sect. 6.4 with that of Ornstein-Uhlenbeck presented here we must remark at once that in the two approaches the symbol D takes two different meanings, so that we will accordingly be obliged to adopt two separate notations:

- in the Einstein-Smoluchowski theory we will dub D_X the diffusion coefficient of the Wiener process $W_X(t)$ that directly represents the position of the Brownian particle; since moreover the variance of such a position linearly grows in time as $D_X t$, we also find that its physical dimensions are

$$[D_X] = \frac{\mathrm{mt}^2}{\mathrm{sec}}$$

  while from (6.76) we know that its value in terms of physical constants is

$$D_X = \frac{kT}{3\pi\eta a}$$

- in the Ornstein-Uhlenbeck approach, instead, we will now label as D_V the diffusion coefficient of the Wiener noise $W_V(t)$ affecting the velocity Eq. (9.4), so that $\beta^2 = D_V/2\alpha$ is the asymptotic velocity variance and the physical dimensions will be

$$[D_V] = \frac{\text{mt}^2}{\text{sec}^3}$$

while from (6.76),(9.5),(9.7) and (9.8) we also know that its value is

$$D_V = 2\alpha\beta^2 = \frac{12\pi\eta akT}{m^2} = \alpha^2 D_X$$

We can then compare the two theories by remarking first of all that in the Ornstein-Uhlenbeck model the position variance is deduced from (9.10) and is

$$\boldsymbol{V}[X(t)] = \frac{\beta^2}{\alpha^2}\left(2\alpha t - 3 + 4e^{-\alpha t} - e^{-2\alpha t}\right)$$

while the Einstein-Smoluchowky result is asymptotically recovered as

$$\boldsymbol{V}[X(t)] \simeq \frac{2\beta^2}{\alpha} t = \frac{D_V}{\alpha^2} t = D_X t \qquad\qquad \alpha t \gg 1$$

This apparently suggests that the Einstein-Smoluchiwski theory should be deemed a good approximation of that of Ornstein-Uhlenbeck either for large times t (after a ***transient delay***) or for large values of the viscous drag coefficient α (***over-damped regime***).

Proposition 9.2 *Within the notations of the Proposition 9.1, if $\alpha \to +\infty$ keeping β^2/α finite, then the Ornstein-Uhlenbeck position process $X(t)$ with initial condition $X(0) = x_0$ converges in distribution—in the sense of the Definition 5.4 – to a Wiener process $W_X(t)$ with diffusion coefficient $D_X = 2\beta^2/\alpha = D_V/\alpha^2$ and $W_X(0) = x_0$.*

Proof From (9.9) and (9.10) we see in fact that, in the over-damped limit $\alpha \to +\infty$ for every fixed s, t, the expectation and the covariance of the position process $X(t)$ converge to

$$\boldsymbol{E}[X(t)] \to x_0 \qquad\qquad \boldsymbol{cov}[X(s), X(t)] \to D_X(s \wedge t)$$

and since $X(t)$ is Gaussian it also converges in distribution to a Wiener process $W_X(t)$ with diffusion coefficient D_X and initial condition $W_X(0) = x_0$. Remark that to suppose an over-damped regime is equivalent to take a very short transient delay. ■

By summarizing, from now on we will take for granted that in an over-damped regime, or anyway after a transient delay $t \gg 1/\alpha$, the position of a free Brownian motion is well described by a Wiener process obeying to the (trivial) *SDE*

$$dX(t) = dW_X(t) \tag{9.17}$$

Remark in particular that the position $X(t)$ diffuses *isotropically* because we see from (9.6) that the initial velocity v_0 is quickly wiped out by the background noise so that, after a short delay, $\boldsymbol{E}[V(t)] \to 0$ for $t \gg 1/\alpha$. The present discussion about the Brownian motion in the over-damped regime will be resumed in a more general setting later on in the Proposition 9.7.

9.3 Ornstein-Uhlenbeck Markovianity

We have seen in the Proposition 9.1 that in the Ornstein-Uhlenbeck theory the velocity $V(t)$ is a Markov process, while the position $X(t)$ is not. From a mathematical standpoint this follows from the fact that $V(t)$ satisfies the Langevin *SDE* (9.4), and hence is Markovian according to the Proposition 8.18, while the relation (9.3) means that $X(t)$ is instead differentiable and satisfies a random differential equation (see Sect. 8.4.2 and [4]). From the discussion of Sect. 7.1.1, however, we also know that it is in general possible to recover a process Markovianity by adding the information needed to this end: typically this means that we should consider vector processes with several components in order to supply all the required additional information. In our discussion a clue comes from the remark that in the Newtonian dynamics the state of the system is not determined by the position $x(t)$ alone, and must instead be described in the phase space by the pair $x(t)$, $v(t)$ of position and velocity. This hints that we should rather consider the phase space vector process

$$\boldsymbol{Z}(t) = \begin{pmatrix} X(t) \\ V(t) \end{pmatrix}$$

so that the system of our two Eqs. (9.3) and (9.4) can be given as a unique vector *SDE*

$$d\boldsymbol{Z}(t) = \boldsymbol{a}(\boldsymbol{Z}(t))\,dt + \mathbb{C}\,d\boldsymbol{W}(t) \tag{9.18}$$

where we took

$$\boldsymbol{a}(z) = \boldsymbol{a}(x, v) = \begin{pmatrix} v \\ -\alpha v \end{pmatrix} \qquad \mathbb{C} = \begin{pmatrix} 0 & 0 \\ 0 & 1 \end{pmatrix} \tag{9.19}$$

while $\boldsymbol{W}(t)$ is now a vector Wiener process with

$$W(t) = \begin{pmatrix} W_X(t) \\ W_V(t) \end{pmatrix}$$

To not overload our discussion we we did not previously mentioned the **vector *SDE*'s** like (9.18), that generally speaking take the form

$$dZ(t) = a(Z(t), t)\, dt + \mathbb{C}(Z(t), t)\, dW(t) \tag{9.20}$$

but we will here give for granted that—with some burdening in the notations—most of the results stated in the previous sections hold even for the *SDE*'s of the type (9.20). The solution of (9.18) with the degenerate initial condition

$$Z(0) = z_0 = \begin{pmatrix} x_0 \\ v_0 \end{pmatrix} \tag{9.21}$$

apparently is the vector $Z(t)$ whose components are the solutions (9.11) and (9.12) previously found, but in this new formulation a new trait comes to the fore that has been neglected in the discussion of the Sect. 9.1: the need to calculate also the *cross-correlation* of the two processes $X(t)$ and $V(t)$, and more generally their *joint distribution* in addition to their respective marginals.

Proposition 9.3 *The cross-covariance of the Ornstein-Uhlenbeck processes $X(t)$ and $V(t)$ is*

$$\boldsymbol{cov}\,[X(s), V(t)] = \frac{\beta^2}{\alpha}\left[1 + \frac{|t-s|}{t-s}\left(e^{-\alpha|t-s|} - 1\right) - 2e^{-\alpha t} + e^{-\alpha(t+s)}\right] \tag{9.22}$$

Proof By using again the centered processes (9.14) and (9.15) we first find that

$$\begin{aligned} \boldsymbol{cov}\,[X(s), V(t)] &= \boldsymbol{E}\left[\widetilde{X}(s)\widetilde{V}(t)\right] = \boldsymbol{E}\left[\widetilde{V}(t)\int_0^s \widetilde{V}(t')\,dt'\right] \\ &= \int_0^s \boldsymbol{E}\left[\widetilde{V}(t)\widetilde{V}(t')\right] dt' = \int_0^s \boldsymbol{cov}\left[V(t), V(t')\right] dt' \end{aligned}$$

and then from (9.7) we can write

$$\boldsymbol{cov}\,[X(s), V(t)] = \beta^2 \int_0^s \left(e^{-\alpha|t-t'|} - e^{-\alpha(t+t')}\right) dt'$$

The result (9.22) finally follows from a boring elementary integration. ■

Proposition 9.4 *The solution $Z(t)$ of the SDE (9.18) with initial conditions (9.21) is a Gaussian vector Markov process; the joint law of its two components at the time t is $\mathfrak{N}(\boldsymbol{b}, \mathbb{A})$ with*

$$\boldsymbol{b} = \begin{pmatrix} \boldsymbol{E}[X(t)] \\ \boldsymbol{E}[V(t)] \end{pmatrix} = \begin{pmatrix} x_0 + v_0\left(1 - e^{-\alpha t}\right)/\alpha \\ v_0\, e^{-\alpha t} \end{pmatrix} \tag{9.23}$$

$$\mathbb{A} = \begin{pmatrix} \boldsymbol{V}[X(t)] & \boldsymbol{cov}[X(t), V(t)] \\ \boldsymbol{cov}[X(t), V(t)] & \boldsymbol{V}[V(t)] \end{pmatrix}$$
$$= \frac{\beta^2}{\alpha^2}\begin{pmatrix} 2\alpha t - 3 + 4e^{-\alpha t} - e^{-2\alpha t} & \alpha\left(1 - 2e^{-\alpha t} + e^{-2\alpha t}\right) \\ \alpha\left(1 - 2e^{-\alpha t} + e^{-2\alpha t}\right) & \alpha^2\left(1 - e^{-2\alpha t}\right) \end{pmatrix} \tag{9.24}$$

We will skip instead for short to provide the explicit form of the joint distribution of the pair $\mathbf{Z}(s)$, $\mathbf{Z}(t)$ *that at any rate can be worked out along similar lines.*

Proof Since $\mathbf{Z}(t)$ satisfies the (9.18), a generalization of the Proposition 8.18 entails that such a solution too is a vector Markov process. As for its distribution, we also know from the Proposition 9.1 that the two components $X(t)$ and $V(t)$ of $\mathbf{Z}(t)$ *individually* are Gaussian processes, but this occurrence—to be sure—is not enough to entail that such components also are *jointly* gaussian. For the time being we will take that conclusion for granted without a proof by postponing to the next proposition the outline of a possible checking procedure, and we will confine ourselves here to remark just that in this event the expressions (9.23) and (9.24) for the vector of the means and the covariance matrix of $\mathbf{Z}(t)$ follow from (9.22), (9.7) and (9.10). ∎

Even the Proposition 8.19 establishing a correspondence between *SDE*'s and Fokker-Planck equations can be suitably generalized to the case of vector processes of the type (9.20), and in this case the Fokker-Planck equation will of course take the form of a multivariate equation like (7.81) whose coefficients—at least when there is only one Wiener noise—are found from the following rules that generalize (8.30)

$$\boldsymbol{A}(\boldsymbol{x}, t) = \boldsymbol{a}(\boldsymbol{x}, t) \qquad \mathbb{B}(\boldsymbol{x}, t) = D\,\mathbb{C}(\boldsymbol{x}, t)\mathbb{C}^{\mathrm{T}}(\boldsymbol{x}, t) \tag{9.25}$$

where $\mathbb{C}^{\mathrm{T}}$ denotes the transposition of $\mathbb{C}$

Proposition 9.5 *The joint pdf's of the r-vec* $\mathbf{Z}(t)$ *solution of the* SDE *(9.18) with initial conditions (9.21) abides by the following phase space Fokker-Planck equation*

$$\partial_t f(x, v, t) = -v\partial_x f(x, v, t) + \alpha\,\partial_v\left[v f(x, v, t)\right] + \frac{D}{2}\,\partial_v^2 f(x, v, t) \tag{9.26}$$
$$f(x, v, 0) = f_0(x, v)$$

Proof It would be enough to write down a bivariate Fokker-Planck (7.81) keeping into account (9.25) and (9.19). Remark that while according to the Proposition 9.1 the velocity $V(t)$ is an Ornstein-Uhlenbeck process and hence its *pdf* satisfies a Fokker-Planck equation of the type (8.47)—that could also be recovered from (9.26) with an x-marginalization—the position $X(t)$ on the contrary is not individually a Markov process and hence its *pdf* is not the solutions of some partial differential equation: in particular a v-marginalization of (9.26) to recover this supposed equation would not lead to any coherent result.

The Eq. (9.26) also enables us to design a procedure to directly check our claim in the Proposition 9.4 that the vector process $\boldsymbol{Z}(t)$ is in fact Gaussian: being $\boldsymbol{Z}(t)$ Markovian it is enough indeed to prove that the transition *pdf* is Gaussian. To this end we could simply write down explicitly (what we did not for short in the previous proposition) the presumed Gaussian bivariate *pdf*'s of $\boldsymbol{Z}(t)$ from (9.23) and (9.24), then the corresponding transition *pdf* and finally verify by direct calculation that it is the solutions of (9.26) with degenerate initial conditions $f(x, v, 0) = \delta(x - x_0)\delta(v - v_0)$. We will neglect however the details of this proof. ■

9.4 Brownian Particle in a Force Field

Let us suppose now that our Brownian particle is embedded in an external force field, so that the system of Eqs. (9.3) and (9.4) becomes

$$dX(t) = V(t)\,dt \tag{9.27}$$

$$dV(t) = \gamma\,(X(t), t)\,dt - \alpha V(t)\,dt + dW_V(t) \tag{9.28}$$

where $\gamma(x, t)$ is a new term with the dimensions of an acceleration brought in to reckon our force field. This new system can again be rephrased as a unique vector *SDE* of the type (9.18) for $\boldsymbol{Z}(t)$ with the following coefficients

$$\boldsymbol{a}(z) = \boldsymbol{a}(x, v) = \begin{pmatrix} v \\ \gamma(x, t) - \alpha v \end{pmatrix} \qquad \mathbb{C} = \begin{pmatrix} 0 & 0 \\ 0 & 1 \end{pmatrix} \tag{9.29}$$

so that $\boldsymbol{Z}(t)$, as a solution of (9.18), still is a vector Markov process, but now the two equations of the system are apparently coupled in a way no longer allowing to solve them individually one after the other. A complete investigation of this problem would consequently put forward more difficulties w.r.t. the previous free case, so that instead of the general solutions we will rather investigate the possibility of extending—under suitable conditions—the approximate approach already presented in the Sect. 9.2. According to the Proposition 9.2 we know indeed that for a free Brownian motion a Wiener process on the configuration space (positions x) under suitable conditions is a good approximation for the position of the vector Markov process $\boldsymbol{Z}(t)$ on the phase space x, v. When instead the Brownian motion occurs in a force field such a Markovian approximation on the configuration space has been found by Smoluchowski and we will outline in the following its main features.

We start first by supposing that the force field i constant

$$\gamma(x, t) = \gamma_0$$

so that the two equations of our system become

$$dX(t) = V(t)\,dt \tag{9.30}$$
$$dV(t) = \left[\gamma_0 - \alpha V(t)\right] dt + dW_V(t) \tag{9.31}$$

and being no longer coupled they can be easily solved as in the free case. The extra constant γ_0 can indeed be reabsorbed with the following redefinition of the velocity process

$$V_\gamma(t) = V(t) - \frac{\gamma_0}{\alpha}$$

that now, instead of (9.31), satisfies the equation

$$dV_\gamma(t) = -\alpha V_\gamma(t)\,dt + dW_V(t)$$

that formally coincides with the Ornstein-Uhlenbeck Eq. (9.4) for $V(t)$ in the free case. The solution $V_\gamma(t)$ is then again of the form (9.12), and hence we deduce from (9.6) that, with an arbitrary initial condition and for times $t \gg 1/\alpha$, $V_\gamma(t)$ will asymptotically vanish, and consequently the velocity $V(t)$ will tend to the constant γ_0/α. We can then conclude that—after a short transient delay—the position $X(t)$ of the vector Markov process $\mathbf{Z}(t)$ will comply with the equation

$$dX(t) = \frac{\gamma_0}{\alpha}\,dt + dW_X(t) \tag{9.32}$$

that generalizes that of the free case (9.17), and whose solution simply is a Wiener process superposed to a constant drift $\gamma_0 t/\alpha$.

The next step consists then in the remark that this discussion hints to an extension of the previous result to the case of a field $\gamma(x,t)$ varying *slowly* w.r.t the time scale $1/\alpha$ characteristic of the model, so that it can be deemed roughly constant. We get in this way the ***Smoluchowski equation***

$$dX(t) = \frac{\gamma(X(t),t)}{\alpha}\,dt + dW_X(t) \tag{9.33}$$

that constitutes the ground for an approximate theory where the dynamics only appears as a drift term in a *SDE*, and the position becomes a Markov process. The Smoluchowski equation defines thus in a configuration space a *dynamical* theory with many important outcomes.

Exemple 9.6 Elastic restoring force: The Smoluchowski approximation provides acceptable solutions when the force field is a linear (elastic) restoring force

$$\gamma(x,t) = -\omega^2 x \tag{9.34}$$

so that the equations of the Ornstein-Uhlenbeck theory are

$$dX(t) = V(t)\,dt \tag{9.35}$$
$$dV(t) = -\omega^2 X(t)\,dt - \alpha V(t)\,dt + dW_V(t) \tag{9.36}$$

that is in a vector notation

$$d\boldsymbol{Z}(t) = \boldsymbol{a}(\boldsymbol{Z}(t), t)\,dt + \mathbb{C}(\boldsymbol{Z}(t), t)\; d\boldsymbol{W}(t)$$
$$\boldsymbol{a}(z) = \boldsymbol{a}(x, v) = \begin{pmatrix} v \\ -\omega^2 x - \alpha v \end{pmatrix} \qquad \mathbb{C} = \begin{pmatrix} 0 & 0 \\ 0 & 1 \end{pmatrix}$$

The solutions of (9.35) and (9.36) can be explicitly calculated (for details see [5]), but they are rather cumbersome and we will skip an explicit description of them. It is instead more interesting to point out that the account provided by the solution of the corresponding Smoluchowski equation

$$dX(t) = -\frac{\omega^2}{\alpha} X(t)\,dt + dW_X(t) \tag{9.37}$$

is indeed rather simple and accurate.[1] The Eq. (9.37)—with a suitable coefficient redefinition—looks in fact again as an Ornstein-Uhlenbeck Eq. (8.46) for the position $X(t)$ that now becomes a Gaussian Markov process with law $\mathfrak{N}(x_0 e^{-\omega^2 t/\alpha}, \beta^2(1 - e^{-2\omega^2 t/\alpha}))$ and with

$$\beta^2 = \frac{\alpha D_X}{2\omega^2} = \frac{kT}{m\omega^2}$$

where k is the Boltzmann constant and T the temperature. This process also has an asymptotic, invariant distribution $\mathfrak{N}(0, \beta^2)$ accounting for a situation where—either after a transient delay, or in an overdamped regime—our particle no longer diffuses endlessly because of the contrast exercised by the binding restoring force.

The scope of the Smoluchowski approximation (9.33) is not confined only to the case of the elastic restoring forces: a more comprehensive formulation, anticipated at the end of the Sect. 9.2, is presented in the next proposition where it has been deemed expedient to define the new velocity field

$$c(x, t) = \frac{\gamma(x, t)}{\alpha}$$

Proposition 9.7 *Under reasonable regularity conditions on* $c(x, t)$, *if* $X(t)$ *and* $V(t)$ *are the solutions of the* SDE *system*

[1] The behavior of a Brownian motion under the effect of an elastic restoring force has also been empirically investigated with a few clever experiments by E. Kappler [6] confirming the idea that the Smoluchowski approximation holds well when the drag α is large.

$$dX(t) = V(t)\,dt \qquad\qquad X(0) = x_0$$
$$dV(t) = \alpha c(X(t), t)\,dt - \alpha V(t)\,dt + \alpha dW(t) \qquad V(0) = v_0$$

while Y(t) is the solution of the SDE

$$dY(t) = c(Y(t), t)\,dt + dW(t) \qquad Y(0) = x_0$$

then, for every given v_0, *we have*

$$\lim_{\alpha\to\infty} X(t) = Y(t) \qquad \boldsymbol{P}\text{-}a.s.$$

uniformly in t in every compact of $[0, +\infty)$

Proof Omitted: see [7] p. 71 ■

9.5 Boltzmann Distribution

In the Smoluchowski equation for a Brownian particle in a force field

$$dX(t) = \frac{\gamma(X(t), t)}{\alpha}\,dt + dW(t) \tag{9.38}$$

the external dynamics embodied by $\gamma(x, t)$ only appears in the form of the drift velocity $c = \gamma/\alpha$, while it is completely missing in the diffusion term $b = 1$. In the present section we will consider the case of time-independent force fields endowed with a *potential energy* $\phi(x)$ such that

$$m\gamma(x) = -\phi'(x) \tag{9.39}$$

As a consequence the Eq. (9.38) becomes

$$dX(t) = -\frac{\phi'(X(t))}{\alpha m}\,dt + dW(t)$$

On the other hand from (6.76) and (9.5) we get

$$\frac{1}{\alpha m} = \frac{D}{2kT} = \frac{D\beta}{2}$$

where the thermodynamic parameter $1/kT$ traditionally designated as β (a notation that we deemed better to maintain here) must not be misinterpreted as the homonym parameter of the Ornstein-Uhlenbeck process of the previous sections. As a consequence the Smoluchowski equation takes the form

$$dX(t) = -\frac{D}{2}\beta\phi'(X(t))\,dt + dW(t)$$

and hence from the Proposition 8.19 with $a(x,t) = -\frac{D}{2}\beta\phi'(x)$, and $b(x,t) = 1$ we find the following Fokker-Planck equation for our Brownian motion in a potential $\phi(x)$

$$\partial_t f(x,t) = \frac{D}{2}\,\partial_x[\beta\phi'(x)f(x,t)] + \frac{D}{2}\,\partial_x^2 f(x,t) \tag{9.40}$$

Proposition 9.8 *When it exists, the stationary solution of the Eq.* (9.40) *is the* ***Boltzmann distribution***

$$f(x) = \frac{e^{-\beta\phi(x)}}{Z(\beta)} \tag{9.41}$$

where the normalization constant

$$Z(\beta) = \int_{-\infty}^{+\infty} e^{-\beta\phi(x)}\,dx \tag{9.42}$$

is also called ***partition function.***

Proof To check first that the Boltzmann distribution is a solution of (9.40) it is enough to remark from (9.41) that $\partial_t f = 0$, and then that $\partial_x f = -\beta\phi' f$. If conversely $f(x)$ is a stationary solution of (9.40), we first have $\partial_t f = 0$ and then from (9.40) we find that $f(x)$ must satisfy the first order equation

$$\beta\phi'(x)f(x) + f'(x) = C$$

where C is an integration constant. Since on the other hand $f(x)$ must be an integrable *pdf*, the function $f(x)$ must vanish for $x \to \pm\infty$. Assuming then that f with its first derivative vanishes at the infinity fast enough to make the left hand side of our equation infinitesimal as a whole for $x \to \pm\infty$, we will get $C = 0$ and hence the stationary f must in fact satisfy the equation

$$\beta\phi'(x)f(x) + f'(x) = 0$$

whose solution can be obtained with elementary methods and, after normalization, coincides with the Boltzmann distribution (9.41). ∎

Exemple 9.9 Elastic restoring force (continuation): Resuming the discussion of the Example 9.6 with γ as in (9.34), we unsurprisingly find from (9.39) that ϕ is the harmonic oscillator potential

$$\phi(x) = \frac{1}{2}m\omega^2 x^2 \tag{9.43}$$

and hence, within the notation adopted in this section, the Smoluchowski Eq. (9.37) becomes

$$dX(t) = -\frac{D}{2}\,\beta m\omega^2 X(t)\,dt + dW(t) \tag{9.44}$$

From the Proposizione 9.8 we then find the Boltzmann distribution

$$Z(\beta) = \sqrt{\frac{2\pi}{\beta m\omega^2}} = \sqrt{\frac{2\pi kT}{m\omega^2}}$$

$$f(x) = \frac{e^{-\frac{1}{2}\beta m\omega^2 x^2}}{\sqrt{\frac{2\pi}{\beta m\omega^2}}} = \frac{e^{-m\omega^2 x^2/2kT}}{\sqrt{\frac{2\pi kT}{m\omega^2}}}$$

that apparently coincide with the stationary solution $\mathfrak{N}\left(0, \frac{kT}{m\omega^2}\right)$ of the Ornstein-Uhlenbeck Eq. (9.37) investigated in the previous section.

Exemple 9.10 Weight: Take now a negative constant acceleration $\gamma(x) = -g$ for a process confined to the positive half-line $x \geq 0$. Supposing that x represent the height of a corpuscle above a floor placed in $x = 0$, this model will describe the distribution of the Brownian particles under the effect of the weight. We thus obtain from (9.39) a potential $\phi(x) = mgx$, while the Smoluchowski Eq. (9.38) becomes

$$dX(t) = -\frac{D}{2}\,\beta mg\,dt + dW(t)$$

namely has constant coefficient, a case already discussed in the Sect. 8.5.1 but for the fact that now we must impose the additional condition $x \geq 0$, that is $f = 0$ for $x < 0$, so that now the solution can no longer be Gaussian. The corresponding Fokker-Planck Eq. (9.40) is

$$\partial_t f(x,t) = \frac{D}{2}\,\beta mg\,\partial_x f(x,t) + \frac{D}{2}\,\partial_x^2 f(x,t) \qquad x \geq 0$$

and from the Proposition 9.8 we get the stationary solution

$$Z(\beta) = \frac{1}{\beta mg} = \frac{kT}{mg}$$

$$f(x) = \beta mg\, e^{-\beta mgx}\vartheta(x) = \frac{mg}{kT}\, e^{-mgx/kT}\vartheta(x)$$

where ϑ is the Heaviside function (2.13): the invariant distribution is then an exponential $\mathfrak{E}(\beta mg) = \mathfrak{E}\left(\frac{mg}{kT}\right)$ accurately accounting for the upward thinning halo of minute particles in a fluid suspension.

The Proposition 9.8 also enables us to solve a simple problem of **reverse engineering**: *find the potential* ϕ *acting on a Brownian particle and resulting in a given Boltzmann stationary distribution* (9.41).

Exemple 9.11 Student distributions: The family $\mathfrak{T}(\beta)$ of Boltzmann *pdf*'s

$$f(x) = \frac{1}{a\,B\left(\frac{1}{2},\,\frac{\beta m\omega^2 a^2 - 1}{2}\right)}\left(\frac{a^2}{a^2+x^2}\right)^{\frac{1}{2}\beta m\omega^2 a^2} = \frac{e^{-\beta\phi(x)}}{Z(\beta)} \tag{9.45}$$

generalizes that of the Student distributions $\mathfrak{T}_n$ introduced in the Sect. 3.5.2: here $a > 0$ is a characteristic length, $\omega > 0$ is a parameter epitomizing the external potential intensity and

$$B(x, y) = \frac{\Gamma(x)\Gamma(y)}{\Gamma(x+y)}$$

is the Riemann beta function. These distributions are well defined when $\beta m\omega^2 a^2 > 1$, namely if $m\omega^2 a^2 > kT$: this points out that our stationary solutions exist only if the said balance between the potential strength ω and the temperature T is conformed to. From (9.45) we see at once that

$$\phi(x) = \frac{1}{2}m\omega^2 a^2 \ln\left(1 + \frac{x^2}{a^2}\right) \qquad Z(\beta) = a\,B\left(\frac{1}{2},\,\frac{\beta m\omega^2 a^2 - 1}{2}\right) \tag{9.46}$$

so that the Smoluchowski equation becomes

$$dX(t) = -\frac{D}{2}\,\beta m\omega^2 X(t)\,\frac{a^2}{a^2 + X^2(t)}\,dt + dW(t) \tag{9.47}$$

It is illuminating to look to analogies and differences between the Student stationary solution of the Smoluchowski Eq. (9.47), and the Gaussian stationary solution of the Smoluchowski Eq. (9.44). In the Fig. 9.1 two examples of these stationary *pdf*'s are portrayed along with the potentials $\phi(x)$ yielding them: in both these instances the parameters a, ω and T have the same values. The two potentials (9.43) and (9.46)

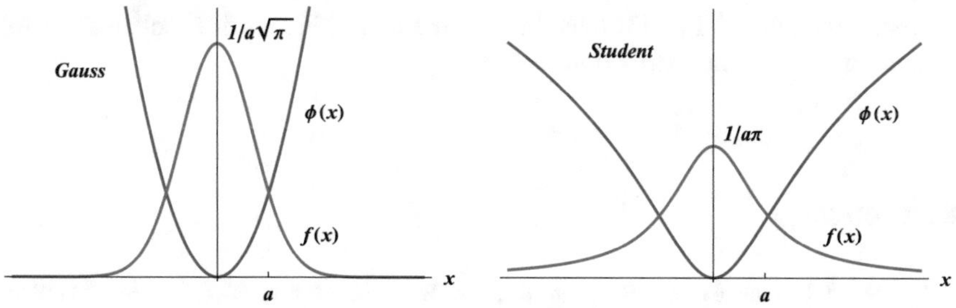

Fig. 9.1 Gauss and student stationary distributions respectively for the Smoluchowski Eqs. (9.44) and (9.47). The temperature T is chosen in such a way that $2kT = m\omega^2 a^2$, entailing in particular that the Student law is in fact a Cauchy. The energy units instead are conventionally fixed in order to make comparable the superposed curves

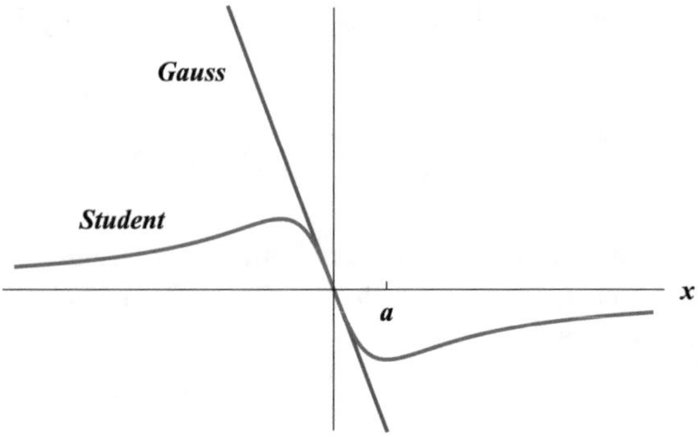

Fig. 9.2 The drift velocities (9.48) respectively for the Smoluchowski Eqs. (9.44) and (9.47). We adopted the same parameter values used in the Fig. 9.1

approximately coincide near to $x = 0$, but for $x \to \pm\infty$ they diverge with distinctly different speed: in the harmonic case the potential (9.43) grows as x^2, while in the Student instance (9.46) it only increases as $\ln x$. From a physical standpoint it is exactly this feature that results in the difference between the two stationary distributions: the harmonic potential (9.43), being more strong and binding, provides indeed Gaussian stationary distributions that visibly are more piled up in $x = 0$ than the Student laws (look also in the Fig. 9.1 at the different behavior of the tails).

Since finally in the Smoluchowski approximation the dynamical effects only appear in the drift velocities $a(x)$, it is also telling to compare their expressions

$$-\frac{D}{2}\beta m\omega^2 x \qquad\qquad -\frac{D}{2}\beta m\omega^2 x\,\frac{a^2}{a^2+x^2} \tag{9.48}$$

respectively derived from the Smoluchowski Eqs. (9.44) and (9.47). We displayed their behaviors in the Fig. 9.2: both the velocity fields drag the Brownian particle toward the center $x = 0$ from every other location on the x axis; while however in the Gaussian case (9.44) the pull is always the same at every distance from $x = 0$, for the Student laws (9.47) it attains a maximum value at a distance a from the center and then asymptotically vanishes. Here again the juxtaposition shows in what sense the harmonic potential (9.43) must be deemed more binding than potential (9.46) producing the Student distributions.

References

1. Ornstein, L.S., Uhlenbeck, G.E.: On the theory of Brownian Motion. Phys. Rev. **36**, 823 (1930)
2. Einstein, A.: Über die von der molekularkinetischen Theorie der Wärme geforderte Bewegung von in ruhenden Flüssigkeiten suspendierten Teilchen. Ann. Phys. **17**, 549 (1905)
3. von Smoluchowski, M.: Zur kinetischen Theorie der Brownschen Molekularbewegung und der Suspensionen. Ann. Phys. **21**, 757 (1906)
4. Neckel, T., Rupp, F.: Random differential equations in scientific computing. Versita, London (2013)

5. Chandrasekhar, S.: Stochastic problems in physics and astronomy. Rev. Mod. Phys. **15**, 1 (1943)
6. Kappler, E.: Versuche zur Messung der Avogadro-Loschmidtschen Zahl aus der Brownschen Bewegung einer Drehwaage. Ann. Phys. **11**, 233 (1931)
7. Nelson, E.: Dynamical Theories of Brownian Motion. Princeton, Princeton UP, (1967)

Chapter 10
Stochastic Mechanics

10.1 Madelung Decomposition

In a paper of 1926 Madelung [1] famously remarked that the Schrödinger equation

$$i\hbar\,\partial_t\psi(\boldsymbol{x},t) = -\frac{\hbar^2}{2m}\nabla^2\psi(\boldsymbol{x},t) + \Phi(\boldsymbol{x})\psi(\boldsymbol{x},t) \tag{10.1}$$

ruling in a 3-dimensional space the evolution of the complex-valued wave function $\psi(\boldsymbol{x},t)$ describing the state of a quantum particle with an external potential $\Phi(\boldsymbol{x})$, can be recast in a new form involving only real functions: if indeed we define $R(\boldsymbol{x},t)$ and $S(\boldsymbol{x},t)$ respectively as the (real) modulus and phase of $\psi(\boldsymbol{x},t)$, namely

$$\psi(\boldsymbol{x},t) = R(\boldsymbol{x},t)\,e^{iS(\boldsymbol{x},t)/\hbar}$$

by separating the real and imaginary parts of (10.1) after a little algebra we find

$$\partial_t S + \frac{m}{2}\left(\frac{\nabla S}{m}\right)^2 + \Phi - \frac{\hbar^2}{2m}\frac{\nabla^2 R}{R} = 0 \tag{10.2}$$

$$\partial_t R^2 + \nabla\left(R^2\,\frac{\nabla S}{m}\right) = 0 \tag{10.3}$$

If now $R^2(\boldsymbol{x},t) = |\psi(\boldsymbol{x},t)|^2$ is interpreted as the density of some fluid and

$$\boldsymbol{v}(\boldsymbol{x},t) = \frac{\nabla S(\boldsymbol{x},t)}{m} \tag{10.4}$$

as its current velocity, then (10.3) can be considered as a *continuity equation* expressing the conservation of the fluid, while the Eq. (10.2) plays the role of a *Hamilton-Jacobi equation* ruling the dynamics of the velocity potential S in the presence of a potential Φ plus an additional *quantum potential* dependent on the wave function

N. Cufaro Petroni, *Probability and Stochastic Processes for Physicists*, UNITEXT for Physics, https://doi.org/10.1007/978-3-030-48408-8_10

$$\Phi_q(\boldsymbol{x}, t) = -\frac{\hbar^2}{2m} \frac{\nabla^2 R(\boldsymbol{x}, t)}{R(\boldsymbol{x}, t)}$$

This ***Madelung decomposition*** has played a leading role in the *causal interpretation* of the quantum mechanics [2, 3]: in that theory a non relativistic particle of mass m, whose wave function obeys the Schrödinger equation (10.1), is considered as a classical object following a continuous and causally defined trajectory with a well defined position and accompanied by a physically real wave field which contributes to determine its motion through the Hamilton-Jacobi equation (10.2). In particular the trajectories are dictated by (10.4) when $\boldsymbol{v}$ is identified with the velocity of the particle passing through $\boldsymbol{x}$ at the time t. Naturally enough the causal interpretation was also obliged to add some randomness to its deterministic description in order to reproduce the statistical predictions of the quantum mechanics: this was done however just by taking the function $|\psi(\boldsymbol{x}, 0)|^2$ as the initial *pdf* of an ensemble of particles, and then letting it to evolve according to the Eqs. (10.3) and (10.2). The uncertainty on the particle position was therefore confined only to that of their initial values with no further perturbation added along their deterministic trajectories, in a way reminiscent of the Liouville systems discussed above in the Sect. 7.2.3.

While this statistical addendum to the causal interpretation looks only as made by hand out of necessity, it is striking to remark that, if we introduce the following drift vector

$$\boldsymbol{A}(\boldsymbol{x}, t) = \frac{\nabla S}{m} + \frac{\hbar}{2m} \frac{\nabla R^2}{R^2}$$

it is easy to see that the continuity equation (10.3) turns out to take the form

$$\partial_t R^2 = -\nabla(\boldsymbol{A}\, R^2) + \frac{\hbar}{2m} \nabla^2 R^2 \tag{10.5}$$

which apparently coincides with a Fokker-Planck equation

$$\partial_t f(\boldsymbol{x}, t) = -\nabla\big[\boldsymbol{A}(\boldsymbol{x}, t) f(\boldsymbol{x}, t)\big] + \frac{D}{2} \nabla^2 f(\boldsymbol{x}, t) \tag{10.6}$$

of the type (7.81) for some hypothetical Markovian vector diffusion process $\boldsymbol{X}(t)$, provided that we adopt as constant diffusion coefficient

$$\mathbb{B} = D = \frac{\hbar}{m}$$

This obviously hints at a possible connection between the density R^2 of the Madelung fluid and the *pdf* of a diffusion process describing the random motion of a classical particle. Taken in itself, however, even this association looks rather awkward: for one thing, while from the Sect. 7.2 we know that with a given $\boldsymbol{A}$ we can find an infinity of solutions of the Fokker-Planck equation (10.6) (one for every initial condition), the quantum mechanics always seems to select just one particular *pdf* among them, namely $f = R^2$ as in (10.5). It must be emphasized indeed that in a Madelung decomposition R^2 and $\boldsymbol{A}$ (or equivalently R and S) are not separate entities: they are both derived from a unique ψ solution of (10.1) and hence they are locked together by their common origin. In other words: not every pair $\boldsymbol{A}, R^2$—with R^2 solution of (10.5)—can be considered as derived from a ψ solution of the Schrödinger equation (10.1). While the mathematical rationale for this binding can of course be traced back to the second, Hamilton-Jacobi equation (10.2) coupled with the Fokker-Planck (10.6) and representing the dynamics of the system, it is apparent at this point that the origin of this dynamics is rather puzzling, so that the whole construction is likely to appear still purely formal.

A new path leading from the classical random phenomena to the Schrödinger wave functions was opened only later by the *stochastic mechanics* pioneered in the 60s by Nelson [4–7]: in this theory the particle flights are those of a Markov process on some probabilistic space with its *pdf*'s ruled by a Fokker-Planck equation, and its trajectories solutions of an Itō *SDE*. Here however, at variance with what has been discussed in the previous chapters, we will deal with processes whose drift velocity is not given a priori, and plays instead the role of a new dynamical variable following an adaptation of the Newton equations of motion allowed by a shrewd definition of its kinematics (velocity and acceleration). The state of the particle being defined now not by the process *pdf* alone, but also by its velocity fields, we will show in the following that a well-suited combination of these elements will provide complex-valued functions satisfying a Schrödinger equation.

Of course we pay this by abandoning the idea of deterministic trajectories peculiar to the causal interpretation, but that notwithstanding the stochastic mechanics still lends us the possibility of looking into the quantum world by keeping intact a description by means of continuous, space-time paths, recovering finally the deterministic trajectories only as averages of the stochastic ones. In a subsequent development [8] this theory has been profitably recast in terms of a stochastic variational principle, but here we will stick for simplicity to the original Nelson formulation with the Newton equations often adopted also in the recent literature (see [9], Sect. 3.8), and—to keep things easy without losing too much of the physical meaning—we will present a simplified account of the results of the stochastic mechanics confining ourselves to the case of *ac*, one-dimensional Markov processes. At variance with the said original treatment, moreover, we will chiefly rely on the use of the Fokker-Planck equations rather than on an application of the stochastic calculus and of the *SDE*'s: the connection between these two possible standpoints, on the other hand, could be easily recovered by taking into account the results summarized in the Sect. 8.4.3.

10.2 Retarded and Advanced Equations

Given a diffusion process $X(t)$ with $t \geq 0$ ruled by an Itō *SDE* (8.26) we know from the Proposition 8.19 that its transition *pdf*'s $f(x, t|y, s)$ in the *retarded* region $0 \leq s \leq t$ satisfies the corresponding Fokker-Planck equation either as a ***forward equation***

$$\partial_t f(x,t|y,s) = -\partial_x\left[A(x,t)f(x,t|y,s)\right] + \frac{1}{2}\,\partial_x^2\left[B(x,t)f(x,t|y,s)\right] \quad (10.7)$$

with initial conditions $f(x,s^+|y,s) = \delta(x-y)$, or as a ***backward equation***

$$\partial_s f(x,t|y,s) = -A(y,s)\partial_y f(x,t|y,s) - \frac{1}{2}\,B(y,s)\partial_y^2 f(x,t|y,s) \quad (10.8)$$

with final conditions $f(x,t|y,t^-) = \delta(x-y)$ as discussed in the Sect. 7.2.2. Since all the *pdf*'s $f(x,t|y,s)$ involved in the Eqs. (10.7) and (10.8) are taken for $s \le t$, namely in the *retarded* region, from now on these equations will both be called ***retarded equations***. The difference between the labels *forward* and *backward* will only point to the fact that in the forward equation (10.7) the variables are the final ones x, t (y, s being only external parameters fixing the *initial* conditions), while in the backward equation (10.8) the variables are the initial y, s (x, t representing now the *final* conditions). In other words the same retarded $f(x,t|y,s)$ is a solution of both (10.7) and (10.8), but only for $s \le t$. It is important to remark at once, on the other hand, that given a retarded solution $f(x,t|y,s)$ for $s \le t$, it would be easy to see from simple examples that it can not be trivially extend into the advanced region $t \le s$ because we would in general find functions which are meaningless as *pdf*'s.

We recall moreover from the Theorem 7.32 that also the *one-time pdf* $f(x,t)$ of our $X(t)$ is a solution of the Fokker-Planck equation

$$\partial_t f(x,t) = -\partial_x\left[A(x,t)f(x,t)\right] + \frac{1}{2}\,\partial_x^2\left[B(x,t)f(x,t)\right] \quad (10.9)$$

This will be called here ***direct equation*** in so far as we look for its solutions from some instant $s \ge 0$ *onward* to $t \ge s$ by selecting an arbitrary *initial condition* $f(x,s) = f_s(x)$. In particular, by choosing $f_s(x) = \delta(x-y)$ we would recover the retarded ($t \ge s \ge 0$) transition *pdf*'s solutions of (10.7), so that the forward Fokker-Planck equation (10.7) for the transition *pdf* turns out to be nothing else than the direct Fokker-Planck equation (10.9) for suitable initial conditions at a time s.

We have already remarked more than once that the knowledge of the *retarded* transition *pdf* $f(x,t|y,s)$ with $t \ge s$ is all we need—along with an arbitrary initial law—to determine the finite dimensional distributions of the process according to the chain rule (7.14). These transition laws, however, do exist also in the *advanced* region $0 \le t \le s$ and their behavior is not without interest, as we will see later in the discussion about the Nelson derivatives of the process. As a matter of fact the *advanced* equations for the *advanced* $f(x,t|y,s)$ with $0 \le t \le s$ could be deduced by reproducing a modified proof of the Theorem 7.32 leading to the retarded equations, but we prefer to follow here a simpler way around. It is easy to understand indeed that the advanced *pdf*'s $f(x,t|y,s)$ for $0 \le t \le s$ can be trivially obtained from the retarded ones via the following continuous version of the ***Bayes formula*** directly resulting from the definition of conditional *pdf*

$$f(x,t|y,s) = \frac{f(y,s|x,t)f(x,t)}{f(y,s)} \tag{10.10}$$

If in fact we take $t \le s$, the calculation of the right hand side only requires the knowledge of the retarded *pdf*'s. Remark however that the advanced *pdf*'s defined in this way inherently depend on $f(x,t)$, and hence on the initial conditions $f_0(x)$ fixed at $t = 0$: a point that will be reconsidered later.

Proposition 10.1 *Take a diffusion process $X(t)$ whose transition pdf $f(x,t|y,s)$ in the retarded region is a solution of the forward Fokker-Planck equation* (10.7) *for $0 \le s \le t$: then, if $f(x,t)$ is the pdf satisfying the Fokker-Planck equation* (10.9) *for a fixed initial condition $f(x,0) = f_0(x)$, the transition pdf in the advanced region $0 \le t \le s$ satisfies the (forward)* ***advanced equation*** *with final condition*

$$\partial_t f(x,t|y,s) = -\partial_x\left[\widehat{A}(x,t)f(x,t|y,s)\right] - \frac{1}{2}\,\partial_x^2\left[B(x,t)f(x,t|y,s)\right] \tag{10.11}$$

$$\widehat{A}(x,t) = A(x,t) - \frac{\partial_x[B(x,t)f(x,t)]}{f(x,t)} \tag{10.12}$$

$$f(x,s^-|y,s) = \delta(x-y) \tag{10.13}$$

Proof In order to deduce the advanced equation (10.11) it will be useful first to rewrite the retarded, backward equation (10.8) by swapping the symbols $s \leftrightarrow t$ and $x \leftrightarrow y$, so that, with $t \le s$, we will have now

$$\partial_t f(y,s|x,t) = -A(x,t)\partial_x f(y,s|x,t) - \frac{1}{2}\,B(x,t)\partial_x^2 f(y,s|x,t) \tag{10.14}$$

Then with $0 \le t \le s$, by using (10.10), (10.14), (10.9) and (10.10) again we get

$$\begin{aligned}
\partial_t f(x,t|y,s) &= \frac{f(x,t)}{f(y,s)}\,\partial_t f(y,s|x,t) + \frac{f(y,s|x,t)}{f(y,s)}\,\partial_t f(x,t)\\
&= \frac{f(x,t)}{f(y,s)}\left\{-A(x,t)\partial_x f(y,s|x,t) - \frac{1}{2}\,B(x,t)\partial_x^2 f(y,s|x,t)\right\}\\
&\quad + \frac{f(y,s|x,t)}{f(y,s)}\left\{-\partial_x\left[A(x,t)f(x,t)\right] + \frac{1}{2}\,\partial_x^2\left[B(x,t)f(x,t)\right]\right\}\\
&= f(x,t)\left\{-A(x,t)\partial_x\frac{f(x,t|y,s)}{f(x,t)} - \frac{1}{2}\,B(x,t)\partial_x^2\frac{f(x,t|y,s)}{f(x,t)}\right\}\\
&\quad + \frac{f(x,t|y,s)}{f(x,t)}\left\{-\partial_x\left[A(x,t)f(x,t)\right] + \frac{1}{2}\,\partial_x^2\left[B(x,t)f(x,t)\right]\right\}\\
&= -\partial_x\left[A(x,t)f(x,t|y,s)\right]\\
&\quad -\frac{1}{2}\,\partial_x\left\{B(x,t)f(x,t)\partial_x\frac{f(x,t|y,s)}{f(x,t)} - \frac{f(x,t|y,s)}{f(x,t)}\partial_x\left[B(x,t)f(x,t)\right]\right\}
\end{aligned}$$

The second term can now be recast in a more telling form: to see that within a simplified notation we first call $f = f(x,t|y,s)$ and $g = f(x,t)$, and then we have

$$
\begin{aligned}
-\frac{1}{2}\partial_x\left[Bg\partial_x\frac{f}{g}-\frac{f}{g}\partial_x(Bg)\right] &= -\frac{1}{2}\partial_x\left[\left(B\partial_x f - Bf\frac{\partial_x g}{g}\right)-\left(Bf\frac{\partial_x g}{g}+f\partial_x B\right)\right]\\
&= -\frac{1}{2}\partial_x\left(B\partial_x f + f\partial_x B - 2f\partial_x B - 2fB\frac{\partial_x g}{g}\right)\\
&= -\frac{1}{2}\partial_x\left[\partial_x(Bf) - 2\left(\partial_x B + B\frac{\partial_x g}{g}\right)f\right]\\
&= -\frac{1}{2}\partial_x^2(Bf) + \partial_x\left[\frac{\partial_x(Bg)}{g}f\right]
\end{aligned}
$$

so that, restoring the complete notation, we finally find the sought extension of the forward equation into the advanced region $0 \le t \le s$:

$$
\partial_t f(x,t|y,s) = -\partial_x\left\{\left[A(x,t) - \frac{\partial_x[B(x,t)f(x,t)]}{f(x,t)}\right]f(x,t|y,s)\right\} - \frac{1}{2}\partial_x^2\left[B(x,t)f(x,t|y,s)\right]
$$

with (10.13) as extremal (final) condition. ■

In short, the complete transition *pdf* $f(x,t|y,s)$ for every $s,t \ge 0$ will obey two different equations—namely (10.7) and (10.11)—respectively in the retarded ($t \ge s$), and in the advanced ($0 \le t \le s$) regions. Remark that it could be deemed preposterous here to call *forward* the Eq. (10.11) since it holds for $t \le s$; we will retain however this name because for us the *forward* qualification refers chiefly to the fact that the variables involved in this equation are the actual variables x, t of the *pdf*, and not the conditioning variables y, s as in the *backward* equations.

In analogy with what it has been done to get the Fokker-Planck equation (10.9) for the one-time *pdf* (see Theorem 7.32) we can also eliminate the conditioning in the Eq. (10.11) multiplying it by $f(y,s)$ and then integrating in dy to get

$$
\partial_t f(x,t) = -\partial_x\left[\widehat{A}(x,t)f(x,t)\right] - \frac{1}{2}\partial_x^2\left[B(x,t)f(x,t)\right] \tag{10.15}
$$

that will now be named ***reverse equation*** because we understand here that it is always solved from some instant $s \ge 0$ *rearward* to $t \le s$ with a *final condition* $f(x,s) = f_s(x)$. Despite its external appearance, however, this equation in fact coincides with (10.9) so that no inconsistency can arise: it is easy to see indeed that, using the definition (10.12) of $\widehat{A}$ and simple algebraic manipulations, we exactly recover the Eq. (10.9) from (10.15). This entails therefore that, given $s \ge 0$, the advanced *pdf* is again a solution of (10.9) (or equivalently of (10.15)) for $t \in [0, s]$, but the new fact is that now we have extremal conditions *at both the endpoints* 0 and s (implementing in so doing a ***stochastic bridge***, see Appendix M): for an advanced, transition *pdf* the natural conditions in s will be $f(x, s^-|y,s) = \delta(x-y)$, while in $t = 0$ the condition entailed by the Bayes theorem (10.10)

$$f(x,0|y,s) = \frac{f(y,s|x,0)f(x,0)}{f(y,s)} = \frac{f(y,s|x,0)f_0(x)}{\int_{\mathbf{R}} f(y,s|z,0)f_0(z)\,dz}$$

is apparently fixed by the choice of an arbitrary initial condition $f_0(x)$. This important point will be discussed in further detail in the examples of the Appendix N.

The previous remarks also emphasize that the right to look into the advanced region of a process always comes with a price: to do that, indeed, we must fix an $f(x,t)$ for every $t \geq 0$—or equivalently an initial condition $f_0(x)$—and in so doing we forfeit the right to choose a different condition in $t = 0$. This in particular implies that the *complete* transition *pdf* $f(x,t|y,s)$ for $s,t \geq 0$ can no longer be used to produce arbitrary process distributions from arbitrary initial conditions through the Chapman-Kolmogorov equation (7.16): it always requires indeed the selection of one initial *pdf* $f_0(x)$. A different f_0 always means a different *complete* transition *pdf*. Its avail is therefore restricted now to the chain rule (7.14) with the aim of finding all the joint laws of the unique process we selected to look at by choosing an $f_0(x)$ in order to have the right to glance into the advanced region.

If however we compare our *reverse* equation (10.15) with the *direct* equation (10.9), we see that in (10.15) the diffusive term (namely that depending on B) displays a *wrong sign* with respect to the sign of the time derivative: as a consequence the reverse equation (10.15) in its present formulation is not a legitimate Fokker-Planck equation. In fact, since a diffusion coefficient must be positive, this difference can not be healed by simply taking $-B(x,t)$ as a new diffusion coefficient for (10.15). By taking instead $-\widehat{A}(x,t)$ as a new drift term, the reverse equation (10.15) becomes

$$-\partial_t f(x,t) = -\partial_x\left[-\widehat{A}(x,t)f(x,t)\right] + \frac{1}{2}\,\partial_x^2\left[B(x,t)f(x,t)\right]$$

which apparently formally is a Fokker-Planck equation with a positive diffusion term, but for the fact that it now displays the *wrong sign* in front of the time derivative in the left hand side. This suggests to operate a ***time reversal***, namely an inversion of the time direction, in order to conclusively recast the reverse equations in the legitimate form of a Fokker-Planck equation similar—but for the revised coefficients—to the direct one: this is an important point that will be resumed in the Sect. 10.4.1 where we will briefly discuss how the stochastic mechanics can manage to have (typically irreversible) diffusion processes reproducing the behavior of the (time-reversal invariant) Schrödinger equation. To perform the proposed time reversal just replace t by $-t$, and define

$$f^*(x,t) \equiv f(x,-t) \tag{10.16}$$

$$A^*(x,t) \equiv -\widehat{A}(x,-t) = -A(x,-t) + \frac{\partial_x[B(x,-t)f(x,-t)]}{f(x,-t)} \tag{10.17}$$

$$B^*(x,t) \equiv B(x,-t) > 0 \tag{10.18}$$

It is easy to see then that in this way the reverse equation (10.15) becomes

$$\partial_t f^*(x,t) = -\partial_x \left[A^*(x,t)f^*(x,t)\right] + \frac{1}{2}\partial_x^2 \left[B^*(x,t)f^*(x,t)\right] \tag{10.19}$$

which in fact is now a full-fledged Fokker-Planck equation that, with an arbitrary $s \geq 0$, is understood to be solved in the advanced region $0 \leq t \leq s$ with two extremal conditions in s and 0, implementing in this way a stochastic bridge as discussed before.

Remark that embedded in the result (10.19) there is our taking into account of both the advanced and the retarded regions of the process. It is crucial indeed in the transformation (10.17) the same knowledge of the *pdf* $f(x,t)$ needed to look into the advanced region: as already remarked in the the present section, for every choice of this *pdf* (namely of an initial condition $f_0(x)$) we have both a different advanced transition *pdf*, and a different form of the drift term in Eq. (10.19). This flexibility of the drift turns out to be paramount to account for the existence of the time-reversal invariant diffusions of the stochastic mechanics as later discussed in the Sect. 10.4.1. By summarizing, the Eq. (10.19) is no longer autonomous from the initial conditions, because its form, in particular that of the coefficient A^* explicitly depends on the choice of $f_0(x)$. In some sense this is not surprising since also the advanced transition *pdf* in Eq. (10.10) depends on the choice of $f_0(x)$. In other words there are many possible advanced extensions (from $t \geq s$ to $0 \leq t \leq s$) of a retarded $f(x,t|y,s)$: one for every possible initial condition $f_0(x)$; and as a consequence we get as many advanced (and time-reversed) equations too. On the other hand, the process distribution being unique, once an initial condition $f_0(x)$ is chosen for the retarded equations we are no longer free to chose another arbitrary condition in $t = 0$ for the time-reversed equation: advanced and retarded equations being but two sides of the same coin, when we fix the details for one representation we also settle the corresponding details for the other.

10.3 Kinematics of a Diffusion Process

10.3.1 Forward and Backward Mean Derivatives

Definition 10.2 The **forward** and **backward mean derivatives** (see [4], Chap. 11) of a Markov process $X(t)$ are the processes defined—when they exist—respectively as the ms limits

$$\vec{\partial} X(t) = \lim_{\Delta t \to 0^+}\text{-}ms\ \boldsymbol{E}\left[\left.\frac{X(t+\Delta t) - X(t)}{\Delta t}\right| X(t)\right] \tag{10.20}$$

$$\overleftarrow{\partial} X(t) = \lim_{\Delta t \to 0^+}\text{-}ms\ \boldsymbol{E}\left[\left.\frac{X(t) - X(t-\Delta t)}{\Delta t}\right| X(t)\right] \tag{10.21}$$

Here and in the following the arrows are not vector labels: they only denote the forward or backward type of the derivatives.

Apparently the existence of these processes is not a foregone conclusion: not only the limits, but also the expectations themselves could not exist. On the other hand, when they turn out to be well defined, these two processes respectively represent the *mean conditional right and left derivatives* of the (usually non derivable) process $X(t)$, and they will play in the following the role of mean ***forward*** and ***backward velocity processes***. It is rather unfortunate that these *forward* and *backward* qualifications partially overlap with that of the equations discussed in the Sect. 10.2: we could have used instead the names of *right* and *left* derivatives and velocities but, in view of the fact that they are known with their names since longtime, we decided to stay true to this tradition. Of course, when they exist, the forward and backward derivatives do not coincide unless the process $X(t)$ is in some sense differentiable.

Moreover, beside the forward and backward velocities, it is also possible to define a ***mean second derivative***

$$\overset{\leftrightarrow}{\partial}{}^2 X(t) \equiv \frac{\overleftarrow{\partial}\overrightarrow{\partial} + \overrightarrow{\partial}\overleftarrow{\partial}}{2} X(t) = \frac{\overleftarrow{\partial}\left(\overrightarrow{\partial} X(t)\right) + \overrightarrow{\partial}\left(\overleftarrow{\partial} X(t)\right)}{2} \tag{10.22}$$

that will play the role of a ***mean acceleration process***. Of course, when the forward and backward derivatives coincide this would be just the average of their second forward and backward derivatives, but it is important to remark that in the general case the expression (10.22) is just one among several possible forms of a mean acceleration that can be surmised by combining two subsequent forward and backward derivatives: the motivations in favor of this particular choice (see [4], Chap. 12) boil down to the argument that this acceleration will allow us to formulate the equilibrium dynamics of a process by means of a Newton equation in the form

$$m\,\overset{\leftrightarrow}{\partial}{}^2 X(t) = F(t) \tag{10.23}$$

where $F(t)$ is a random *external* force, in the sense that now a *free* stochastic motion—*noisy* environment notwithstanding—will (at least asymptotically when non stationary: see later a few explicit calculations in the Appendix N) obey the equation

$$\overset{\leftrightarrow}{\partial}{}^2 X(t) = 0$$

There is indeed a precise conceptual interest in making the average acceleration (10.22) exactly zero for a free (namely not acted upon by any *external* force) stationary stochastic systems: the stochastic Newton equation (10.23) is indeed appealing in so far as the effect of the environment (stochastic fluctuations and friction) no longer appear explicitly in the right-hand side, so that F represents only the true *external* force—namely it is the manifestation of the fundamental interactions of nature—and not the aggregate description of some underlying hidden background noise, that is instead summarized in the new acceleration (10.22). On the other hand, taking for instance in consideration the systems of *SDE*'s (9.27) and (9.28) describing the simplest stochastic generalizations of the classical dynamics, the said result could not be achieved neither by simply averaging these *SDE*'s (we will not eliminate the deterministic friction terms), nor by reducing it to a Smoluchowski approximation (we will get only a first order equation of motion rather than a second order, dynamical Newton equation: see [10], Chap. 1 for a discussion about this point). It seems reasonable therefore to surmise that—at least in the asymptotic, stationary conditions—only within the present formulation of the dynamical equations all the environmental effects could be accounted for by (and hence subsumed into) a suitable kinematical term provided by an acceleration of the form (10.22) as it will be shown by discussing a few explicit examples in the Appendix N.

10.3.2 Mean Velocities

Taking into account the discussion on the conditional expectations of Sect. 3.4.2, the definitions (10.20) and (10.21) entail that our mean derivatives are functions of $X(t)$ in such a way that, defining the ***forward*** and ***backward velocity functions***

$$\overrightarrow{v}(x,t) \equiv \lim_{\Delta t\to 0^+} \boldsymbol{E}\left[\frac{X(t+\Delta t)-X(t)}{\Delta t}\,\middle|\, X(t)=x\right] \tag{10.24}$$

$$\overleftarrow{v}(x,t) \equiv \lim_{\Delta t\to 0^+} \boldsymbol{E}\left[\frac{X(t)-X(t-\Delta t)}{\Delta t}\,\middle|\, X(t)=x\right] \tag{10.25}$$

the two derivatives (10.20) and (10.21) can also be written as

$$\overrightarrow{\partial} X(t) = \overrightarrow{v}(X(t),t) \qquad\qquad \overleftarrow{\partial} X(t) = \overleftarrow{v}(X(t),t) \tag{10.26}$$

For our purposes it will then be relevant to explicitly calculate the velocities (10.24) and (10.25) of diffusive Markov processes, namely

$$\overrightarrow{v}(x,t) = \lim_{\Delta t\to 0^+} \frac{1}{\Delta t}\int_{\boldsymbol{R}} (y-x) f(y,t+\Delta t|x,t)\,dy$$

$$\overleftarrow{v}(x,t) = \lim_{\Delta t\to 0^+} \frac{1}{\Delta t}\int_{\boldsymbol{R}} (x-y) f(y,t-\Delta t|x,t)\,dy$$

and of course to do that we will need to take into account both the retarded and the advanced equations of the Sect. 10.2.

Proposition 10.3 *The mean velocities for an ac diffusion process $X(t)$ obeying the Eqs.* (10.7) *and* (10.11) *with a given initial condition, are*

$$\overrightarrow{v}(x,t) = A(x,t) \tag{10.27}$$

$$\overleftarrow{v}(x,t) = \widehat{A}(x,t) = A(x,t) - \frac{\partial_x[B(x,t)f(x,t)]}{f(x,t)} \tag{10.28}$$

We moreover have

$$\boldsymbol{E}\left[\overrightarrow{\partial} X(t)\right] = \boldsymbol{E}\left[\overleftarrow{\partial} X(t)\right] \tag{10.29}$$

Proof Supposing that the conditions to exchange limits and integrals are always met, for the forward velocity (10.27) we have from the definition (10.24)

$$\begin{aligned}
\overrightarrow{v}(x,t) &= \lim_{\Delta t\to 0^+} \frac{\boldsymbol{E}\left[X(t+\Delta t)|X(t)=x\right]-x}{\Delta t} \\
&= \lim_{\Delta t\to 0^+} \frac{1}{\Delta t}\int_{\boldsymbol{R}} y\left[f(y,t+\Delta t|x,t)-\delta(y-x)\right]dy \\
&= \lim_{\Delta t\to 0^+} \int_{\boldsymbol{R}} y\,\frac{f(y,t+\Delta t|x,t)-f(y,t^+|x,t)}{\Delta t}\,dy = \int_{\boldsymbol{R}} y\,\left[\partial_s f(y,s|x,t)\right]_{s=t^+}dy
\end{aligned}$$

On the other hand, by exchanging the roles of x, t and y, s the retarded equation (10.7) can be written as

$$\partial_s f(y,s|x,t) = -\partial_y\left[A(y,s)f(y,s|x,t)\right] + \frac{1}{2}\,\partial_y^2\left[B(y,s)f(y,s|x,t)\right]$$

so that from the properties of the Dirac δ we have

$$\begin{aligned}
\left[\partial_s f(y,s|x,t)\right]_{s=t^+} &= -\partial_y\left[A(y,t)\delta(y-x)\right] + \frac{1}{2}\,\partial_y^2\left[B(y,t)\delta(y-x)\right] \\
&= -A(x,t)\delta'(y-x) + \frac{1}{2}\,B(x,t)\delta''(y-x)
\end{aligned}$$

and by taking into account the well known relations

$$\int_{\boldsymbol{R}} h(y)\delta'(y-x)\,dy = -h'(x) \qquad\qquad \int_{\boldsymbol{R}} h(y)\delta''(y-x)\,dy = h''(x) \tag{10.30}$$

we get the first result (10.27)

$$\overrightarrow{v}(x,t) = -A(x,t)\int_{\boldsymbol{R}} y\,\delta'(y-x)\,dy + \frac{1}{2}\,B(x,t)\int_{\boldsymbol{R}} y\,\delta''(y-x)\,dy = A(x,t)$$

As for the backward velocity (10.25) the argument runs along the same path, but for the fact that we must now take into account the equations for the *advanced pdf*'s. We first remark that now

$$\begin{aligned}\overleftarrow{v}(x,t) &= \lim_{\Delta t\to 0^+}\frac{x - \boldsymbol{E}\left[X(t-\Delta t)|X(t)=x\right]}{\Delta t}\\ &= \lim_{\Delta t\to 0^+}\frac{1}{\Delta t}\int_{\boldsymbol{R}} y\left[\delta(y-x) - f(y,t-\Delta t|x,t)\right]dy\\ &= \lim_{\Delta t\to 0^+}\int_{\boldsymbol{R}} y\,\frac{f(y,t^-|x,t) - f(y,t-\Delta t|x,t)}{\Delta t}\,dy = \int_{\boldsymbol{R}} y\,\left[\partial_s f(y,s|x,t)\right]_{s=t^-}\,dy\end{aligned}$$

and then that, always by exchanging the roles of x, t and y, s, the advanced equation (10.11) becomes

$$\begin{aligned}\partial_s f(y,s|x,t) = -\partial_y\left\{\left[A(y,s) - \frac{\partial_y[B(y,s)f(y,s)]}{f(y,s)}\right]f(y,s|x,t)\right\}\\ -\frac{1}{2}\,\partial_y^2\left[B(y,s)f(y,s|x,t)\right]\end{aligned}$$

so that we now have

$$\left[\partial_s f(y,s|x,t)\right]_{s=t^-} = -\left[A(x,t) - \frac{\partial_x[B(x,t)f(x,t)]}{f(x,t)}\right]\delta'(y-x) - \frac{1}{2}\,B(x,t)\delta''(y-x)$$

and hence again from (10.30) we easily get also the second result (10.28). Finally, for (10.29) we remark first of all that from (10.27)

$$\boldsymbol{E}\left[\overrightarrow{\partial}X(t)\right] = \boldsymbol{E}\left[\overrightarrow{v}\,(X(t),t)\right] = \int_{\boldsymbol{R}}\overrightarrow{v}(x,t)f(x,t)\,dx = \int_{\boldsymbol{R}} A(x,t)f(x,t)\,dx$$

and then that from (10.28), (10.12) and the boundary conditions $f(\pm\infty, t) = 0$

$$\begin{aligned}\boldsymbol{E}\left[\overleftarrow{\partial}X(t)\right] &= \int_{\boldsymbol{R}}\widehat{A}(x,t)f(x,t)\,dx = \int_{\boldsymbol{R}} A(x,t)f(x,t)\,dx - \int_{\boldsymbol{R}}\partial_x\big[B(x,t)f(x,t)\big]\,dx\\ &= \int_{\boldsymbol{R}} A(x,t)f(x,t)\,dx - \big[B(x,t)f(x,t)\big]_{-\infty}^{+\infty} = \int_{\boldsymbol{R}} A(x,t)f(x,t)\,dx\end{aligned}$$

so that (10.29) follows at once. ■

For later convenience we now define

$$V(t) = \frac{\overrightarrow{\partial}X(t) + \overleftarrow{\partial}X(t)}{2} = v\big(X(t),t\big) \qquad v(x,t) \equiv \frac{\overrightarrow{v}(x,t) + \overleftarrow{v}(x,t)}{2} \tag{10.31}$$

$$U(t) = \frac{\overrightarrow{\partial}X(t) - \overleftarrow{\partial}X(t)}{2} = u\big(X(t),t\big) \qquad u(x,t) \equiv \frac{\overrightarrow{v}(x,t) - \overleftarrow{v}(x,t)}{2} \tag{10.32}$$

that will be called respectively ***current*** and ***osmotic velocities***: it is easy to see then that from (10.27) and (10.28) we have

$$v(x,t) = \frac{A(x,t) + \widehat{A}(x,t)}{2} = A(x,t) - \frac{1}{2}\frac{\partial_x[B(x,t)f(x,t)]}{f(x,t)} \quad (10.33)$$

$$u(x,t) = \frac{A(x,t) - \widehat{A}(x,t)}{2} = \frac{1}{2}\frac{\partial_x[B(x,t)f(x,t)]}{f(x,t)} \quad (10.34)$$

while from (10.29) we find

$$\boldsymbol{E}[V(t)] = \boldsymbol{E}\left[\overrightarrow{\partial} X(t)\right] = \boldsymbol{E}\left[\overleftarrow{\partial} X(t)\right] \qquad \boldsymbol{E}[U(t)] = 0 \quad (10.35)$$

The name of current velocity for $v(x,t)$ is justified by remarking that by adding the two direct and reverse Fokker-Planck equation (10.9) and (10.15) we find from (10.33) the ***continuity equation*** in the form

$$\partial_t f(x,t) = -\partial_x[v(x,t)f(x,t)] \quad (10.36)$$

while the name of osmotic velocity for $u(x,t)$ in (10.34) is explained by its unique association with the diffusion coefficient $B(x,t)$ of the process.

10.3.3 Mean Acceleration

From the definition (10.22), from (10.26) and by taking into account the Proposition 10.3 we now see that the mean *acceleration* of our diffusion process takes the form

$$\overleftrightarrow{\partial}^2 X(t) = \frac{\overleftarrow{\partial}\overrightarrow{\partial} + \overrightarrow{\partial}\overleftarrow{\partial}}{2} X(t) = \frac{\overleftarrow{\partial}\left[A(X(t),t)\right] + \overrightarrow{\partial}\left[\widehat{A}(X(t),t)\right]}{2} \quad (10.37)$$

so that, in order to write it down explicitly, we need to know how to calculate the mean derivatives

$$\overrightarrow{\partial}\left[h(X(t),t)\right] \qquad \overleftarrow{\partial}\left[h(X(t),t)\right]$$

of a process $Y(t) = h(X(t),t)$ for an arbitrary, well behaved $h(x,t)$. By adopting the usual notations for the partial derivatives of a smooth function $h(x,t)$

$$h_t(x,t) = \partial_t h(x,t) \qquad h_x(x,t) = \partial_x h(x,t) \qquad h_{xx}(x,t) = \partial_x^2 h(x,t)$$

we will now generalize the results of Proposition 10.3 to $Y(t) = h(X(t),t)$, and we will find again that the mean derivatives of $Y(t)$ will be explicitly given in terms of the coefficients $A(x,t)$ and $B(x,t)$.

Proposition 10.4 *Given a diffusion process $X(t)$ and a smooth function $h(x,t)$, the mean forward and backward derivatives of $Y(t) = h(X(t),t)$ can be written as*

$$\overrightarrow{\partial} Y(t) = \overrightarrow{\mathcal{D}} h(X(t), t) = \overrightarrow{h}(X(t), t) \qquad \overleftarrow{\partial} Y(t) = \overleftarrow{\mathcal{D}} h(X(t), t) = \overleftarrow{h}(X(t), t) \tag{10.38}$$

where, for a process obeying the Eqs. (10.7) *and* (10.11), *we used the* ***mean forward*** *and* ***backward streams***

$$\overrightarrow{h}(x, t) = \overrightarrow{\mathcal{D}} h(x, t) = h_t(x, t) + A(x, t) h_x(x, t) + \frac{1}{2} B(x, t) h_{xx}(x, t) \tag{10.39}$$

$$\overleftarrow{h}(x, t) = \overleftarrow{\mathcal{D}} h(x, t) = h_t(x, t) + \widehat{A}(x, t) h_x(x, t) - \frac{1}{2} B(x, t) h_{xx}(x, t) \tag{10.40}$$

that are defined by means of the ***forward*** *and* ***backward substantial operators*** *(for further details see also Appendix O)*

$$\overrightarrow{\mathcal{D}} \equiv \partial_t + A\, \partial_x + \frac{1}{2} B\, \partial_x^2 \qquad \overleftarrow{\mathcal{D}} \equiv \partial_t + \widehat{A}\, \partial_x - \frac{1}{2} B\, \partial_x^2 \tag{10.41}$$

Proof The streams (10.39) and (10.40) are a generalization of the velocities (10.27) and (10.28): it would be enough indeed to take $h(x, t) = x$ to find that $\overrightarrow{h}(x, t)$ and $\overleftarrow{h}(x, t)$ of (10.39) and (10.40) become respectively $\overrightarrow{v}(x, t)$ and $\overleftarrow{v}(x, t)$ of (10.27) and (10.28). As a consequence the proof follows the same lines of the Proposition 10.3 and we will omit it here referring to the Appendix O for further details. ∎

The *forward and backward substantial operators* effectively reduce the mean derivatives action on the processes $Y(t)$ to operations on the ordinary functions $h(x, t)$: remark indeed that in our notation the symbols $\overrightarrow{\partial}$, $\overleftarrow{\partial}$ act on the processes according to (10.20) and (10.21), while the symbols $\overrightarrow{\mathcal{D}}$, $\overleftarrow{\mathcal{D}}$ operate on the ordinary functions according to (10.41). We stress again that if we choose $h(x, t) = x$ it is easy to check from (10.27), (10.28) and (10.41) that in analogy with (10.26) we have

$$\overrightarrow{\mathcal{D}} x = A(x, t) = \overrightarrow{v}(x, t) \qquad \overleftarrow{\mathcal{D}} x = \widehat{A}(x, t) = \overleftarrow{v}(x, t)$$

while from (10.39) and (10.40), we have $\overrightarrow{h}(x, t) = \overrightarrow{v}(x, t)$ and $\overleftarrow{h}(x, t) = \overleftarrow{v}(x, t)$. As a consequence the results of the previous Proposition 10.3 can again be deduced as a particular case of Proposition 10.4: when indeed $h(x, t) = x$ the Eq. (10.38) for the processes are simply brought back to the Eq. (10.26). Another easy corollary of the Propositions 10.3 and 10.4 is that when $h(x, t) = \alpha x + \beta$, with α and β arbitrary constants, it is

$$\overrightarrow{\mathcal{D}}(\alpha x + \beta) = \alpha \overrightarrow{\mathcal{D}} x = \alpha \overrightarrow{v}(x, t) \qquad \overleftarrow{\mathcal{D}}(\alpha x + \beta) = \alpha \overleftarrow{\mathcal{D}} x = \alpha \overleftarrow{v}(x, t)$$

which apparently reproduce the *linearity properties* of the usual derivatives, while on the other hand for the process trajectories these relations read as

$$\vec{\partial}\big(\alpha X(t)+\beta\big)=\alpha\,\vec{\partial}X(t)\qquad\qquad \overleftarrow{\partial}\big(\alpha X(t)+\beta\big)=\alpha\,\overleftarrow{\partial}X(t) \tag{10.42}$$

As a matter of fact the knowledge of both the velocities $u(x,t),\ v(x,t)$, and of the diffusion coefficient $B(x,t)$ is tantamount to the knowledge of the global law of the process, namely of the state of our system. From the current and osmotic velocities $u(x,t),\ v(x,t)$ and from (10.33), (10.34) we indeed find first that

$$A(x,t)=v(x,t)+u(x,t)$$

and then, with a little algebra, that the one-time *pdf* is (N is a normalization constant)

$$f(x,t)=N\,\frac{e^{2\int\frac{u(x,t)}{B(x,t)}\,dx}}{B(x,t)}$$

Since now we know $A(x,t)$ and $B(x,t)$ we can solve the Fokker-Planck equation (10.7) to have the retarded transition *pdf*, and finally with the one-time *pdf* $f(x,t)$ we can find all the joint laws of the process from the chain rule (7.14). While on the other hand the diffusion coefficient B can be considered as given a priori in order to describe the random background, it is apparent that it would be useful to be able to find $u(x,t),\ v(x,t)$ from some dynamical condition, as for instance the stochastic form of the the Newton equation (10.23), and that to this end we should first find an explicit form for the mean acceleration appearing therein. The Proposition 10.4 turns out therefore to be instrumental in obtaining this outcome because, within the notations introduced in the present section, we have from (10.37)

$$\overleftrightarrow{\partial}^2 X(t)=\overleftrightarrow{a}(X(t),t)\qquad\qquad \overleftrightarrow{a}(x,t)\equiv\frac{\overleftarrow{D}A(x,t)+\vec{D}\widehat{A}(x,t)}{2} \tag{10.43}$$

where the new function $\overleftrightarrow{a}(x,t)$ will be called ***mean acceleration***, and therefore we are reduced to calculate substantial derivatives of the velocities A and $\widehat{A}$.

Proposition 10.5 *The mean acceleration* (10.43) *for an ac diffusion* $X(t)$ *with current and osmotic velocities* $v(x,t)$, $u(x,t)$ *and diffusion coefficient* $B(x,t)$, *takes the form*

$$\overleftrightarrow{a}(x,t)=\partial_t v(x,t)+\partial_x\left[\frac{v^2(x,t)-u^2(x,t)}{2}\right]-\frac{B(x,t)}{2}\,\partial_x^2 u(x,t) \tag{10.44}$$

Proof From (10.43), (10.41), (10.33) and (10.34) it follows that

$$\begin{aligned}\overset{\leftrightarrow}{a} &= \frac{\vec{D}\widehat{A} + \overleftarrow{D}A}{2} = \frac{1}{2}\left(\partial_t \widehat{A} + A\partial_x\widehat{A} + \frac{B}{2}\,\partial_x^2\widehat{A} + \partial_t A + \widehat{A}\partial_x A - \frac{B}{2}\,\partial_x^2 A\right)\\ &= \partial_t\left(\frac{A+\widehat{A}}{2}\right) + \frac{A\partial_x\widehat{A} + \widehat{A}\partial_x A}{2} - \frac{B}{2}\,\partial_x^2\left(\frac{A-\widehat{A}}{2}\right)\\ &= \partial_t v + \frac{A\partial_x\widehat{A} + \widehat{A}\partial_x A}{2} - \frac{B}{2}\,\partial_x^2 u\end{aligned}$$

and since from $A = v + u,\ \widehat{A} = v - u$ it is immediate to check that

$$A\partial_x\widehat{A} + \widehat{A}\partial_x A = 2(v\partial_x v - u\partial_x u) = \partial_x(v^2 - u^2)$$

we at once get the result (10.44). ■

10.4 Dynamics of a Diffusion Process

We confine now our analysis to the *ac* diffusion processes $X(t)$ ruled by a Fokker-Planck equation with a ***constant diffusion coefficient*** $B(x,t) = D$. This requirement on the other hand is not really restricting: it is possible to show indeed that with a suitable transformation of the process we can always meet this condition (for more details see Appendix P). In particular the examples of Gaussian processes that are discussed in the Appendix N are already fully within this framework. As a consequence, for given D, $A(x,t)$ and $f(x,t)$, we first of all find that (10.27) and (10.28) become

$$\vec{v}(x,t) = A(x,t) \tag{10.45}$$

$$\overleftarrow{v}(x,t) = \widehat{A}(x,t) \;=\; A(x,t) - D\,\frac{\partial_x f(x,t)}{f(x,t)} \tag{10.46}$$

while the forward and backward substantial operators are now reduced to

$$\vec{D} = \partial_t + A\,\partial_x + \frac{D}{2}\,\partial_x^2 \qquad\qquad \overleftarrow{D} = \partial_t + \widehat{A}\,\partial_x - \frac{D}{2}\,\partial_x^2 \tag{10.47}$$

It is apparent on the other hand that under these hypotheses the direct and reverse equations (10.9) and (10.15) are reduced to the form

$$\partial_t f(x,t) = -\,\partial_x\big[A(x,t)f(x,t)\big] + \frac{D}{2}\,\partial_x^2 f(x,t) \tag{10.48}$$

$$\partial_t f(x,t) = -\,\partial_x\big[\widehat{A}(x,t)f(x,t)\big] - \frac{D}{2}\,\partial_x^2 f(x,t) \tag{10.49}$$

By taking moreover into account (10.45) and (10.46) also the current and osmotic velocities (10.31) and (10.32) take the form

$$v(x,t) = \frac{A(x,t) + \widehat{A}(x,t)}{2} = A(x,t) - \frac{D}{2}\frac{\partial_x f(x,t)}{f(x,t)} \tag{10.50}$$

$$u(x,t) = \frac{A(x,t) - \widehat{A}(x,t)}{2} = \frac{D}{2}\frac{\partial_x f(x,t)}{f(x,t)} \tag{10.51}$$

We reiterate once more that the knowledge of D and of the two functions u and v (or equivalently of $\overrightarrow{v}$ and $\overleftarrow{v}$) is tantamount to the knowledge of the law of the process $X(t)$ because now

$$A(x,t) = v(x,t) + u(x,t) \qquad f(x,t) = N\, e^{\frac{2}{D}\int u(x,t)\,dx} \tag{10.52}$$

where $\int u(x,t)\,dx$ is a suitable x-primitive of $u(x,t)$.

Proposition 10.6 *Within our previous notations the osmotic and current velocities obey the following coupled system of non linear partial differential equations*

$$\partial_t v(x,t) = \overleftrightarrow{a}(x,t) + \partial_x\left[\frac{D}{2}\,\partial_x u(x,t) + \frac{u^2(x,t) - v^2(x,t)}{2}\right] \tag{10.53}$$

$$\partial_t u(x,t) = -\,\partial_x\left[\frac{D}{2}\,\partial_x v(x,t) + u(x,t)v(x,t)\right] \tag{10.54}$$

where $\overleftrightarrow{a}(x,t)$ is the acceleration (10.43).

Proof Equation (10.54) easily follows from (10.51) and from the continuity equation (10.36):

$$\begin{aligned}\partial_t u &= \frac{D}{2}\,\partial_x\left(\frac{\partial_t f}{f}\right) = -\frac{D}{2}\,\partial_x\left(\frac{\partial_x(vf)}{f}\right) = -\frac{D}{2}\,\partial_x\left(v\,\frac{\partial_x f}{f} + \partial_x v\right)\\ &= -\,\partial_x\left(uv + \frac{D}{2}\,\partial_x v\right)\end{aligned}$$

Equation (10.53) instead immediately proceeds from the Proposition 10.5 taking $B(x,t) = D$ in (10.44) and rearranging its terms. ■

If now $\overleftrightarrow{\partial}^2 X(t)$ is the acceleration of a particle of mass m acted upon by some external force $F(t)$ derivable from a potential $\Phi(x)$ according to

$$F(t) = -\partial_x\Phi(X(t))$$

we will have from the Newton equation (10.23) and (10.43)

$$m\overleftrightarrow{a}(x,t) = -\,\partial_x\Phi(x) \tag{10.55}$$

while from its definition (10.51) the osmotic velocity can always be written as

$$u(x,t) = D\,\frac{\partial_x R(x,t)}{R(x,t)} \qquad\qquad R(x,t) \equiv \sqrt{f(x,t)} \tag{10.56}$$

We make moreover the additional hypothesis that there is some function $S(x,t)$ such that the current velocity can also be written as

$$v(x.t) = \frac{\partial_x S(x,t)}{m} \tag{10.57}$$

Apparently this is a rather trivial claim in our one-dimensional setting (you will always find an x-primitive of v), but in the general formulation this amounts to suppose that the *vector* velocity field v can be derived as the *gradient* of some *scalar* function S, and this is apparently not trivial at all for an arbitrary vector v. We must remark however that this point has subsequently been totally clarified in the more general framework of the variational reformulation of the stochastic mechanics [8] where the condition (10.57) is no longer a hypothesis, but rather a result of the overall theory.

Theorem 10.7 Schrödinger equation: *Within the previous notations, the complex valued wave function*

$$\psi(x,t) = R(x,t)\,e^{iS(x,t)/mD} \tag{10.58}$$

is a solution of the linear equation

$$imD\,\partial_t\psi(x,t) = -\frac{mD^2}{2}\,\partial_x^2\psi(x,t) + \Phi(x)\psi(x,t) \tag{10.59}$$

if and only if $u(x,t)$ *and* $v(x,t)$ *from* (10.56) *and* (10.57) *are solutions of the non linear system* (10.53) *and* (10.54). *In particular, by taking*

$$D = \frac{\hbar}{m} \tag{10.60}$$

where $\hbar$ *is the Planck constant, the Eq.* (10.59) *coincides with the one-dimensional version of the Schrödinger equation* (10.1).

Proof By making use of (10.55) we first reduce the Eqs. (10.53) and (10.54) to

$$\partial_t v = \partial_x\left(\frac{u^2 - v^2}{2} + \frac{D}{2}\,\partial_x u - \frac{\Phi}{m}\right)$$
$$\partial_t u = -\,\partial_x\left(uv + \frac{D}{2}\,\partial_x v\right)$$

and then, taking also (10.56) and (10.57) into account, we find

$$\partial_x \left[\partial_t S + \frac{(\partial_x S)^2}{2m} - \frac{mD^2}{2} \frac{\partial_x^2 R}{R} + \Phi \right] = 0$$

$$\partial_x \left[\frac{\partial_t R}{R} + \frac{1}{2m} \left(2\, \partial_x S \, \frac{\partial_x R}{R} + \partial_x^2 S \right) \right] = 0$$

With $c_S(t)$ and $c_R(t)$ arbitrary functions of t only we therefore have

$$\partial_t S + \frac{(\partial_x S)^2}{2m} - \frac{mD^2}{2} \frac{\partial_x^2 R}{R} + \Phi = c_S(t)$$

$$\frac{\partial_t R}{R} + \frac{1}{2m} \left(2\, \partial_x S \, \frac{\partial_x R}{R} + \partial_x^2 S \right) = c_R(t)$$

Since however in (10.56) and (10.57) R and S are determined by u and v up to arbitrary constants (namely arbitrary functions of t only), c_R and c_S can always be disposed of by suitably redefining our quantities, and therefore we can always manage to take $c_R(t) = c_S(t) = 0$ so that finally, by rearranging the terms, we have

$$\partial_t S + \frac{m}{2} \left(\frac{\partial_x S}{m} \right)^2 + \Phi - \frac{mD^2}{2} \frac{\partial_x^2 R}{R} = 0 \tag{10.61}$$

$$\partial_t R^2 + \partial_x \left(R^2 \frac{\partial_x S}{m} \right) = 0 \tag{10.62}$$

that apparently turn out to be the one-dimensional analog of (10.2) and (10.3). We resort therefore to the Madelung decomposition introduced in the Sect. 10.1: by plugging (10.58) into (10.59) and by separating the real and the imaginary parts we indeed easily recover the Eqs. (10.61) and (10.62) that turn out then to be equivalent to (10.59). ■

This derivation of the Schrödinger equation first of all entails that the Born rule—according to which $R^2 = |\psi|^2$ is the *pdf* of the random position of a quantum particle—is no longer a postulate, but naturally springs from (10.56), namely from the fact that now $R^2 = f$ is in fact exactly the *pdf* of a legitimate diffusion process describing the position of the particle. This in particular also means that a natural and straightforward particle interpretation of the Madelung fluid, without the *ad hoc* introduction of a quantum potential, is indeed possible, but only by allowing a random character to the underlying trajectories. In the semiclassical limit $\hbar = mD \to 0$ the randomness disappears and the trajectories become those of the classical theory, while the Madelung fluid, through the vanishing of the quantum potential, reduces to the classical Hamilton-Jacobi fluid. In so doing the stochastic mechanics provides a simple general and unified formulation of quantum and classical mechanics based on the theory of stochastic processes, with random trajectories that become deterministic in the classical limit. On the other hand, since at every instant the *pdf*'s of stochastic mechanics keep the same values as in quantum mechanics, the two theories cannot be distinguished at the operational level: they are just different pictures of a unique theory describing quantum phenomena, as has also been confirmed in the subsequent developments of the theory. For the time being, however, this only means that a particle picture with (random) trajectories is not forbidden, but does not directly imply that particles and trajectories really exist in the physical sense. Since in any case our starting point is the Schrödinger equation which does not involve trajectories, it is clear indeed that the particle picture with trajectories of the stochastic mechanics is only a possibility needing some new information of physical nature to be confirmed or excluded.

It is important to remark finally that the nature of a possible background, non-relativistic *Zitterbewegung* that could account for the random character of the quantum particle trajectories is a question still far to be settled. This is a point that was initially underlined by Nelson himself by introducing his hypothesis of an *universal Brownian motion* [5], but despite extensive research and several innovative proposals (from electromagnetic [11] to gravitational [12] fluctuations) this is an issue that remains a relevant open problem.

10.4.1 Time Reversal Invariance

A preliminary objection usually leveled to the theory presented in this chapter is that the stochastic processes are typical expressions of diffusion phenomena sharing an inherent irreversible character, while the Schrödinger equation is time-reversal invariant and therefore cannot have anything to do with the stochastic processes. While we can not embark here in a complete discussion of this point, it is essential to remark at least that the time irreversible character of the phenomenological *SDE*'s (as for instance the Ornstein-Uhlenbeck equations of the Chap. 9) is strictly related to the fact that the drift coefficients appearing in them have a well defined, prearranged form depending on the physical environment, in particular on the external fields acting on the system. This in fact prevents the possibility of transforming these *SDE*'s under time-reversal while retaining their qualitative physical meaning. It is easy to show, indeed, by simple examples that under time-reversal the transformed solutions would acquire unacceptable (anti-diffusive) forms, and this apparently entails the irreversible character of the diffusion equations of the dissipative processes. Take for instance the Fokker-Planck equation (7.92) of a simple Wiener process with its solution

$$f(x,t) = \frac{e^{-\frac{x^2}{2(a^2+Dt)}}}{\sqrt{2\pi(a^2+Dt)}} \sim \mathfrak{N}(0, a^2+Dt)$$

corresponding to an initial condition $f_0 \sim \mathfrak{N}(0, a^2)$ with an arbitrary variance a^2: a straightforward time reversal $t \to -t$ would produce the equation

$$\partial_t f(x,t) = -\frac{D}{2}\partial_x^2 f(x,t)$$

with a negative diffusion coefficient, while $f(x,t)$ would be transformed in the function

$$\frac{e^{-\frac{x^2}{2(a^2-Dt)}}}{\sqrt{2\pi(a^2-Dt)}}$$

that will preposterously shrink its variance from a^2 to 0 when t goes from 0 to a^2/D, while utterly losing every physical meaning as soon as t exceeds a^2/D.

On the contrary, as it is briefly sketched in the end of the Sect. 10.2, the drift coefficients of the Fokker-Planck equations of stochastic mechanics have a genuine dynamical meaning, and then they acceptably transform under time reversal: there is no preordained, natural form for them so that the time inverted ones are acceptable too. This is the reason why generally speaking a stochastic scheme can share ***time-reversal invariance*** properties, as happens in particular to the processes discussed here. To show that by means of a simple example consider the free version ($\Phi = 0$) of the Schrödinger equation (10.59)

$$-i\partial_t\psi(x,t) = \frac{D}{2}\,\partial_x^2\psi(x,t)$$

It is well known first of all that the modifications produced by a time reversal $t \to -t$ can be immediately reabsorbed by taking the complex conjugate of the equation, and that this transformation does not change the *pdf* of the state because its value $|\psi|^2$ is left unchanged by a complex conjugation. To see that in further detail consider (in analogy with the previous example) the centered wave packet of minimal uncertainty solution of our Schrödinger equation with an arbitrary initial variance a^2

$$\psi(x,t) = \sqrt[4]{\frac{2a^2}{\pi}}\,\frac{e^{-\frac{x^2}{2(2a^2+iDt)}}}{\sqrt{2a^2+iDt}} \qquad\qquad \psi_0(x) = \psi(x,0) = \sqrt[4]{\frac{2a^2}{\pi}}\,\frac{e^{-\frac{x^2}{4a^2}}}{\sqrt{2a^2}}$$

it is easy to see indeed that

$$|\psi_0(x)|^2 = \frac{e^{-\frac{x^2}{2a^2}}}{\sqrt{2\pi a^2}} \sim \mathfrak{N}(0,a^2)$$

A little algebra would show now from (10.58) that the modulus and phase of our wave function $\psi(x,t)$ are

$$R(x,t) = \sqrt[4]{\frac{2a^2}{\pi}}\,\frac{e^{-\frac{a^2x^2}{4a^4+D^2t^2}}}{\sqrt[4]{4a^4+D^2t^2}} \qquad\qquad S(x,t) = \frac{mD}{2}\left(\frac{Dtx^2}{4a^4+D^2t^2} - \arctan\frac{Dt}{2a^2}\right)$$

and hence from (10.56) the *pdf* of the particle position is the time dependent, centered Gaussian

$$f(x,t) = R^2(x,t) = \sqrt{\frac{2a^2}{\pi}}\,\frac{e^{-\frac{2a^2x^2}{4a^4+D^2t^2}}}{\sqrt{4a^4+D^2t^2}} \sim \mathfrak{N}\left(0, a^2 + \frac{D^2t^2}{4a^2}\right)$$

while from (10.57) and (10.56) the current and osmotic velocities are

$$v(x,t) = \frac{D^2t\,x}{4a^4+D^2t^2} \qquad u(x,t) = -\frac{2Da^2\,x}{4a^4+D^2t^2}$$

As a consequence from (10.50) and (10.51) the drift coefficients of the diffusion (or equivalently from (10.45) and (10.46) its forward and backward velocities $\overrightarrow{v}$ and $\overleftarrow{v}$) are

$$A(x,t) = \overrightarrow{v}(x,t) = v(x,t) + u(x,t) = \frac{D^2t - 2Da^2}{4a^4+D^2t^2}\,x$$
$$\widehat{A}(x,t) = \overleftarrow{v}(x,t) = v(x,t) - u(x,t) = \frac{D^2t + 2Da^2}{4a^4+D^2t^2}\,x$$

and hence, according to (10.9), the *pdf* $f(x,t)$ of the process $X(t)$ associated to $\psi(x,t)$ by the stochastic mechanics satisfies the following direct Fokker-Planck equation

$$\partial_t f(x,t) = -\partial_x\left[\frac{D^2t - 2Da^2}{4a^4+D^2t^2}\,x\,f(x,t)\right] + \frac{D}{2}\,\partial_x^2 f(x,t)$$

which by the way is no longer that of a simple Wiener process. As for the time reversed equation (10.19) moreover we see from the discussion concluding the Sect. 10.2, and in particular from the transformation (10.17) of the drift coefficient, that here

$$A^*(x,t) = -\widehat{A}(x,-t) = \frac{D^2t - 2Da^2}{4a^4+D^2t^2}\,x = A(s,t)$$

and therefore that under time reversal the Fokker-Planck equation becomes

$$\partial_t f^*(x,t) = -\partial_x\left[\frac{D^2t - 2Da^2}{4a^4+D^2t^2}\,x\,f^*(x,t)\right] + \frac{D}{2}\,\partial_x^2 f^*(x,t)$$

which formally coincides (but for the meaning of the symbols) with the previous one, and validates in so doing its time reversal invariance. This is also confirmed by the remark that even the time reversed *pdf* (10.16) would coincide with the original $f(x,t)$ that is a time function of t^2 only and hence is unaffected by a change of the t sign. By summarizing, the diffusion process associated by the stochastic mechanics to our minimal uncertainty wave function $\psi(x,t)$ is perfectly time reversal invariant in the sense that the time reversed process is again a totally legitimate diffusion.

References

1. Madelung, E.: Eine anschauliche Deutung der Gleichung von Schrödinger. Z. Physik **40**, 332 (1926)
2. Bohm, D.: A suggested interpretation of the quantum theory in terms of "hidden" variables I–II. Phys. Rev. **85**, 166, 180 (1952)
3. Holland, P.R.: The Quantum Theory of Motion. Cambridge UP, Cambridge (1993)
4. Nelson, E.: Dynamical Theories of Brownian Motion. Princeton UP, Princeton (1967)
5. Nelson, E.: Derivation of the Schrödinger equation from Newtonian mechanics. Phys. Rev. **150**, 1079 (1966)
6. Nelson, E.: Quantum Fluctuations. Princeton UP, Princeton (1985)
7. Guerra, F.: Structural aspects of stochastic mechanics and stochastic field theory. Phys. Rep. **77**, 263 (1981)
8. Guerra, F., Morato, L.M.: Quantization of dynamical systems and stochastic control theory. Phys. Rev. D **27**, 1774 (1983)
9. Paul, W., Baschnagel, J.: Stochastic Processes: From Physics to Finance. Springer, Heidelberg (2013)
10. McCauley, J.L.: Classical Mechanics. Cambridge UP, Cambridge (1997)
11. de la Peña, L., Cetto, A.M.: The Quantum Dice. Springer, Berlin (1996)
12. Calogero, F.: Cosmic origin of quantization. Phys. Lett. A **228**, 335 (1997)

Appendix A
Consistency (Sect. 2.3.4)

Consistency conditions are instrumental in the two Kolmogorov Theorems 2.35 and 2.37, but they are also crucial in the supposedly more elementary discussion about copulas at the end of the Sect. 2.3.4. In the following we will show that compliance with these conditions is not at all a foregone conclusion, even in the very simple context we will restrict to: that of discrete distributions on finite sets of integer numbers.

Take first a trivariate, discrete distribution on the set $\{0, 1\} \times \{0, 1\} \times \{0, 1\}$ of the 0 - 1 triples that (with a notation taken from the Sect. 2.1) will be denoted as

$$p_{ijk} = \boldsymbol{P}\{i, j, k\} \qquad i, j, k \in \{0, 1\}$$

Such a distribution always is well define provided that

$$0 \leq p_{ijk} \leq 1 \qquad \sum_{i,j,k} p_{ijk} = 1 \tag{A.1}$$

Here and in the following it will be understood that the summation indices always take the values 0 and 1. From p_{ijk} it is then possible to deduce—as in the Sect. 2.3.3—the three bivariate, marginal distributions on $\{0, 1\} \times \{0, 1\}$

$$p_{jk}^{(1)} = \sum_i p_{ijk} \qquad p_{ik}^{(2)} = \sum_j p_{ijk} \qquad p_{ij}^{(3)} = \sum_k p_{ijk}$$

and the three univariate, marginal (Bernoulli) distributions on $\{0, 1\}$

$$p_k^{(1,2)} = \sum_{i,j} p_{ijk} \qquad p_i^{(2,3)} = \sum_{j,k} p_{ijk} \qquad p_j^{(1,3)} = \sum_{i,k} p_{ijk}$$

N. Cufaro Petroni, *Probability and Stochastic Processes for Physicists*, UNITEXT for Physics, https://doi.org/10.1007/978-3-030-48408-8

Apparently this procedure also entails by construction the *consistency* of the three levels of distributions because the extra marginalization relations

$$p_k^{(1,2)} = \sum_j p_{jk}^{(1)} = \sum_i p_{ik}^{(2)}$$
$$p_i^{(2,3)} = \sum_k p_{ik}^{(2)} = \sum_j p_{ij}^{(3)}$$
$$p_j^{(1,3)} = \sum_k p_{jk}^{(1)} = \sum_i p_{ij}^{(3)}$$

are always trivially satisfied. We are interested now in finding to what extent—if at all—this consistency can be preserved when we start instead backward from the lowest level, namely from some univariate distributions.

Start then now with three arbitrary, univariate Bernoulli distributions on $\{0, 1\}$ (upper indices are now gone, because we are no longer supposing *a priori* to have deduced them from some other given multivariate distribution)

$$p_i = \begin{cases} P & i = 1 \\ 1 - P & i = 0 \end{cases} \qquad 0 \leq P \leq 1$$
$$q_j = \begin{cases} Q & j = 1 \\ 1 - Q & j = 0 \end{cases} \qquad 0 \leq Q \leq 1$$
$$r_k = \begin{cases} R & k = 1 \\ 1 - R & k = 0 \end{cases} \qquad 0 \leq R \leq 1$$

and ask first if it would be possible to find three bivariate distributions p_{ij}, q_{jk} e r_{ik} having the given Bernoulli as their marginals in the sense that

$$\sum_j p_{ij} = \sum_k r_{ik} = p_i \qquad \sum_i p_{ij} = \sum_k q_{jk} = q_j \qquad \sum_j q_{jk} = \sum_i r_{ik} = r_k \tag{A.2}$$

This is a linear system of 12 equations in the 12 unknowns p_{ij}, q_{jk} and r_{ik}, but we should also remember that in order to be acceptable our solutions must take values in $[0, 1]$ in compliance with the conditions

$$\sum_{ij} p_{ij} = \sum_{jk} q_{jk} = \sum_{ik} r_{ik} = 1$$

Only 9 among the 12 equations (A.2) are however linearly independent,[1] so that in general we expect ∞^3 solutions, with three free parameters p, q, r to be chosen—if possible—in a way giving rise to acceptable solutions. It is easy to check now that, for given P, Q, R of the initial distributions, the solutions can be put in the form

[1]The rank of the coefficient matrix is indeed 9, and it coincides with the rank of the same matrix augmented with the column of the constant terms.

$$\begin{cases} p_{11} = p \\ p_{10} = P - p \\ p_{01} = Q - p \\ p_{00} = 1 - P - Q + p \end{cases} \qquad \begin{cases} q_{11} = q \\ q_{10} = Q - q \\ q_{01} = R - q \\ q_{00} = 1 - R - Q + q \end{cases} \qquad \begin{cases} r_{11} = r \\ r_{10} = P - r \\ r_{01} = R - r \\ r_{00} = 1 - P - R + r \end{cases}$$

and that in their turn they are acceptable distributions provided that P, Q, R, p, q, r comply with the following restrictions

$$0 \le P \le 1 \qquad 0 \le Q \le 1 \qquad 0 \le R \le 1 \tag{A.3}$$
$$0 \le p \le P \wedge Q \qquad 0 \le q \le Q \wedge R \qquad 0 \le r \le P \wedge R \tag{A.4}$$

which can always be easily met (here $x \wedge y = \min\{x, y\}$). In conclusion, however taken the numbers P, Q, R in $[0, 1]$ (namely, for every choice of the initial univariate distributions), we can always find (infinite) bivariate distributions consistent with the given univariate.

Go on now to the next level: take 6 numbers P, Q, R, p, q, r in compliance with the conditions (A.3) and (A.4) (namely: take arbitrary, but consistent univariate and e bivariate distributions p_i, q_j, r_k and p_{ij}, q_{jk}, r_{ik}) and ask if it is always possible to find also a trivariate distribution p_{ijk} which turns out to be consistent with these given univariate and bivariate. In short ask if we can always find 8 numbers p_{ijk} in compliance with the limitations (A.1), and satisfying the 12 equations

$$\sum_k p_{ijk} = p_{ij} \qquad \sum_i p_{ijk} = q_{jk} \qquad \sum_j p_{ijk} = r_{ik} \tag{A.5}$$

The system (A.5) apparently is overdetermined (12 equations and 8 unknowns), but we could check that both the coefficient matrix, and that augmented with the column of the constant terms

$$\begin{pmatrix} 1\,1\,0\,0\,0\,0\,0\,0 \\ 0\,0\,1\,1\,0\,0\,0\,0 \\ 0\,0\,0\,0\,1\,1\,0\,0 \\ 0\,0\,0\,0\,0\,0\,1\,1 \\ 1\,0\,0\,0\,1\,0\,0\,0 \\ 0\,1\,0\,0\,0\,1\,0\,0 \\ 0\,0\,1\,0\,0\,0\,1\,0 \\ 0\,0\,0\,1\,0\,0\,0\,1 \\ 1\,0\,1\,0\,0\,0\,0\,0 \\ 0\,1\,0\,1\,0\,0\,0\,0 \\ 0\,0\,0\,0\,1\,0\,1\,0 \\ 0\,0\,0\,0\,0\,1\,0\,1 \end{pmatrix} \qquad \begin{pmatrix} 1\,1\,0\,0\,0\,0\,0\,0 & p \\ 0\,0\,1\,1\,0\,0\,0\,0 & P-p \\ 0\,0\,0\,0\,1\,1\,0\,0 & Q-p \\ 0\,0\,0\,0\,0\,0\,1\,1 & 1-P-Q+p \\ 1\,0\,0\,0\,1\,0\,0\,0 & q \\ 0\,1\,0\,0\,0\,1\,0\,0 & Q-q \\ 0\,0\,1\,0\,0\,0\,1\,0 & R-q \\ 0\,0\,0\,1\,0\,0\,0\,1 & 1-Q-R+q \\ 1\,0\,1\,0\,0\,0\,0\,0 & r \\ 0\,1\,0\,1\,0\,0\,0\,0 & R-r \\ 0\,0\,0\,0\,1\,0\,1\,0 & P-r \\ 0\,0\,0\,0\,0\,1\,0\,1 & 1-P-R+r \end{pmatrix}$$

have the same rank 7. Hence—according to the Rouché-Capelli theorem—the system (A.5), albeit overdetermined, turns out to be compatible, and in fact has infinite solutions with one free parameter s. It is possible to show then that the solutions of the system (A.5) take the form

$$\begin{cases} p_{111} = 1 - P - Q - R + p + q + r - s \\ p_{110} = P + Q + R - 1 - q - r + s \\ p_{101} = P + Q + R - 1 - p - q + s \\ p_{100} = 1 - Q - R + q - s \\ p_{011} = P + Q + R - 1 - p - r + s \\ p_{010} = 1 - P - R + r - s \\ p_{001} = 1 - P - Q + p - s \\ p_{000} = s \end{cases} \tag{A.6}$$

and we must ask now if—for every choice of the numbers P, Q, R, p, q, r in compliance with the conditions (A.3) and (A.4)—is it possible to find some $s \in [0, 1]$ such that (A.6) are acceptable according to the limitations (A.1). Surprisingly enough the answer to this question is in the negative, and we will show that by means of a counterexample.

Since it would be easy to check that the sum of the p_{ijk} in (A.6) always adds up to 1, we are left with the problem of looking if all these 8 can be in [0, 1], at least for some choice of s. Suppose then—in compliance with the condition (A.3) and (A.4)—to take in particular

$$P = Q = R = \frac{1}{2} \qquad p = q = \frac{2 + \sqrt{2}}{8} \approx 0.426777 \qquad r = \frac{1}{4} \tag{A.7}$$

namely the following consistent family of bivariate and univariate distributions

$$\begin{cases} p_{11} = \frac{2+\sqrt{2}}{8} \\ p_{10} = \frac{2-\sqrt{2}}{8} \\ p_{01} = \frac{2-\sqrt{2}}{8} \\ p_{00} = \frac{2+\sqrt{2}}{8} \end{cases} \qquad \begin{cases} q_{11} = \frac{2+\sqrt{2}}{8} \\ q_{10} = \frac{2-\sqrt{2}}{8} \\ q_{01} = \frac{2-\sqrt{2}}{8} \\ q_{00} = \frac{2+\sqrt{2}}{8} \end{cases} \qquad \begin{cases} r_{11} = {}^1/_4 \\ r_{10} = {}^1/_4 \\ r_{01} = {}^1/_4 \\ r_{00} = {}^1/_4 \end{cases} \tag{A.8}$$

$$p_1 = p_0 = {}^1/_2 \qquad q_1 = q_0 = {}^1/_2 \qquad r_1 = r_0 = {}^1/_2$$

With this choice the (A.6) become

$$\begin{cases} p_{111} = \frac{1+\sqrt{2}}{4} - s \approx 0.603553 - s \\ p_{110} = -\frac{\sqrt{2}}{8} + s \approx -0.176777 + s \\ p_{101} = -\frac{\sqrt{2}}{4} + s \approx -0.353553 + s \\ p_{100} = \frac{2+\sqrt{2}}{8} - s \approx 0.426777 - s \\ p_{011} = -\frac{\sqrt{2}}{8} + s \approx -0.176777 + s \\ p_{010} = \frac{1}{4} - s = 0.25 - s \\ p_{001} = \frac{2+\sqrt{2}}{8} - s \approx 0.426777 - s \\ p_{000} = s \end{cases}$$

and it is easy to see that there exists no value of $s \in [0, 1]$ such that all the p_{ijk} lie in $[0, 1]$: to this end it is enough to remark that we should choose $s \geq 0.353553$ in order to have $p_{101} \geq 0$, and that in this case it would be $p_{010} \leq 0.25 - 0.353553 = -0.103553$. In short: *there are consistent families of univariate and bivariate distributions not allowing a consistent trivariate one.*

For later convenience, it is useful to remark here that the same conclusions could have been drawn for a given set of univariate and *conditional distributions*, instead of *joint, bivariate distributions*. It is easy to understand indeed that, from a formal point of view to give the set (A.8) it is equivalent to give the set of the univariate and conditional probabilities $p_{i|j} = p_{ij}/p_j$, $q_{j|k} = q_{jk}/q_k$, $r_{i|k} = r_{ik}/r_k$, namely

$$\begin{cases} p_{1|1} = \frac{2+\sqrt{2}}{4} \\ p_{1|0} = \frac{2-\sqrt{2}}{4} \\ p_{0|1} = \frac{2-\sqrt{2}}{4} \\ p_{0|0} = \frac{2+\sqrt{2}}{4} \end{cases} \qquad \begin{cases} q_{1|1} = \frac{2+\sqrt{2}}{4} \\ q_{1|0} = \frac{2-\sqrt{2}}{4} \\ q_{0|1} = \frac{2-\sqrt{2}}{4} \\ q_{0|0} = \frac{2+\sqrt{2}}{4} \end{cases} \qquad \begin{cases} r_{1|1} = {}^1/_2 \\ r_{1|0} = {}^1/_2 \\ r_{0|1} = {}^1/_2 \\ r_{0|0} = {}^1/_2 \end{cases} \tag{A.9}$$
$$p_1 = p_0 = {}^1/_2 \qquad q_1 = q_0 = {}^1/_2 \qquad r_1 = r_0 = {}^1/_2$$

that again can fit no trivariate distribution in a unique probability space.

It is crucial to point out moreover that the previously underlined circumstance does not pertain to the nature of the *probability spaces*, but it is rather a feature of the *families of distributions*. If indeed we would suppose *a priori* to be inside a given, *unique* probability space $(\Omega, \mathcal{F}, \boldsymbol{P})$, and if we take only the distributions defined from triples of events $A, B, C \in \mathcal{F}$ through relations such as

$$\begin{aligned} &p_{111} = \boldsymbol{P}\{ABC\} \quad p_{110} = \boldsymbol{P}\{AB\overline{C}\} \quad \dots \\ &p_{11} = \boldsymbol{P}\{AB\} \quad p_{10} = \boldsymbol{P}\{A\overline{B}\} \quad \dots \quad q_{11} = \boldsymbol{P}\{BC\} \quad \dots \\ &p_1 = \boldsymbol{P}\{A\} \quad p_0 = \boldsymbol{P}\{\overline{A}\} \quad q_1 = \boldsymbol{P}\{B\} \quad \dots \end{aligned}$$

it would be easy to show that they would always be perfectly consistent. The pointed out impossibility of finding trivariate laws consistent with arbitrary given bivariate and univariate ones appears instead only when we consider families of distributions *without a priori connecting them with a unique probability space*. From the example produced in the present appendix we can say indeed that families of univariate and bivariate laws with parameters of the type (A.7), while perfectly consistent among them, are not derivable as marginals of a unique trivariate, and hence can not be described as probabilities of events in a unique probability space. On the other hand it would be useful to remember that, while laws and distributions are directly connected with empirical observations, the probability spaces (albeit very important to give rigor to the theory) are theoretical constructs introduced with the aim of describing how the probabilities are combined: and in principle the model for these combinations could be different from that of the probability spaces defined in the Chap. 1.

The relevance of this last remark is better understood, however, if we consider a point which has been so far left in the background: it is all too natural indeed to ask why should we worry about the paradoxical behavior of a family of distributions so carefully tailored to be baffling as that in (A.8) or (A.9): has ever been observed in the reality some physical system displaying such an awkward behavior? Could not be this just an anomalous, but practically irrelevant case? Even the answer to this question, however, is rather surprising: the distributions (A.8), or (A.9) have not at all been chosen in a captious or malicious way, and are instead of a considerable conceptual interest. We will show now indeed that the conditional distributions[2] (A.9) are the *quantum mechanical distributions* (calculated with the usual procedures based on the square modulus of scalar products) of the possible values of the three observables $\hat{\boldsymbol{\alpha}} \cdot \boldsymbol{S}$, $\hat{\boldsymbol{\beta}} \cdot \boldsymbol{S}$, and $\hat{\boldsymbol{\gamma}} \cdot \boldsymbol{S}$ projecting the spin $\boldsymbol{S} = (\sigma_x, \sigma_y, \sigma_z)$ of the Pauli matrices on three versors $\hat{\boldsymbol{\alpha}}$, $\hat{\boldsymbol{\beta}}$ and $\hat{\boldsymbol{\gamma}}$ lying in the x, z plane at angles 0, $^{\pi}/_{4}$, $^{\pi}/_{2}$ with the z axis, in an initial eigenstate of σ_y (this example has been first pointed out in [1]).

The Cartesian components $\hat{\boldsymbol{v}} = (\sin\theta\cos\phi, \sin\theta\sin\phi, \cos\theta)$ of a versor in a three-dimensional space depend on both the angle $\theta \in [0, \pi]$ between $\hat{\boldsymbol{v}}$ and the z axis, and the angle $\phi \in [0, 2\pi]$ between its projection ont the x-y plane and the x axis. As a consequence the versors of our example have the following components

$$\hat{\boldsymbol{\alpha}} = (0, 0, 1) \qquad \hat{\boldsymbol{\beta}} = \left(^{\sqrt{2}}/_{2},\, 0,\, ^{\sqrt{2}}/_{2}\right) \qquad \hat{\boldsymbol{\gamma}} = (1, 0, 0)$$

the spin projections are

$$\begin{aligned}
\hat{\boldsymbol{\alpha}} \cdot \boldsymbol{S} &= \sigma_z = \begin{pmatrix} 1 & 0 \\ 0 & -1 \end{pmatrix} \\
\hat{\boldsymbol{\beta}} \cdot \boldsymbol{S} &= \frac{\sqrt{2}}{2}(\sigma_x + \sigma_z) = \frac{\sqrt{2}}{2}\begin{pmatrix} 1 & 1 \\ 1 & -1 \end{pmatrix} \\
\hat{\boldsymbol{\gamma}} \cdot \boldsymbol{S} &= \sigma_x = \begin{pmatrix} 0 & 1 \\ 1 & 0 \end{pmatrix}
\end{aligned}$$

while the system is supposed to be in an eigenstate of

$$\sigma_y = \begin{pmatrix} 0 & -i \\ i & 0 \end{pmatrix}$$

It is easy to check now that the previous four observables have eigenvalues ± 1, that the orthonormal systems of eigenvectors of the spin projections are

$$\begin{aligned}
|\,\alpha+\rangle &= \begin{pmatrix} 1 \\ 0 \end{pmatrix} & |\,\beta+\rangle &= \tfrac{\sqrt{2-\sqrt{2}}}{2}\begin{pmatrix} 1 \\ \sqrt{2}-1 \end{pmatrix} & |\,\gamma+\rangle &= \tfrac{1}{\sqrt{2}}\begin{pmatrix} 1 \\ 1 \end{pmatrix} \\
|\,\alpha-\rangle &= \begin{pmatrix} 0 \\ 1 \end{pmatrix} & |\,\beta-\rangle &= \tfrac{\sqrt{2-\sqrt{2}}}{2}\begin{pmatrix} 1 \\ -\sqrt{2}-1 \end{pmatrix} & |\,\gamma-\rangle &= \tfrac{1}{\sqrt{2}}\begin{pmatrix} 1 \\ -1 \end{pmatrix}
\end{aligned}$$

[2] It is expedient here to use the conditional distributions (A.9) rather than the joint bivariate distributions (A.8) because in quantum mechanics we can not calculate *joint* distributions when the observables do not commute, while the corresponding *conditional* distributions are always available.

and finally that the two orthonormal eigenvectors of σ_y (possible states of our system) are

$$|\,y+\rangle = \frac{1}{\sqrt{2}}\begin{pmatrix}1\\ i\end{pmatrix} \qquad |\,y-\rangle = \frac{1}{\sqrt{2}}\begin{pmatrix}1\\ -i\end{pmatrix}$$

Take now $|\,y+\rangle$ as the system state: if we call p_i, q_j e r_k the distributions respectively of $\hat{\boldsymbol{\alpha}}\cdot\boldsymbol{S}$, $\hat{\boldsymbol{\beta}}\cdot\boldsymbol{S}$ and $\hat{\boldsymbol{\gamma}}\cdot\boldsymbol{S}$, we first find in agreement with (A.9)

$$\begin{array}{lll} p_1 = |\langle\alpha+|y+\rangle|^2 = {}^1/_2 & q_1 = |\langle\beta+|y+\rangle|^2 = {}^1/_2 & r_1 = |\langle\gamma+|y+\rangle|^2 = {}^1/_2 \\ p_0 = |\langle\alpha-|y+\rangle|^2 = {}^1/_2 & q_0 = |\langle\beta-|y+\rangle|^2 = {}^1/_2 & r_0 = |\langle\gamma-|y+\rangle|^2 = {}^1/_2 \end{array}$$

As for the *conditional distributions* $p_{i|j}$, $q_{j|k}$, $r_{i|k}$ they will be then calculated from the usual quantum procedure as $|\langle\alpha\pm|\beta\pm\rangle|^2$, $|\langle\beta\pm|\gamma\pm\rangle|^2$, $|\langle\alpha\pm|\gamma\pm\rangle|^2$, so that, by using the explicit form of our eigenvectors, we get the following conditional probabilities

$$\begin{cases} p_{1|1} = |\langle\alpha+|\beta+\rangle|^2 = \frac{2+\sqrt{2}}{4} \\ p_{1|0} = |\langle\alpha+|\beta-\rangle|^2 = \frac{2-\sqrt{2}}{4} \\ p_{0|1} = |\langle\alpha-|\beta+\rangle|^2 = \frac{2-\sqrt{2}}{4} \\ p_{0|0} = |\langle\alpha-|\beta-\rangle|^2 = \frac{2+\sqrt{2}}{4} \end{cases}$$
$$\begin{cases} q_{1|1} = |\langle\beta+|\gamma+\rangle|^2 = \frac{2+\sqrt{2}}{4} \\ q_{1|0} = |\langle\beta+|\gamma-\rangle|^2 = \frac{2-\sqrt{2}}{4} \\ q_{0|1} = |\langle\beta-|\gamma+\rangle|^2 = \frac{2-\sqrt{2}}{4} \\ q_{0|0} = |\langle\beta-|\gamma-\rangle|^2 = \frac{2+\sqrt{2}}{4} \end{cases}$$
$$\begin{cases} r_{1|1} = |\langle\alpha+|\gamma+\rangle|^2 = {}^1/_2 \\ r_{1|0} = |\langle\alpha+|\gamma-\rangle|^2 = {}^1/_2 \\ r_{0|1} = |\langle\alpha-|\gamma+\rangle|^2 = {}^1/_2 \\ r_{0|0} = |\langle\alpha-|\gamma-\rangle|^2 = {}^1/_2 \end{cases}$$

which are nothing else than (A.9), and hence could not possibly fit any trivariate distribution in a unique probability space. In other words, the systems of (univariate and conditional) distributions coming from quantum mechanics can not in generale be coherently shoehorned into a (unique) classical probabilistic model.

In short, our example shows that there are quantum systems that do not allow a coherent description within the framework of a unique probability space, and consequently brings to the fore *the probabilistic roots of the quantum paradoxes*. It is well known, on the other hand, that the probabilistic models of the quantum mechanics are not centered around a probability space $(\Omega, \mathcal{F}, \boldsymbol{P})$, but are rather related to states as vectors in some Hilbert space with all the aftereffects we know. The discussion in the present appendix, however, hints also that having conditional distributions not consistent with a unique probability space is an open possibility even independently from quantum models (albeit these seem today to be the only available concrete examples). In other words, there is more in the multivariate families of laws than there is within the framework of Kolmogorov probability spaces, so that the possibility of having conditional distributions which behave in a *quantum* way is already allowed in the usual probability if we drop any reference to probability spaces. The inconsistencies recalled here are indeed known since longtime and have motivated many inquiries to find general conditions for the existence of Kolmogorovian models for given families of laws: in this perspective the celebrated *Bell inequalities* (proved in the 60's within a discussion about the Einstein-Podolski-Rosen paradox: see [2]) can be considered as an example of such conditions that apparently are not always satisfied by the quantum systems.

Appendix B
Inequalities (Sect. 3.3.2)

In the present appendix we will draw attention on a few important integral inequalities that—in their present probabilistic formulation—will be used in the text.

Proposition B.1 Jensen inequality: *If $g(x)$ is a convex (downward) Borel function, and if X is an integrable rv, it is*

$$g\left(\boldsymbol{E}\left[X\right]\right) \leq \boldsymbol{E}\left[g(X)\right]$$

Proof Jensen inequality is a rather general property instrumental in the proof of the subsequent propositions. If $g(x)$ is downward convex, for every $x_0 \in \boldsymbol{R}$ it exists a number $\lambda(x_0)$ such that

$$g(x) \geq g(x_0) + (x - x_0)\lambda(x_0)\,, \qquad \forall x \in \boldsymbol{R}$$

By replacing then x_0 with $\boldsymbol{E}\left[X\right]$, and computing the functions in X we get

$$g(X) \geq g(\boldsymbol{E}\left[X\right]) + (X - \boldsymbol{E}\left[X\right])\lambda(\boldsymbol{E}\left[X\right])$$

and the result follows by taking the expectation of both sides of this equation ■

Corollary B.1 Lyapunov inequality: *If X is a rv we have*

$$\boldsymbol{E}\left[|X|^s\right]^{1/s} \leq \boldsymbol{E}\left[|X|^t\right]^{1/t} \qquad 0 < s \leq t$$

In particular it is

$$\boldsymbol{E}\left[|X|\right] \leq \boldsymbol{E}\left[|X|^2\right]^{1/2} \leq \cdots \leq \boldsymbol{E}\left[|X|^n\right]^{1/n} \leq \cdots$$

Proof Take $r = t/s \geq 1$ and $Y = |X|^s$ and then use Jensen inequality with the convex function $g(x) = |x|^r$ to get$|\boldsymbol{E}\left[Y\right]|^r \leq \boldsymbol{E}\left[|Y|^r\right]$, namely

N. Cufaro Petroni, *Probability and Stochastic Processes for Physicists*, UNITEXT for Physics, https://doi.org/10.1007/978-3-030-48408-8

$$\boldsymbol{E}\left[|X|^s\right]^{t/s} \leq \boldsymbol{E}\left[|X|^t\right]$$

and the result follows at once. The subsequent inequality chain is just a particular case. A relevant implication of the Lyapunov inequality is that if a *rv* X has a finite absolute moment of order r ($\boldsymbol{E}\left[|X|^r\right] < +\infty$), then all the absolute moments of an order lesser than r are also finite; this instead is not true in general for the absolute moments of order larger than r ■

Proposition B.2 Hölder inequality: *Take two numbers* p, q *with*

$$1 < p < +\infty \qquad 1 < q < +\infty \qquad \frac{1}{p} + \frac{1}{q} = 1$$

and the rv's X, Y *with* $\boldsymbol{E}\left[|X|^p\right] < +\infty$ *and* $\boldsymbol{E}\left[|Y|^q\right] < +\infty$*: then the product* XY *is also integrable, and we have*

$$\boldsymbol{E}\left[|XY|\right] \leq \boldsymbol{E}\left[|X|^p\right]^{1/p} \boldsymbol{E}\left[|Y|^q\right]^{1/q}$$

Remark that the well known ***Schwarz inequality***

$$\boldsymbol{E}\left[|XY|\right]^2 \leq \boldsymbol{E}\left[X^2\right] \boldsymbol{E}\left[Y^2\right]$$

is a particular case of the Hölder inequality for $p = q = 2$.

Proof Omitted: see [3] p. 193. We will recall here just the proof of the Schwarz inequality. Consider first the case $\boldsymbol{E}\left[X^2\right] \neq 0$ and $\boldsymbol{E}\left[Y^2\right] \neq 0$ and take

$$\tilde{X} = \frac{X}{\sqrt{\boldsymbol{E}\left[X^2\right]}} \qquad \tilde{Y} = \frac{Y}{\sqrt{\boldsymbol{E}\left[Y^2\right]}}$$

so that $\boldsymbol{E}\left[\tilde{X}^2\right] = 1$ and $\boldsymbol{E}\left[\tilde{Y}^2\right] = 1$. Since $\left(|\tilde{X}| - |\tilde{Y}|\right)^2 \geq 0$, and hence

$$2\,|\tilde{X}\,\tilde{Y}| \leq \tilde{X}^2 + \tilde{Y}^2$$

we have

$$2\,\boldsymbol{E}\left[|\tilde{X}\,\tilde{Y}|\right] \leq \boldsymbol{E}\left[\tilde{X}^2\right] + \boldsymbol{E}\left[\tilde{Y}^2\right] = 2$$

namely

$$\boldsymbol{E}\left[|\tilde{X}\,\tilde{Y}|\right]^2 \leq 1 = \boldsymbol{E}\left[\tilde{X}^2\right] \cdot \boldsymbol{E}\left[\tilde{Y}^2\right]$$

and the result follows by making use of the definitions of $\tilde{X}$ and $\tilde{Y}$ in terms of X and Y. When instead at least one of the expectations vanishes, for instance if $\boldsymbol{E}\left[X^2\right] = 0$, from 5 of Proposition 3.26 we get $X = 0$ $\boldsymbol{P}$-a.s., and hence from 3 of the same proposition we have also $\boldsymbol{E}\left[|XY|\right] = 0$. It is straightforward then to see how the result follows even in this case ■

Proposition B.3 Minkowski inequality: *Given the number p with*

$$1 \leq p < +\infty$$

and two rv's X, Y *such that* $\boldsymbol{E}\left[|X|^p\right] < +\infty$ *and* $\boldsymbol{E}\left[|Y|^p\right] < +\infty$, *then also* $\boldsymbol{E}\left[|X+Y|^p\right] < +\infty$ *and we have*

$$\boldsymbol{E}\left[|X+Y|^p\right]^{1/p} \leq \boldsymbol{E}\left[|X|^p\right]^{1/p} + \boldsymbol{E}\left[|Y|^p\right]^{1/p}$$

Proof Omitted: see [3] p. 194 ■

Appendix C
Bertrand's Paradox (Sect. 3.5.1)

In the first chapter of his classic book [4] Joseph Bertrand dwells for a while on the definition of probability, and in particular he remarks that the random models with an *uncountable* number of possible results are prone to particularly insidious misunderstandings. If for example we ask what is the probability that a real number chosen at random between 0 and 100 is larger than 50, our natural answer is $\frac{1}{2}$. Since however the real numbers between 0 and 100 are also bijectively associated to their squares between 0 and 10 000, we also feel that our question should be equivalent to ask for the probability that our random number turns out to be larger than $50^2 = 2\,500$. If however we take at random a number between 0 and 10 000, intuitively again the probability of exceeding 2 500 would now be $\frac{3}{4}$ instead of $\frac{1}{2}$. The two problems look equivalent, but their two answers (apparently both legitimates) are different: what is the root of this paradox? Bertrand states—correctly—that the two questions are fallacious because the locution *at random* is too careless, as a few other examples could show: he listed many telling cases, but we will linger for a while only on the following one which is widely acknowledged as the *Bertrand paradox*.

Looking at the Fig. C.1, take *at random* a chord on the radius 1 circle Γ: what is the probability that its length exceeds that of the edge of an inscribed equilateral triangle (namely $\sqrt{3}$)? Three acceptable answers are possible, but they are all numerically different (in the following we will always make reference to the Fig. C.1):

1. To take a chord at random is equivalent to choose the location of its middle point (its orientation would be an aftermath), and to get the chord longer than the triangle edge it is necessary and sufficient to take this middle point inside the concentric circle γ with radius $\frac{1}{2}$ inscribed in the triangle. The required probability is then the ratio between the area $\frac{\pi}{4}$ of γ and the area π of Γ, and consequently we have $p_1 = \frac{1}{4}$
2. By symmetry the position of one chord endpoint along the circle is immaterial to our calculations: then, for a given endpoint, the chord length will only be contingent on the angle (between 0 and π) with the tangent line τ in the chosen endpoint. If then we draw the triangle with one vertex in the chosen endpoint, the

N. Cufaro Petroni, *Probability and Stochastic Processes for Physicists*, UNITEXT for Physics, https://doi.org/10.1007/978-3-030-48408-8

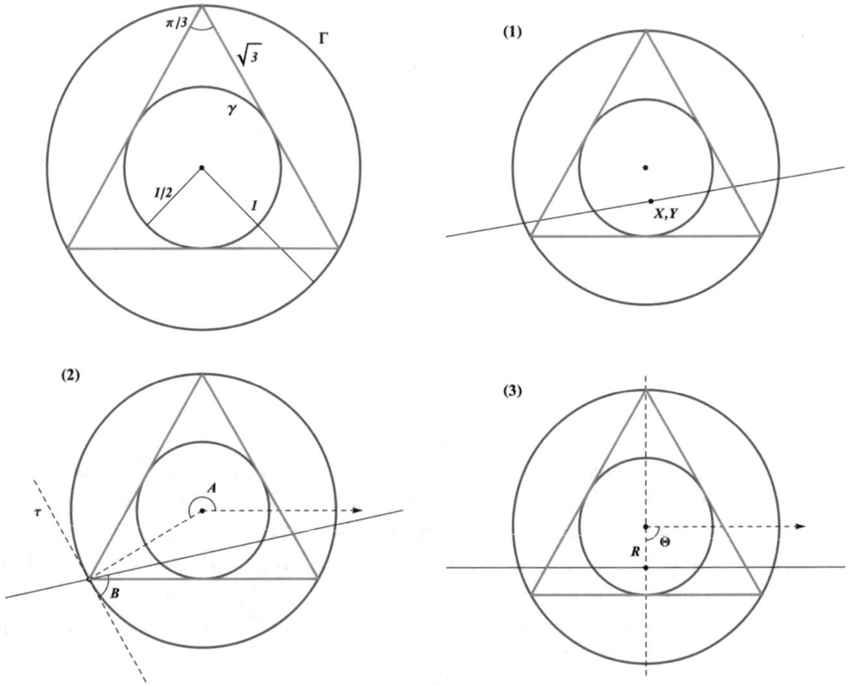

Fig. C.1 Bertrand's paradox

chord at random will exceed its edge if the angle with the tangent falls between $\frac{\pi}{3}$ and $\frac{2\pi}{3}$, and the corresponding probability will be $p_2 = \frac{1}{3}$

3. Always by symmetry, also the random chord direction does not affect the required probability. Fix then such a direction, and remark that the chord will exceed $\sqrt{3}$ if its intersection with the orthogonal diameter falls within a distance from the center smaller than $\frac{1}{2}$: this happens with probability $p_3 = \frac{1}{2}$.

To find our paradox origin we must remember that taking a number *at random* usually means that this number is *uniformly* distributed in some interval. It is possible to show however that what is considered as uniformly distributed in every one of the three proposed solutions can not be at the same time uniformly distributed in the other two: in other words, in our three solutions—by differently choosing what is uniformly distributed—we surreptitiously adopt three different probability measures, and consequently it is not astonishing that the three answers mutually disagree.

To be more precise let us define (see Fig. C.1) the three *rv* pairs representing the coordinates describing the position of our chord in the three proposed solutions:

1. the Cartesian coordinates (X, Y) of the chord middle point
2. the angles (A, B) respectively giving the position of the fixed endpoint and the chord orientation w.r.t. the tangent
3. the polar coordinates (R, Θ) of the chord-diameter intersection.

In every instance however there is the concealed (namely not explicitly acknowledged) hypothesis that the corresponding pair of coordinates is uniformly distributed, but these three assumptions are not mutually consistent, as we will see at once, because they require three different probability measures on the probability space where all our *rv*'s are defined. In particular the three solutions respectively assume the following uniform, joint distributions (here $\chi_{[a,b]}(x)$ is an indicator):

1. the joint, uniform *pdf* on $\boldsymbol{R}^2$

$$f_{XY}(x.y) = \frac{1}{\pi}\,\chi_{[0,1]}(x^2+y^2) \tag{C.1}$$

   of the pair (X, Y): here the two *rv*'s are *not* independent
2. the joint, uniform *pdf* on $\boldsymbol{R}^2$

$$f_{AB}(\alpha,\beta) = \frac{1}{2\pi^2}\,\chi_{[0,2\pi]}(\alpha)\chi_{[0,\pi]}(\beta) \tag{C.2}$$

   of the pair (A, B) with independent components
3. and finally the joint, uniform *pdf* on $\boldsymbol{R}^2$

$$f_{R\Theta}(r,\theta) = \frac{1}{2\pi}\,\chi_{[0,1]}(r)\chi_{[-\pi,\pi]}(\theta) \tag{C.3}$$

   of the pair (R, Θ) again with independent components.

Surely enough if we would adopt a unique probability space for our three solutions, the three numerical results would be exactly coincident, but in this case only one of the three *rv* pairs could be uniformly distributed, while the other joint distributions should be derived from the results of the Sect. 3.47 for the functions of *rv*'s. The crucial point is that there are indeed a few precise transformations allowing to go from a pair of our *rv*'s to the other: by using these transformations we can show that if a pair is jointly uniform, then the other two can not have the same property.

Without going into the details of every possible combination we will confine ourselves to discuss just the relations between the solutions (1) and (3). The transformations between the Cartesian coordinates (X, Y) and the polar ones (R, Θ) are well known:

$$\begin{cases} x = r\cos\theta \\ y = r\sin\theta \end{cases} \qquad \begin{cases} r = \sqrt{x^2+y^2} \\ \theta = \arctan\frac{y}{x} \end{cases} \qquad \begin{matrix} r > 0 \\ -\pi < \theta \le \pi \end{matrix}$$

with a Jacobian determinant

$$J(r,\theta) = \begin{vmatrix} \frac{\partial r}{\partial x} & \frac{\partial r}{\partial y} \\ \frac{\partial \theta}{\partial x} & \frac{\partial \theta}{\partial y} \end{vmatrix} = \begin{vmatrix} \cos\theta & \sin\theta \\ -\frac{1}{r}\sin\theta & \frac{1}{r}\cos\theta \end{vmatrix} = \frac{1}{r}$$

As a consequence, if (X, Y) have the jointly uniform *pdf* (C.1), then the joint law of the pair (R, Θ) must be deduced from (3.63) and will not be uniform: it will have instead the *pdf*

$$f^{(1)}_{R\Theta}(r, \theta) = \frac{r}{\pi}\,\chi_{[0,1]}(r)\chi_{[-\pi,\pi]}(\theta)$$

apparently different from the $f_{R\Theta}$ in (C.3). By taking advantage of this distribution $f^{(1)}_{R\Theta}$ it is easy to see now that also the probability in the framework of the solution (3) would be

$$p_3 = \int_0^{\frac{1}{2}} \frac{r}{\pi}\,dr \int_{-\pi}^{\pi} d\theta = \frac{1}{4}$$

in perfect agreement with the solution (1).

It is important to remark in conclusion that—as already pointed out at the beginning of this appendix—the Bertrand-type paradoxes arise only when we consider probability measures on uncountable sets. To clarify this last point it would be enough to resume our initial problem of calculating the probability $p^{(1)}$ that a real number X taken *at random* in $[0, 100]$ exceeds 50: this we would readily concede to be $p^{(1)} = \frac{1}{2}$. The paradox appears when we try to calculate the probability $p^{(2)}$ that the square of our real number X^2 taken *at random* in $[0, 10\,000]$ exceeds $50^2 = 2\,500$, because in this case we are spontaneously bent to think that it should now be $p^{(2)} = \frac{1}{4}$. But the fact is—as in the previous examples—that if X is uniform in $[0, 100]$, then X^2 can not be uniform in $[0, 10\,000]$, and vice-versa. In this case however it is easy to see that the paradox does not show up when we ask for the probability ($p^{(1)} = \frac{1}{2}$) of choosing at random an *integer number* larger than 50 among the (equiprobable) numbers from 1 to 100. In this case in fact we would have the same answer ($p^{(2)} = \frac{1}{2}$) also for the question of calculating the probability of choosing *at random* a number larger than 2 500 among the squared integers $1, 4, 9, \dots, 10\,000$, because now our set is again constituted of just 100 equiprobable integers.

Appendix D
L^p Spaces of *rv*'s (Sect. 4.1)

The symbol $L^p(\Omega, \mathcal{F}, \boldsymbol{P})$, or even L^p, denotes the set of *rv*'s defined on $(\Omega, \mathcal{F}, \boldsymbol{P})$ with $\boldsymbol{E}\left[|X|^p\right] < +\infty$ and $p > 0$. These sets can be equipped of geometric structures especially suitable for the applications. Remark first that,for every $p > 0$, we can always give them a ***metric***, namely a distance between two *rv*'s defined as

$$\boldsymbol{d}(X, Y) = \boldsymbol{E}\left[|X - Y|^p\right]^{1/p}$$

In this case L^p is a *metric space*. If moreover $p \geq 1$, the Minkowski inequality (Proposition B.3) enables us to state that L^p is also a ***vector space*** such that linear combinations of its elements again are in L^p. On these vector spaces L^p it is also possible to define a ***norm***, namely a length of the vectors $X \in L^p$ defined as

$$\|X\|_p = \boldsymbol{E}\left[|X|^p\right]^{1/p}$$

and hence also the convergence toward X of the sequences $(X_n)_{n \in \boldsymbol{N}}$ as the numerical convergence toward zero $\|X_n - X\|_p \to 0$. since these *normed spaces* are also *complete*,[3] they are ***Banach spaces***, where the distance is implemented through the norm as

$$\boldsymbol{d}(X, Y) = \|X - Y\|_p$$

Remark that from the Lyapunov inequality (Corollary B.1) we immediately conclude that

[3]In a normed space $(\mathcal{E}, \|\cdot\|)$ a sequence $(x_n)_{n \in \boldsymbol{N}}$ is a *Cauchy sequence* when

$$\lim_{n,m} \|x_n - x_m\| = 0$$

A normed space is said to be *complete* if every Cauchy sequence of elements of $\mathcal{E}$ converges toward another element of $\mathcal{E}$. In this case $(\mathcal{E}, \|\cdot\|)$ is also called a *Banach space*.

N. Cufaro Petroni, *Probability and Stochastic Processes for Physicists*, UNITEXT for Physics, https://doi.org/10.1007/978-3-030-48408-8

$$\|X\|_1 \leq \|X\|_p \leq \|X\|_q \qquad 1 \leq p \leq q < +\infty$$

As a consequence, if $1 \leq p \leq q$ and $X \in L^q$, then also $X \in L^p$, and therefore

$$L^1 \supseteq L^p \supseteq L^q \qquad 1 \leq p \leq q < +\infty$$

Among the Banach spaces L^p with $p \geq 1$, an especially relevant role is played by then case $p = 2$, namely by the space $L^2(\Omega, \mathcal{F}, \boldsymbol{P})$: it is easy to show in fact that in this case the norm $\|\cdot\|_2$ can be implemented through a ***scalar product***

$$\langle X, Y \rangle = \boldsymbol{E}[XY]$$

in the sense that in L^2 we have

$$\|X\|_2 = \sqrt{\langle X, X \rangle} = \sqrt{\boldsymbol{E}\left[X^2\right]}$$

The spaces equipped with a scalar product, when they are also complete, take the name of ***Hilbert spaces***. The existence of a scalar product in a probability space allows not only to use of functional analysis methods, but also to extend notions borrowed from the geometry. We will say for instance that two *rv*'s $X, Y \in L^2$ are ***orthogonal*** when $\langle X, Y \rangle = \boldsymbol{E}[XY] = 0$, and we will say that a set of *rv*'s in L^2 is an ***orthogonal system*** when however taken among them two different *rv*'s they are orthogonal. If moreover the elements of an orthogonal system are also normalized, that is $\|X\|_2 = 1$ for every element, then the set constitutes an ***orthonormal system***. Remark finally that, if two *rv*'s are not correlated we find

$$\langle X, Y \rangle = \boldsymbol{E}[XY] = \boldsymbol{E}[X]\,\boldsymbol{E}[Y]$$

so that they are orthogonal *iff* at least one has a vanishing expectation.

Appendix E
Moments and Cumulants (Sect. 4.2.1)

If all the moments $m_n = \boldsymbol{E}[X^n]$ of a *rv* X exist and are finite, the Theorem 4.11 states that we can write down the power expansion of the *chf* $\varphi(u)$ of X; moreover the Theorems 4.12 and 4.13 say that the *chf* $\varphi(t)$ uniquely determines the *pdf* $f(x)$ of X (that for simplicity's sake we suppose to be *ac*). It makes then sense to ask the following question known as ***moments problem***: can we trace back in a unique way the *pdf* $f(x)$ of a *rv* X from the knowledge of its moments $(m_n)_{n\in N}$? In particular the problem of uniqueness can be stated as follows: given two *pdf*'s $f(x)$ and $g(x)$ such that

$$\int_{-\infty}^{+\infty} x^n f(x)\,dx = \int_{-\infty}^{+\infty} x^n g(x)\,dx\,, \qquad n \geq 1$$

can we conclude that $f(x) = g(x)$ for every x? As a matter of fact it is possible to show with counterexamples (see [3] p. 294) that in general the answer is in the negative: it is possible indeed to explicitly produce different distributions that have the same sequence of momenta. it will therefore be important to establish under what sufficient conditions the moment problem admits one, and only one solution.

Theorem E.1 *Take a rv X and its moments* $m_n = \boldsymbol{E}[X^n]$ *and* $\mu_n = \boldsymbol{E}[|X|^n]$*: if all the absolute moments* μ_n *are finite and if*

$$\overline{\lim_n}\,\frac{\mu_n^{1/n}}{n} < +\infty$$

then the moments m_n *prescribe in a unique way the law of X. These sufficient conditions are in particular definitely met when the distribution of X is concentrated in a limited interval.*

Proof Omitted: see [3] p. 295 ■

The formula 4.18 of the Theorem 4.11 about the series expansion of the *chf* of a *rv* X can moreover be extended to the *chf* $\varphi(\boldsymbol{u})$ of *r-vec* $\boldsymbol{X} = (X_1, \ldots, X_n)$ taking the form

N. Cufaro Petroni, *Probability and Stochastic Processes for Physicists*, UNITEXT for Physics, https://doi.org/10.1007/978-3-030-48408-8

$$\varphi(u_1, \ldots, u_n) = \sum_{\{\boldsymbol{k}\}} \frac{i^{|\boldsymbol{k}|}}{k_1! \ldots k_n!} u_1^{k_1} \cdot \ldots \cdot u_n^{k_n} m_n(k_1, \ldots, k_n)$$

where for short we have set $\{\boldsymbol{k}\} = \{k_1, \ldots, k_n\}$ and $|\boldsymbol{k}| = k_1 + \ldots + k_n$, while

$$m_n(\boldsymbol{u}) = m_n(k_1, \ldots, k_n) = \boldsymbol{E}\left[X_1^{k_1} \cdot \ldots \cdot X_n^{k_n}\right]$$

are the *mixed moments* of the components of $\boldsymbol{X}$. Also this expansion is of course cut down to a finite sum with an infinitesimal remainder (Taylor formula) if the moments do not exist from a certain order onward.

It is helpful now to define also the *logarithmic characteristic* of the *rv* X

$$\eta(u_1, \ldots, u_n) = \ln \varphi(u_1, \ldots, u_n)$$

that is sometimes used instead of the *chf*. This is indeed often easier to handle than the φ and its properties can be more straightforward to study. For example for a Gaussian *rv* $\mathfrak{N}(b, a^2)$ it is

$$\eta(u) = ibu - \frac{a^2u^2}{2}$$

while for a Cauchy $\mathfrak{C}(a, b)$ it is

$$\eta(u) = ibu - a|u|$$

and for a Poisson $\mathfrak{P}(\alpha)$ we have

$$\eta(u) = \alpha(e^{iu} - 1)$$

Also a logarithmic characteristic of a *r-vec* admits (with the required clarifications on the existence of the moments) a series expansion of the type

$$\eta(u_1, \ldots, u_n) = \sum_{\{\boldsymbol{k}\}} \frac{i^{|\boldsymbol{k}|}}{k_1! \ldots k_n!} u_1^{k_1} \cdot \ldots \cdot u_n^{k_n} c_n(k_1, \ldots, k_n)$$

but its coefficients $c_n(k_1, \ldots, k_n)$, called ***cumulants***, no longer are just the expectation values of *rv*'s products. By comparing the two expansions it is however possible to deduce the relations between the cumulants and the mixed moments of the components of $\boldsymbol{X}$: for instance we find (here the choice of the non zero indices is arbitrary and only illustrative)

$$\begin{aligned}
c_n(1, 0, 0, \ldots, 0) &= m_n(1, 0, 0, \ldots, 0) \\
c_n(1, 1, 0, \ldots, 0) &= m_n(1, 1, 0, \ldots, 0) - m_n(1, 0, 0, \ldots, 0)m_n(0, 1, 0, \ldots, 0) \\
c_n(1, 1, 1, \ldots, 0) &= m_n(1, 1, 1, \ldots, 0) - m_n(1, 1, 0, \ldots, 0)m_n(0, 0, 1, \ldots, 0) \\
&\quad - m_n(1, 0, 1, \ldots, 0)m_n(0, 1, 0, \ldots, 0) \\
&\quad - m_n(0, 1, 1, \ldots, 0)m_n(1, 0, 0, \ldots, 0) \\
&\quad + 2m_n(1, 0, 0, \ldots, 0)m_n(0, 1, 0, \ldots, 0)m_n(0, 0, 1, \ldots, 0)
\end{aligned}$$

The complete relations are rather involuted and we will ignore them (for details see [3] p. 290-1 and [5] p. 34), but we will remark that the value of the cumulants with more than one non zero index is a measure of the correlation between the corresponding components X_k. If for instance the X_k are all independent the *chf* is factorized and hence $\eta(u_1, \ldots, u_n)$ is the sum of n terms, each dependent on one u_k only. In this case it is easy to see from the cumulant expansion that the c_n with more than one non zero index identically vanish.

Finally, while—because of Lyapunov inequality $\boldsymbol{E}\left[X^n\right]^2 \leq \boldsymbol{E}\left[X^{2n}\right]$—the moments can not be all zero from a certain order onward (all the moments contain relevant information), for the cumulants this is possible at least in special cases. It is possible to show in particular (see [3] p. 288) that if $\eta(\boldsymbol{u})$ is a polynomial, its degree can not exceed 2: see for example the logarithmic characteristic of $\mathfrak{N}(b, a^2)$. As a consequence either all the cumulants vanish except the first two, or the number of non zero cumulants is infinite.

Appendix F
Binomial Limit Theorems (Sect. 4.3)

The earliest versions of the limit theorems (beginning of the XVIII century) basically pertained to sequences of binomial *rv*'s and were proved by exploiting the analytic properties of these particular distributions. The modern variants discussed in the Chap. 4 instead, while validating substantially the same results, cover much more general contexts and use more advanced demonstration techniques. In this appendix we will briefly summarize some of the said archaic forms of the limit theorems that still retain their suggestive power.

The oldest theorem due to J. Bernoulli [6] starts by remarking that if the *rv*'s of the sequence $(X_n)_{n\in N}$ are *iid* $\mathfrak{B}(1; p)$—they may represent the results of white and black ball drawings according to the Bernoulli model of the Sects. 2.1.2 and 3.2.4—the sums $S_n = X_1 + \cdots + X_n$ are binomial $\mathfrak{B}(n; p)$: as a consequence we know that

$$\boldsymbol{E}[S_n] = np \qquad \boldsymbol{V}[S_n] = np(1-p)$$

This leads to the remark that the expectation of the *rv empirical frequency* S_n/n also coincide with the *probabilità* p of drawing a white ball in every single trial:

$$\boldsymbol{E}\left[\frac{S_n}{n}\right] = p$$

The frequency S_n/n however is a *rv*, not a number as p is, and hence its random value will not in general coincide with p in a single n-tuple of drawings. It is important then to assess how far the *rv frequency* S_n/n deviates from its expectation (that is from the *probability* p) in order to appraise the confidence level of a possible estimation of p (in general not known) through the empirical value of the frequency S_n/n. It is apparent indeed that the unique quantity available to the empirical observations is a frequency counting, and not the value p of an *a priori* probability. We could say that the foundational problem of the ***statistics*** is to determine under what conditions a measurement of the empirical frequency S_n/n allows a *reliable estimation* of p. We

N. Cufaro Petroni, *Probability and Stochastic Processes for Physicists*, UNITEXT for Physics, https://doi.org/10.1007/978-3-030-48408-8

will show now in what sense the difference between frequency and a priori probability can be deemed to be small when n is large enough.

Theorem F.1 Bernoulli Law of Large Numbers: *Take a sequence S_n binomial rv's $\mathfrak{B}(n; p)$: then it is*

$$\frac{S_n}{n} \xrightarrow{P} p$$

Proof From the Chebyshev inequality (3.42), and from the properties of the binomial *rv's* $\mathfrak{B}(n; p)$ we have

$$\boldsymbol{P}\left\{\left|\frac{S_n}{n} - p\right| \geq \epsilon\right\} \leq \frac{1}{\epsilon^2}\,\boldsymbol{V}\left[\frac{S_n}{n}\right] = \frac{1}{n^2\epsilon^2}\,\boldsymbol{V}[S_n] = \frac{np(1-p)}{n^2\epsilon^2} = \frac{p(1-p)}{n\epsilon^2} \leq \frac{1}{4n\epsilon^2}$$

that immediately leads to the required result according to the Definition 4.1 ■

Also the original de Moivre [7] version of the ***Central Limit Theorem*** was confined to sequences of binomial *rv*'s $\mathfrak{B}\left(n; \frac{1}{2}\right)$, and even the subsequent Laplace [8] variants still exploited the properties of sequences of $\mathfrak{B}(n; p)$ *rv*'s with $0 < p < 1$. These limit theorems were presented under multiple guises, but here we will restrict ourselves to the most popular only. Take a sequence of *iid* Bernoulli *rv*'s $X_n \sim \mathfrak{B}(1; p)$: we know that $S_n = X_1, \ldots, X_n \sim \mathfrak{B}(n; p)$, and that from (3.35) and (3.36) the standardized sums

$$S_n^* = \frac{S_n - np}{\sqrt{npq}} \tag{F.1}$$

will take the $n + 1$ (non integer) values

$$x_k = \frac{k - np}{\sqrt{npq}} \qquad k = 0, 1, \ldots, n$$

We have then from from (2.1)

$$\boldsymbol{P}\{S_n^* = x_k\} = \boldsymbol{P}\{S_n = k\} = p_n(k) = \binom{n}{k} p^k q^{n-k} = p_n\big(np + x_k\sqrt{npq}\big)$$

The classical formulation of the binomial limit theorems in point consists in asymptotical ($n \to \infty$) results that allow to express the probabilities of the *rv* S_n^* in terms of Gauss functions. We will not give them in their rigorous form that is rather tortuous (for details see [3] p. 55–63), but we will summarize only the essential results.

A first result known as ***Local Limit Theorem (LLT)*** is the rigorous formulation of the statement that, for large values of n, the values $p_n(k) = p_n\big(np + x_k\sqrt{npq}\big)$ of the binomial distribution are well approximated by a Gaussian function

$$\frac{e^{-(k-np)^2/2npq}}{\sqrt{2\pi npq}} = \frac{1}{\sqrt{npq}}\,\frac{e^{-x_k^2/2}}{\sqrt{2\pi}}$$

It must be said however that this approximation is good only if k is not too far from the expectation np of S_n, namely if x_k is not too far from 0. More precisely the *LLT* states that, for large values of n, there exist two sequences of positive numbers A_n and B_n such that

$$\boldsymbol{P}\{S_n = k\} \simeq \frac{e^{-(k-np)^2/2npq}}{\sqrt{2\pi npq}} \qquad \text{if } |k - np| \le A_n$$

$$\boldsymbol{P}\{S_n^* = x_k\} \simeq \frac{1}{\sqrt{npq}} \frac{e^{-x_k^2/2}}{\sqrt{2\pi}} \qquad \text{if } |x_k| \le B_n$$

The approximation instead is not so good if we move away from the center toward the tails of the distribution, namely if k is too far from np and x_k is too far from 0.

To remove these restrictions we move on to a second formulation known as ***Integral Limit Theorem (ILT)***. To this end remark first that, for given p and n, the numbers x_k are equidistant with

$$\Delta x_k = x_{k+1} - x_k = \frac{1}{\sqrt{npq}}$$

For $n \to \infty$ and x_k not too far from 0, the *LLT* entitles us to write

$$\boldsymbol{P}\{S_n^* = x_k\} = p_n\big(np + x_k\sqrt{npq}\big) \simeq \frac{e^{-x_k^2/2}}{\sqrt{2\pi}} \Delta x_k$$

Since $\Delta x_k \to 0$ for $n \to \infty$, the set of points x_k tend to cover all the real line, and hence, in a suitable sense, for large n and arbitrary $a < b$, we could expect that the value of

$$\boldsymbol{P}\{a < S_n^* \le b\} = \sum_{k:a<x_k\le b} p_n\big(np + x_k\sqrt{npq}\big) \simeq \sum_{k:a<x_k\le b} \frac{e^{-x_k^2/2}}{\sqrt{2\pi}} \Delta x_k$$

is well approximated by the integral

$$\int_a^b \frac{e^{-x^2/2}}{\sqrt{2\pi}} dx = \Phi(b) - \Phi(a)$$

where $\Phi(x)$ is the standard error function (2.16). The *ILT* states indeed that, for every $-\infty \le a < b \le +\infty$, and for $n \to \infty$ we always find

$$\boldsymbol{P}\{a < S_n^* \le b\} \to \int_a^b \frac{e^{-x^2/2}}{\sqrt{2\pi}} dx = \Phi(b) - \Phi(a)$$

that is, with $\alpha = np + a\sqrt{npq}$ and $\beta = np + b\sqrt{npq}$,

$$\boldsymbol{P}\{\alpha < S_n \le \beta\} \to \Phi\left(\frac{\beta - np}{\sqrt{npq}}\right) - \Phi\left(\frac{\alpha - np}{\sqrt{npq}}\right)$$

From a technical standpoint the difference between the two formulations of the binomial limit theorems is that, while in the *LLT* we compare the individual values of the a (discrete) standardized binomial distribution with those of a (continuous) standard normal function, in the *ILT* we compare sums of the said binomial with integrals of the standard Gaussian *pdf* on arbitrary intervals: this has the effect of making relatively negligible the local tail effects and hence of producing an unqualified convergence.

The usual proofs of these two theorems resort to rather convoluted analytical argumentations that we will neglect (see [3] pp. 55–63): we will instead once more highlight the advantages of the *chf*'s by giving an undemanding proof of the convergence in distribution of the standard binomials S_n^* in (F.1) to a standard normal $\mathfrak{N}(0, 1)$. If indeed $X_1, \dots, X_n$ are *iid* Bernoulli *rv*'s $\mathfrak{B}(1; p)$, taken

$$Y_k = \frac{X_k - p}{\sqrt{npq}} = \frac{X_k}{\sqrt{npq}} - \sqrt{\frac{p}{nq}}$$

we can write

$$S_n^* = \sum_{k=1}^{n} Y_k$$

and since from (4.3) and (4.8) we find

$$\varphi_{Y_k}(u) = \boldsymbol{E}\left[e^{iuY_k}\right] = e^{-iu\sqrt{p/nq}}\varphi_{X_k}\left(\frac{u}{\sqrt{npq}}\right) = p\,e^{iu\sqrt{q/np}} + q\,e^{-iu\sqrt{p/nq}}$$

the S_n^* *chf* turns out to be

$$\varphi_{S_n^*}(u) = \boldsymbol{E}\left[e^{iuS_n^*}\right] = \prod_{k=1}^{n} \boldsymbol{E}\left[e^{iuY_k}\right] = \left(p\,e^{iu\sqrt{q/np}} + q\,e^{-iu\sqrt{p/nq}}\right)^n$$

From a power expansion of the exponentials we then have

$$\begin{aligned}\varphi_{S_n^*}(u) &= \left[p\left(1 + iu\sqrt{\frac{q}{np}} - \frac{u^2}{2}\frac{q}{np}\right) + q\left(1 - iu\sqrt{\frac{p}{nq}} - \frac{u^2}{2}\frac{p}{nq}\right) + o\left(\frac{1}{n}\right)\right]^n \\ &= \left[1 - \frac{u^2}{2n} + o\left(\frac{1}{n}\right)\right]^n \xrightarrow{n} e^{-u^2/2}\end{aligned}$$

and hence from the Lévy Theorem 4.16 we get $S_n^* \xrightarrow{d} \mathfrak{N}(0, 1)$.

Appendix G
Non Uniform Point Processes (Sect. 6.1.1)

In the limiting procedure adopted to define the point processes in the Sect. 6.1.1 we have supposed the point distributions on every finite interval $\mathfrak{U}\,[-^T/_2, {}^T/_2]$ to be always uniform. This assumption however is not unavoidable and could be suitably revised imagining that the intensity of the dots shower may vary according to the place.

To scrutinize this idea remember first that, in the uniform case considered up to now the *rv* N enumerating the points falling in a given interval of width $\Delta t > 0$ turns out to be distributed according a Poisson law $\mathfrak{P}(\alpha)$ with $\alpha = \lambda\Delta t$, so that $\boldsymbol{E}[N] = \alpha = \lambda\Delta t$. Keeping then into account that

$$\alpha = \boldsymbol{E}[N] \to 0 \qquad \text{when} \qquad \Delta t \to 0$$

and adopting the notation $\Delta\nu = \alpha = \boldsymbol{E}[N] =$ *average number of points falling in an interval of width* Δt, we can also write

$$\lambda = \frac{\Delta\nu}{\Delta t} \longrightarrow \frac{d\nu}{dt} \qquad\qquad \Delta t \to 0$$

in compliance with the idea that λ represents the *average number of points per unit time*. A constant λ, as previously supposed, would embody the idea of a uniform points density, but we are also free to suppose that $\lambda(t)$ is in fact a time dependent density, so that

$$\lambda(t) = \frac{d\nu(t)}{dt} \qquad \text{namely} \qquad d\nu(t) = \lambda(t)\,dt$$

and hence

$$\alpha = \Delta\nu = \int_t^{t+\Delta t} \lambda(s)\,ds \tag{G.1}$$

N. Cufaro Petroni, *Probability and Stochastic Processes for Physicists*, UNITEXT for Physics, https://doi.org/10.1007/978-3-030-48408-8

If now N is the number of random points falling into $[t, t + \Delta t]$, retracing the same steps previously trodden for the uniform case we could show once again (see [9] p. 59, 291) that N is distributed according to a Poisson law $\mathfrak{P}(\alpha)$, but for the fact that now the value of α will be (G.1) and will be contingent not only on the interval width Δt, but also on its time location t. This entails in particular that—at variance with those of a simple Poisson process—the increments of a non uniform counting process (also known in the literature with the name of *non-homogeneous* counting processes) are no longer stationary because their distribution depends not only on their width Δt but also on their location t. Remark finally that $\lambda(t)$ is a density (measuring the average number of points per unit time), but it is not a *pdf*. Typically we find indeed that

$$\int_{-\infty}^{+\infty} \lambda(t)\, dt = +\infty$$

in agreement with the fact that such an integral represents the total (infinite) number of the points thrown on the *entire* time axis. In the main text we always suppose that a constant intensity λ, but the possible generalizations can always be easily elaborated by adopting the previous remarks as a stepping stone.

Appendix H
Stochastic Calculus Paradoxes (Sect. 6.4.2)

To show the mistakes one can incur by carelessly enforcing the usual rules of the calculus when dealing with stochastic processes, let us try to extend the Langevin heuristic procedure outlined in the Sect. 6.4.2 to a slightly different problem: the shot noise produced in the vacuum tubes by the random arrivals of individual electrons (see [5] pp. 11–15).

The random current $I(t)$ produced by the electrons will be modeled here as a shot noise with $h(t) = \vartheta(t)qe^{-at}$, so that, by keeping into account (6.66) with a Poisson white noise of intensity λ, our process will be

$$
\begin{aligned}
I(t) &= \sum_{k=1}^{\infty} h(t - T_k) = [h * \dot{N}](t) = \int_{-\infty}^{+\infty} h(t-s)\dot{N}(s)\,ds \\
&= qe^{-at}\int_{-\infty}^{t} e^{as}\dot{N}(s)\,ds
\end{aligned}
$$

By making use also of the white noise (6.65) derived from the compensated Poisson process in the Example 6.21, from the usual differentiation rules we then get

$$
\begin{aligned}
\dot{I}(t) &= -qae^{-at}\int_{-\infty}^{t} e^{as}\dot{N}(s)\,ds + q\dot{N}(t) \\
&= -aI(t) + q\dot{N}(t) = [\lambda q - aI(t)] + q\dot{\widetilde{N}}(t) \qquad \text{(H.1)}
\end{aligned}
$$

This is now a first order differential equation akin to that of Langevin (6.78), where however the role of the zero average fluctuating force $B(t)$ is played by $q\dot{\widetilde{N}}(t)$, a process that again will be supposed uncorrelated with $I(t)$. To study the $I(t)$ fluctuations we will look at the behavior of its variance

$$
\boldsymbol{V}[I(t)] = \boldsymbol{E}\left[I^2(t)\right] - \boldsymbol{E}[I(t)]^2
$$

and to do that we take the expectation of (H.1)

N. Cufaro Petroni, *Probability and Stochastic Processes for Physicists*, UNITEXT for Physics, https://doi.org/10.1007/978-3-030-48408-8

$$\frac{d}{dt}\boldsymbol{E}\left[I(t)\right] = \lambda q - a\boldsymbol{E}\left[I(t)\right]$$

so that we have

$$\boldsymbol{E}\left[I(t)\right] = \frac{\lambda q}{a} + Ce^{-at} \tag{H.2}$$

where C is an integration constant. To get the variance we must now calculate $\boldsymbol{E}\left[I^2(t)\right]$: multiplying (H.1) by $I(t)$, from the usual calculus rules we find first

$$\frac{1}{2}\frac{dI^2(t)}{dt} = I(t)\dot{I}(t) = \lambda q I(t) - aI^2(t) + qI(t)\dot{\tilde{N}}(t) \tag{H.3}$$

and then taking the expectation

$$\frac{1}{2}\frac{d}{dt}\boldsymbol{E}\left[I^2(t)\right] = \lambda q\boldsymbol{E}\left[I(t)\right]) - a\boldsymbol{E}\left[I^2(t)\right] \tag{H.4}$$

From (H.2) we thus have

$$\frac{d}{dt}\boldsymbol{E}\left[I^2(t)\right] + 2a\boldsymbol{E}\left[I^2(t)\right] = 2\lambda q\boldsymbol{E}\left[I(t)\right] = 2\lambda q\left(\frac{\lambda q}{a} + Ce^{-at}\right)$$

and with another integration constant A

$$\boldsymbol{E}\left[I^2(t)\right] = \left(\frac{\lambda q}{a}\right)^2 + C\frac{2\lambda q}{a}e^{-at} + Ae^{-2at}$$

The variance of our random current will finally be

$$\boldsymbol{V}\left[I(t)\right] = \boldsymbol{E}\left[I^2(t)\right] - \boldsymbol{E}\left[I(t)\right]^2 = \left(\frac{\lambda q}{a}\right)^2 + C\frac{2\lambda q}{a}e^{-at} + Ae^{-2at} - \left(\frac{\lambda q}{a} + Ce^{-at}\right)^2$$

and therefore asymptotically in time we paradoxically find

$$\lim_{t\to+\infty}\boldsymbol{V}\left[I(t)\right] = 0 \tag{H.5}$$

namely, after a transient delay, the fluctuations just vanish, while we could have reasonably expected a convergence toward some constant non-zero variance. Let us scrutinize this baffling result in more detail.

Take again the—seemingly undisputable—relation adopted in (H.3):

$$\frac{dI^2(t)}{dt} = 2I(t)\dot{I}(t)$$

and, in the light of the discussion of Sect. 6.3, retrace its usual justification. Habitually with an infinitesimal dt we write

$$d\left[I^2(t)\right] = I^2(t+dt) - I^2(t) = \left[I(t) + dI(t)\right]^2 - I^2(t) = 2I(t)dI(t) + \left[dI(t)\right]^2 \tag{H.6}$$

and then, assuming that $dI(t) = \dot{I}(t)dt$, we just neglect the second order term $\left[\dot{I}(t)dt\right]^2$ to attain the result. Here however—since $I(t)$ is not differentiable—we are no longer entitled to say that $\left[dI(t)\right]^2$ coincides with some $\left[\dot{I}(t)dt\right]^2$, that is with an infinitesimal of order larger than dt. We must rather go back to the Eq. (H.1) sidestepping the utilization of derivatives

$$dI(t) = \lambda q dt - aI(t)dt + q d\widetilde{N}(t) \tag{H.7}$$

plug that into (H.6)

$$\begin{aligned} d\left[I^2(t)\right] &= 2I(t)\left[\lambda q dt - aI(t)dt + q d\widetilde{N}(t)\right] + \left[\lambda q dt - aI(t)dt + q d\widetilde{N}(t)\right]^2 \\ &= \left[2\lambda q I(t) - 2aI^2(t)\right]dt + \left[\lambda q - aI(t)\right]^2 (dt)^2 \\ &\quad + 2qI(t)\, d\widetilde{N}(t) + 2q\left[\lambda q - aI(t)\right] d\widetilde{N}(t)\, dt + q^2\left[d\widetilde{N}(t)\right]^2 \end{aligned}$$

and finally, taking the expectations, neglect the higher order terms in dt (remember that according to (6.72) $\boldsymbol{E}\left[d\widetilde{N}^2\right] = \lambda dt$ is of the first order in dt) to find

$$d\boldsymbol{E}\left[I^2(t)\right] = \left(2\lambda q \boldsymbol{E}\left[I(t)\right] - 2a\boldsymbol{E}\left[I^2(t)\right]\right) dt + \lambda q^2 dt$$

Instead of (H.4) we therefore have

$$\frac{1}{2}\frac{d}{dt}\boldsymbol{E}\left[I^2(t)\right] = \lambda q \boldsymbol{E}\left[I(t)\right] - a\boldsymbol{E}\left[I^2(t)\right] + \frac{\lambda q^2}{2}$$

with the new additional term $\lambda q^2/2$, so that from (H.2), retracing the steps leading to the puzzling result (H.5) we now attain a solution with the right asymptotic behaviors for $t \to +\infty$

$$\begin{aligned} \boldsymbol{E}\left[I(t)\right] &= \frac{\lambda q}{a} + Ce^{-at} \longrightarrow \frac{\lambda q}{a} \\ \boldsymbol{E}\left[I^2(t)\right] &= \left(\frac{\lambda q}{a}\right)^2 + \frac{\lambda q^2}{2a} + C\frac{2\lambda q}{a}e^{-at} + Ae^{-2at} \longrightarrow \left(\frac{\lambda q}{a}\right)^2 + \frac{\lambda q^2}{2a} \\ \boldsymbol{V}\left[I(t)\right] &= \boldsymbol{E}\left[I^2(t)\right] - \boldsymbol{E}\left[I(t)\right]^2 \longrightarrow \frac{\lambda q^2}{2a} > 0 \end{aligned}$$

This shows that our remarks about the stochastic infinitesimals discussed in the Sect. 6.3—even if inaccurate and intuitive—play a pivotal role to get an acceptable result.

The way Langevin—even relying on a non-rigorous mathematical formulation—managed to avoid the previous mistakes and to get the correct results deserves some scrutiny. It is interesting to remark indeed that, at variance with (H.3), the two relations (6.79) and (6.80) for the position process $X(t)$, even if only symbolic, are basically correct: to show that we notice first that the dissimilarity between the two formulations (6.77) and (6.78) of the dynamical equations conceals indeed a few important details. Their diversity rests in fact on the idea that $X(t)$ is differentiable, namely that a process $\dot{X}(t) = V(t)$ exists such that

$$X(t) = \int_0^t V(s)\, ds$$

and then that the Newton equation

$$m\ddot{X}(t) = -6\pi\eta a\dot{X}(t) + B(t) \tag{H.8}$$

is equivalent to the system

$$\begin{aligned} \dot{X}(t) &= V(t) \\ m\dot{V}(t) &= -6\pi\eta a V(t) + B(t) \end{aligned} \tag{H.9}$$

Here however, since the random force $B(t)$ directly affects $V(t)$ only, the velocity process—at variance with $X(t)$—will turn out to be not differentiable, so that the Langevin equation (H.9) (with $I(t)$ replaced by $V(t)$) will have the same form of the Eq. (H.1) adopted for the shot noise. This apparently entails first that if Langevin had used (H.9) along with the formula

$$\frac{dV^2(t)}{dt} = 2V(t)\dot{V}(t) \tag{H.10}$$

he would have reached about $V(t)$ the same paradoxical conclusions drawn from (H.5) for the shot noise: after a transient delay the Brownian particle would have stopped, with $V(t) = 0$ not only on average, but even $\boldsymbol{P}$-a.s. namely with a zero variance. His argument starts instead from the Newton equation (H.8) for the position process $X(t)$, and avails himself of the—symbolic, but essentially error-free—relations

$$\frac{d}{dt}\left[X^2(t)\right] = 2X(t)\dot{X}(t) \tag{H.11}$$

$$\frac{d^2}{dt^2}\left[X^2(t)\right] = 2\dot{X}^2(t) + 2X(t)\ddot{X}(t) = 2V^2(t) + 2X(t)\ddot{X}(t) \tag{H.12}$$

in order to find the Einstein result (6.82). We have then to explain why the Eqs. (H.11) and (H.12) may be rather safely used, while (H.10), as we have seen, would lead to paradoxes.

First of all let us remark that, being $X(t)$ differentiable, it is $dX(t) = \dot{X}(t)dt = V(t)dt$, so that the infinitesimal $dX(t)$ is of the first order in dt, and hrnce (H.11) holds allowing us to write

$$\frac{d}{dt}\big[X^2(t)\big] = 2X(t)\dot{X}(t) = 2X(t)V(t) \tag{H.13}$$

The Eq. (H.12), instead, while basically correct, remains purely symbolic because it involves a derivative $\ddot{X}(t) = \dot{V}(t)$ that does not exist. To understand then why this is nonetheless acceptable we must remark that

$$\begin{aligned} d\big[X(t)V(t)\big] &= X(t+dt)V(t+dt) - X(t)V(t) \\ &= \big[X(t) + dX(t)\big]\,\big[V(t) + dV(t)\big] - X(t)V(t) \\ &= \big[X(t) + V(t)dt\big]\,\big[V(t) + dV(t)\big] - X(t)V(t) \\ &= V^2(t)dt + X(t)dV(t) + V(t)dV(t)dt \end{aligned}$$

On the other hand $dV(t)$ is an infinitesimal of the order $O\left(dt^{1/2}\right)$ because (as we have seen in the Sect. 8.1), the fluctuating force $B(t)$ of the Langevin equation (H.9) is a Wienerian white noise $\dot{W}(t)$, so that putting (H.9) in the form

$$m dV(t) = -6\pi\eta a V(t)dt + dW(t)$$

we find that $dV(t)$ is an infinitesimal of the same order of $dW(t)$, namely $O\left(dt^{1/2}\right)$. We can therefore safely maintain that $dV(t)dt$ is an infinitesimal of higher order, more precisely $O\left(dt^{3/2}\right)$, so that at first order we can write

$$d\big[X(t)V(t)\big] = V^2(t)dt + X(t)dV(t)$$

and hence, symbolically at least and not wrongly, we can state that

$$\frac{d}{dt}\big[X(t)V(t)\big] = V^2(t) + X(t)\dot{V}(t) = V^2(t) + X(t)\ddot{X}(t)$$

so that (H.12) will be fully vindicated through (H.13).

Appendix I
Pseudo-Markovian Processes (Sect. 7.1.2)

We will provide here a simple example[4] of a non-Markovian process whose transition probabilities nevertheless abide by the Chapman–Kolmogorov condition: processes of this kind are also called pseudo-Markovian. Consider a process defined on a discrete and finite time span ($t = 1, 2, 3$) and taking only two values (0 and 1): it will be represented then just as a finite sequence $X = (X_1, X_2, X_3)$ of three $0/1$ *rv*'s. The trajectories of this rudimentary (but legitimate) process are reduced to the $8 = 2^3$ possible triplets of 0, 1 symbols, and its distribution can be given by choosing in a consistent way the probabilities allotted to these 8 samples. By adopting the shorthand notations

$$
\begin{aligned}
&p_{1,2,3}(x_1, x_2, x_3) = \boldsymbol{P}\{X_1 = x_1, X_2 = x_2, X_3 = x_3\}\\
&p_{1,2}(x_1, x_2) = \boldsymbol{P}\{X_1 = x_1, X_2 = x_2\} \quad \dots \qquad p_1(x_1) = \boldsymbol{P}\{X_1 = x_1\} \quad \dots\\
&p_{3|2,1}(x_3|x_2, x_1) = \boldsymbol{P}\{X_3 = x_3 \mid X_2 = x_2, X_1 = x_1\} \quad \dots\\
&p_{3|2}(x_3|x_2) = \boldsymbol{P}\{X_3 = x_3 \mid X_2 = x_2\} \quad \dots
\end{aligned}
$$

we will therefore specify in the Table I.1 the joint distribution $p_{1,2,3}(x_1, x_2, x_3)$ of our process by simply assigning a probability to every single sample This completely defines in fact the law of the process because all the other lower-order marginal distributions can then be deduced from the Table I.1 as in the following examples

$$
\begin{aligned}
p_{1,2}(0, 0) &= p_{1,2,3}(0, 0, 0) + p_{1,2,3}(0, 0, 1) = {}^1/_4\\
p_{1,2}(1, 0) &= p_{1,2,3}(1, 0, 0) + p_{1,2,3}(1, 0, 1) = {}^1/_4\\
p_{2,3}(0, 0) &= p_{1,2,3}(0, 0, 0) + p_{1,2,3}(1, 0, 0) = {}^1/_4\\
p_2(0) &= p_{1,2,3}(0, 0, 0) + p_{1,2,3}(1, 0, 0) + p_{1,2,3}(0, 0, 1) + p_{1,2,3}(1, 0, 1) = {}^1/_2\\
p_2(1) &= p_{1,2,3}(0, 1, 0) + p_{1,2,3}(1, 1, 0) + p_{1,2,3}(0, 1, 1) + p_{1,2,3}(1, 1, 1) = {}^1/_2
\end{aligned}
$$

[4]This example is presented by N.G. van Kampen as an exercise in [10] p. 79, but its origin goes back to a P. Lévy note [11] taken up again first by W. Feller [12] and then by E. Parzen [13] p. 203.

N. Cufaro Petroni, *Probability and Stochastic Processes for Physicists*, UNITEXT for Physics, https://doi.org/10.1007/978-3-030-48408-8

Table I.1 Probabilities attributed to the 8 samples of the process $X = (X_1, X_2, X_3)$

X_1	X_2	X_3	$p_{1,2,3}$
0	0	0	0
0	0	1	$^1/_4$
0	1	0	$^1/_4$
0	1	1	0
1	0	0	$^1/_4$
1	0	1	0
1	1	0	0
1	1	1	$^1/_4$

The marginal distributions resulting from this procedure are collected in the Table I.2 We are able now to calculate also the conditional distributions and to check first of all that our process X is not Markovian: we have indeed

$$p_{3|2,1}(0\,|\,0,0) = \frac{p_{1,2,3}(0,0,0)}{p_{1,2}(0,0)} = \frac{0}{^1/_4} = 0$$
$$p_{3|2}(0\,|\,0) = \frac{p_{2,3}(0,0)}{p_2(0)} = \frac{^1/_4}{^1/_2} = {^1/_2}$$

so that, at least in one instance, it is $p_{3|2,1} \neq p_{3|2}$, and hence the process is not Markovian. That notwithstanding it is also easy to see that the transition probabilities $p_{2|1}$, $p_{3|2}$ and $p_{3|1}$ satisfy the Chapman–Kolmogorov equations, namely—according to the Definition 7.6—that they are *Markovian transition probabilities*. We can indeed deduce from the Table I.2 that in any event it is

$$p_{2|1}(x_2\,|\,x_1) = p_{3|2}(x_3\,|\,x_2) = p_{3|1}(x_3\,|\,x_1) = {^1/_2} \qquad x_1, x_2, x_3 = 0, 1$$

and hence that the Chapman–Kolmogorov equations always hold:

$$\sum_{x_2=0}^{1} p_{3|2}(x_3\,|\,x_2)p_{2|1}(x_2\,|\,x_1) = {^1/_2}\cdot{^1/_2} + {^1/_2}\cdot{^1/_2} = {^1/_4} + {^1/_4} = {^1/_2} = p_{3|1}(x_3\,|\,x_1)$$

This surprising result is discussed in further detail in the Sect. 7.1.2, where it is pointed out that it can be understood by remembering that—on the same sample trajectory space—a process could possibly be endowed with several global distributions, all different but sharing the same family of Markovian transition laws: among these processes however only one—if any—can exhibit the Markov property. In the present example—on the space of the 8 sample trajectories (x_1, x_2, x_3)—the process X with the law specified in the Table I.1 has not the Markov property, but its transition distributions are Markovian: this enables us then, through the chain rule of the

Table I.2 Bivariate and univariate marginal distributions of $X = (X_1, X_2, X_3)$ deduced from the Table I.1

X_1	X_2	$p_{1,2}$
0	0	$^1/_4$
0	1	$^1/_4$
1	0	$^1/_4$
1	1	$^1/_4$

X_2	X_3	$p_{2,3}$
0	0	$^1/_4$
0	1	$^1/_4$
1	0	$^1/_4$
1	1	$^1/_4$

X_1	X_3	$p_{1,3}$
0	0	$^1/_4$
0	1	$^1/_4$
1	0	$^1/_4$
1	1	$^1/_4$

X_1	p_1
0	$^1/_2$
1	$^1/_2$

X_2	p_2
0	$^1/_2$
1	$^1/_2$

X_3	p_3
0	$^1/_2$
1	$^1/_2$

Table I.3 Distribution of the Markov process $\widetilde{X}$ sharing its transition probabilities with X

X_1	X_2	X_3	$\widetilde{p}_{1,2,3}$
0	0	0	$^1/_8$
0	0	1	$^1/_8$
0	1	0	$^1/_8$
0	1	1	$^1/_8$
1	0	0	$^1/_8$
1	0	1	$^1/_8$
1	1	0	$^1/_8$
1	1	1	$^1/_8$

Proposition 7.4, to define another process $\widetilde{X}$ (on the same trajectories, but with a different distribution $\widetilde{p}$) that will turn out to be Markovian. The joint distribution of this new process $\widetilde{X}$ for every value of the triplet $x_1, x_2, x_3 = 0, 1$ is indeed calculated from

$$\widetilde{p}_{1,2,3}(x_1, x_2, x_3) = p_{3|2}(x_3 \,|\, x_2)\, p_{2|1}(x_2 \,|\, x_1)\, p_1(x_1) = {}^1/_2 \cdot {}^1/_2 \cdot {}^1/_2 = {}^1/_8$$

and is summarized in the Table I.3, while its bivariate and univariate distributions stay unchanged w.r.t. those of the initial process X.

Appendix J
Fractional Brownian Motion (Sect. 7.1.10)

It has been pointed out in the text that there are instances of—even rather elementary—transition *pdf*'s that are not Markovian in the sense that they do not satisfy the Chapman–Kolmogorov equation (7.17). In this case this transition *pdf*–while possibly being the legitimate conditional *pdf* of some conjectural stochastic process—in no way can play the role of the transition *pdf* of a Markov process: in other words we are not entitled to use the chain rule in order to retrieve the global law of the process from this transition *pdf* alone. We also remarked however in the Sect. 7.1.10 that to find the process distribution we can possibly make up for the lack of Markovianity by means of Gaussianity. Let us remember then that a relevant case of a Gaussian, non Markovian process is the so-called ***fractional Brownian motion*** $Y(t)$ that in some respects can be considered as a generalization of the usual Wiener process: for further details see [14, 15] Chap. 7, and the tutorial in [16].

Starting from the remark that it is easy to check with a direct calculation that for $s, t > 0$ it is

$$\min\{s, t\} = \frac{t + s - |t - s|}{2} = \frac{|t| + |s| - |t - s|}{2}$$

we recall first that the autocovariance (6.47) of a Wiener process $W(t)$ is

$$C_W(s, t) = D \min\{s, t\} = D\,\frac{|t| + |s| - |t - s|}{2}$$

and that all the joint laws of $W(t)$ could also be deduced by plugging this autocovariance into the characteristic function (7.64). In order to define the Gaussian laws of the fractional Brownian motion $Y(t)$ we then generalize the previous setting taking $m_Y(t) = 0$ and the new autocovariance function

$$C_Y(s, t) = \frac{D}{2}\left(|t|^{2H} + |s|^{2H} - |t - s|^{2H}\right)$$

N. Cufaro Petroni, *Probability and Stochastic Processes for Physicists*, UNITEXT for Physics, https://doi.org/10.1007/978-3-030-48408-8

where H, called *Hurst index*, is a real number with $0 < H < 1$. It is therefore apparent that the autocovariance of the usual Wiener process is retrieved when in particular $H = {}^1/_2$, while for all the other values of H it is possible to prove that C_Y is non-negative definite so that it can legitimately be used to define the distribution of the Gaussian process $Y(t)$ called fractional Brownian motion. Of course the particular properties of the process $Y(t)$ change according to the value of H and can be examined in detail—even if we will neglect to do it—because all the finite, joint distributions of the process are explicitly known. In particular the value of the Hurst index H is associated to the correlation of the $Y(t)$ increments, and hence also to the regularity of the trajectories. It is possible to show indeed that—omitting the Wiener case $H = {}^1/_2$ that comes up with independent increments—a value $H > {}^1/_2$ entails a *positive correlation* among the increments, while a value $H < {}^1/_2$ hints to a *negative correlation*. From an intuitive standpoint we could say that the former behavior (either increasing or decreasing) of the trajectory affects the latter one: when $H > {}^1/_2$ the positive correlation entails more regular trajectories (the former behavior tends to be confirmed), while if $H < {}^1/_2$ the negative correlation produces the opposite effect (the former behavior is contradicted by the latter one) giving rise to reinforced chaos.

Appendix K
Ornstein-Uhlenbeck Equations (Sect. 7.2.4)

We will give here an explicit derivation of the coefficients of the forward equation for an Ornstein-Uhlenbeck process $X(t)$, and we will show that its *pdf*'s are solutions of the Fokker–Planck equation (7.95) put forward in the Proposition 7.40. Remark also that the sample continuity of the Ornstein-Uhlenbeck processes directly follows from the vanishing of the jump term.

We will start precisely by proving that the jump term (7.65) vanishes so that $X(t)$ is sample continuous. Remark indeed that from (7.57) we have

$$
\begin{aligned}
\frac{1}{\Delta t} f(x, t+\Delta t \mid y, t) &= \frac{e^{-\frac{[(x-y)+y(1-e^{-\alpha\Delta t})]^2}{2\beta^2(1-e^{-2\alpha\Delta t})}}}{\Delta t\sqrt{2\pi\beta^2(1-e^{-2\alpha\Delta t})}} \\
&= \frac{\alpha\, e^{-\frac{(x-y)^2}{2\beta^2(1-e^{-2\alpha\Delta t})}}}{\alpha\Delta t\sqrt{2\pi\beta^2(1-e^{-2\alpha\Delta t})}}\, e^{-\frac{y^2(1-e^{-\alpha\Delta t})^2+2y(x-y)(1-e^{-\alpha\Delta t})}{2\beta^2(1-e^{-2\alpha\Delta t})}} \\
&= \frac{e^{-\frac{(x-y)^2\alpha\Delta t}{2\beta^2(1-e^{-2\alpha\Delta t})}\frac{1}{\alpha\Delta t}}}{(\alpha\Delta t)^{\frac{3}{2}}\sqrt{2\pi\beta^2\frac{1-e^{-2\alpha\Delta t}}{\alpha\Delta t}}}\,\alpha\, e^{-\frac{y^2(1-e^{-\alpha\Delta t})+2y(x-y)}{2\beta^2(1+e^{-\alpha\Delta t})}}
\end{aligned}
$$

and since it is easy to see that

$$
\lim_{\Delta t\to 0}\alpha\, e^{-\frac{y^2(1-e^{-\alpha\Delta t})+2y(x-y)}{2\beta^2(1+e^{-\alpha\Delta t})}} = \alpha e^{-\frac{y(x-y)}{2\beta^2}} \qquad \lim_{\Delta t\to 0}\frac{1-e^{-2\alpha\Delta t}}{2\alpha\Delta t} = 1
$$

we can carry out the limit in two steps sopping first at the halfway expression

$$
\ell(x|y,t) = \alpha e^{-\frac{y(x-y)}{2\beta^2}}\lim_{\Delta t\to 0}\frac{e^{-\frac{(x-y)^2}{4\beta^2}\frac{1}{\alpha\Delta t}}}{(\alpha\Delta t)^{\frac{3}{2}}\sqrt{4\pi\beta^2}}
$$

and then, for $z = \frac{1}{\alpha\Delta t} \to +\infty$, performing the elementary limit

N. Cufaro Petroni, *Probability and Stochastic Processes for Physicists*, UNITEXT for Physics, https://doi.org/10.1007/978-3-030-48408-8

$$\ell(x|y,t) = \frac{\alpha e^{-\frac{y(x-y)}{2\beta^2}}}{\sqrt{4\pi\beta^2}} \lim_{z\to+\infty} z^{\frac{3}{2}} e^{-\frac{(x-y)^2}{4\beta^2}z} = 0$$

As a consequence the Ornstein-Uhlenbeck equation will be of the Fokker-Planck type and we are left only with the task of calculating its coefficients A and B. Within a slightly changed notation, from (7.66) we first have that

$$A(x,t) = \lim_{\epsilon\to 0^+} \lim_{\Delta t\to 0^+} \frac{1}{\Delta t} \int_{|z-x|<\epsilon} (z-x) f(z, t+\Delta t \,|\, x, t)\, dz$$

an then, by taking

$$y = \frac{z - xe^{-\alpha\Delta t}}{\sqrt{\beta^2(1-e^{-2\alpha\Delta t})}} \qquad\qquad a_\pm = \frac{x(1-e^{-\alpha\Delta t}) \pm \epsilon}{\sqrt{\beta^2(1-e^{-2\alpha\Delta t})}}$$

from (7.57) it follows

$$A(x,t) = \lim_{\epsilon\to 0^+} \lim_{\Delta t\to 0^+} \frac{1}{\Delta t} \left[\sqrt{\beta^2(1-e^{-2\alpha\Delta t})} \int_{a_-}^{a_+} \frac{ye^{-\frac{y^2}{2}}}{\sqrt{2\pi}}\, dy \right.$$
$$\left. - x(1-e^{-\alpha\Delta t}) \int_{a_-}^{a_+} \frac{e^{-\frac{y^2}{2}}}{\sqrt{2\pi}}\, dy \right]$$

Taking now into account the Gaussian primitive functions

$$\int \frac{e^{-\frac{y^2}{2}}}{\sqrt{2\pi}}\, dy = \Phi(y) + const \qquad\qquad \int \frac{ye^{-\frac{y^2}{2}}}{\sqrt{2\pi}}\, dy = -e^{-\frac{y^2}{2}} + const$$

where $\Phi(x)$ is the error function (2.16), we also get

$$A(x,t) = \lim_{\epsilon\to 0^+} \lim_{\Delta t\to 0^+} \left[\sqrt{\frac{\alpha\beta^2}{\pi} \frac{1-e^{-2\alpha\Delta t}}{2\alpha\Delta t}}\, \frac{e^{-\frac{a_-^2}{2}} - e^{-\frac{a_+^2}{2}}}{\sqrt{\Delta t}} \right.$$
$$\left. - \alpha x \frac{1-e^{-\alpha\Delta t}}{\alpha\Delta t} \left(\Phi(a_+) - \Phi(a_-)\right) \right]$$

Since on the other hand for every $\epsilon > 0$ it is

$$\lim_{\Delta t\to 0^+} a_\pm = \pm\infty \qquad \lim_{\Delta t\to 0^+} \left(\Phi(a_+) - \Phi(a_-)\right) = 1 \qquad \lim_{u\to 0} \frac{1-e^{-u}}{u} = 1$$

we will have

$$\lim_{\Delta t\to 0^+} \frac{1-e^{-\alpha\Delta t}}{\alpha\Delta t}\left(\Phi(a_+)-\Phi(a_-)\right)=1 \qquad \lim_{\Delta t\to 0^+}\sqrt{\frac{\alpha\beta^2}{\pi}\frac{1-e^{-2\alpha\Delta t}}{2\alpha\Delta t}}=\sqrt{\frac{\alpha\beta^2}{\pi}}$$

while the other term takes the form

$$\frac{e^{-\frac{a_-^2}{2}}-e^{-\frac{a_+^2}{2}}}{\sqrt{\Delta t}} = e^{-\frac{x^2(1-e^{-\alpha\Delta t})^2}{2\beta^2(1-e^{-2\alpha\Delta t})}}\,\frac{e^{-\frac{\epsilon^2}{2\beta^2(1-e^{-2\alpha\Delta t})}}}{\sqrt{\Delta t}}\left(e^{\frac{\epsilon x(1-e^{-\alpha\Delta t})}{\beta^2(1-e^{-2\alpha\Delta t})}}-e^{-\frac{\epsilon x(1-e^{-\alpha\Delta t})}{\beta^2(1-e^{-2\alpha\Delta t})}}\right)$$

$$= e^{-\frac{x^2(1-e^{-\alpha\Delta t})}{2\beta^2(1+e^{-\alpha\Delta t})}}\,\frac{e^{-\frac{\epsilon^2}{4\beta^2\alpha\Delta t}\frac{2\alpha\Delta t}{1-e^{-2\alpha\Delta t}}}}{\sqrt{\Delta t}}\left(e^{\frac{\epsilon x}{\beta^2(1+e^{-\alpha\Delta t})}}-e^{-\frac{\epsilon x}{\beta^2(1+e^{-\alpha\Delta t})}}\right)$$

so that with $u=\frac{1}{\Delta t}$ the limits are

$$\lim_{\Delta t\to 0^+} e^{-\frac{x^2(1-e^{-\alpha\Delta t})}{2\beta^2(1+e^{-\alpha\Delta t})}}\left(e^{\frac{\epsilon x}{\beta^2(1+e^{-\alpha\Delta t})}}-e^{-\frac{\epsilon x}{\beta^2(1+e^{-\alpha\Delta t})}}\right)=e^{\frac{\epsilon x}{2\beta^2}}-e^{-\frac{\epsilon x}{2\beta^2}}$$

$$\lim_{\Delta t\to 0^+}\frac{2\alpha\Delta t}{1-e^{-2\alpha\Delta t}}=1$$

$$\lim_{\Delta t\to 0^+}\frac{e^{-\frac{\epsilon^2}{4\beta^2\alpha\Delta t}}}{\sqrt{\Delta t}}=\lim_{u\to+\infty} u^{\frac{1}{2}}e^{-\frac{\epsilon^2 u}{4\alpha\beta^2}}=0$$

and hence for every $\epsilon>0$ it follows

$$\lim_{\Delta t\to 0^+}\frac{e^{-\frac{a_-^2}{2}}-e^{-\frac{a_+^2}{2}}}{\sqrt{\Delta t}}=0$$

Collecting then all the factors we finally find

$$A(x,t)=-\alpha x$$

A similar approach, whose details we will neglect here for short, leads finally to establish that the diffusion coefficient B is indeed constant: more precisely, with the notation $D=2\alpha\beta^2$, we have

$$B(x,t)=D=2\alpha\beta^2$$

so that on the whole the Fokker-Planck equation of the Ornstein-Uhlenbeck process takes the form (7.95)

$$\partial_t f(x,t)=\alpha\partial_x[xf(x,t)]+\alpha\beta^2\partial_x^2 f(x,t)=\alpha\partial_x[xf(x,t)]+\frac{D}{2}\partial_x^2 f(x,t)$$

In particular the transition *pdf* (7.57) will be the solution associated to the condition $f(x,t)=\delta(x-y)$ at the time t.

Appendix L
Stratonovich Integral (Sect. 8.2.2)

It is important to recall that there are several definitions of stochastic integral different from that of Itō, and in particular the approach due to R.L. Stratonovich deserves a few clarifications (for details see [5] p. 86, 98 and [17] Sect. 4.5.2). This alternative definition is chiefly based on a Riemann procedure of the type (8.8), where however the values of the integrand $Y(t)$ are taken in the *midpoind* of the intervals $[t_j, t_{j+1}]$, instead than in their left ends t_j, according to the prescription

$$\int_a^b Y(t) \circ dW(t) = \lim_{n,\delta\to 0}\text{-}ms \sum_{j=0}^{n-1} Y\left(\frac{t_j + t_{j+1}}{2}\right)\left[W(t_{j+1}) - W(t_j)\right]$$

Remark the new notation "∘" introduced here to tell apart this integral from the analogous Itō integral. The main appeal of this definition lies in the fact that by its adoption the usual rules of the calculus remain unchanged, and this is likely to be the reason why the *Stratonovich integral* has long been very popular among the physicists. Unfortunately however in no way it enjoys the same properties of its Itō counterpart: rather its convergence and mathematical consistence are not without problems so that its inherent qualities remain quite uncertain. Moreover there is no general rule that allows passing from one definition to another, except for the following result.

Proposition L.1 *(E. Wong, M. Zakai - 1969) If $X(t)$ is a solution of the Itō EDS (8.26), and if $g(x, t)$ is a continuous differentiable function, within a few regularity assumptions that we will neglect here, the Itō and the Stratonovich integrals $\boldsymbol{P}$-a.s. verify the following relation*

$$\int_a^b g(X(t), t) \circ dW(t) = \int_a^b g(X(t), t)\, dW(t) + \frac{1}{2}\int_a^b g_x(X(t), t)\, b(X(t), t)\, dt$$

in the sense that the l.h.s. exists iff the r.h.s. exists, and in this case the two coincide. Here of course $b(x, t)$ is the diffusion coefficient of the Itō SDE *(8.26).*

N. Cufaro Petroni, *Probability and Stochastic Processes for Physicists*, UNITEXT for Physics, https://doi.org/10.1007/978-3-030-48408-8

Proof Omitted: for details see [17] p. 159 ■

Remark in particular that, according to the previous proposition, the Itō and the Stratonovich integrals coincide if $g(x,t) = g(t)$ is x-independent. When moreover $X(t)$ is a solution of the Itō *SDE* (8.26) (according to (8.27)), taking $h(x,t) = b(x,t)$ it is easy to see from the Proposition L.1 that $X(t)$ also is (always in an integral sense) a solution of the following *Stratonovich SDE*

$$dX(t) = \widetilde{a}(X(t),t)\,dt + b(X(t),t) \circ dW(t)$$
$$\widetilde{a}(x,t) = a(x,t) - \frac{1}{2}\,b(x,t)\,b_x(x,t)$$

The previous results may apparently be used to give—at least in these particular instances—a consistent definition of the Stratonovich integral (and *SDE*) relying on the corresponding Itō definitions that, as for them, are well posed. We will keep away however from going along this path and we will always base our considerations on the Itō integral—calculated from the procedure explained in the Sect. 8.2.2—with all its resulting modifications about the calculus rules.

Appendix M
Stochastic Bridges (Sect. 10.2)

Brownian bridge *SDE*'s are stochastic versions of *ODE*'s (Ordinary Differential Equations) for trajectories interpolating two, or more fixed points (see [18] p. 358 and [19] vol. II, p. 172). In general the non-random interpolating trajectories, coinciding with the expectation of the corresponding random bridges, are supposed to be *linear functions* of the time t, but we will argue here that there is no really compelling reason for this choice.

Let us start with a trajectory $z(t)$ connecting two possible values x and y at the endpoints of a time interval $[0, s]$, namely

$$z(t) = x\, g\big({}^{t}/_{s}\big) + y\, h\big({}^{t}/_{s}\big) \qquad 0 \le t \le s \tag{M.1}$$

where, by supposing that $g(u)$ and $h(u)$ satisfy the conditions

$$\begin{cases} g(0) = 1 \\ g(1) = 0 \end{cases} \qquad \begin{cases} h(0) = 0 \\ h(1) = 1 \end{cases} \tag{M.2}$$

we trivially get that

$$z(0) = x \qquad z(s) = y \tag{M.3}$$

In fact every possible, given function $z(t)$ satisfying the extremal conditions (M.3) can always be cast in the form (M.1): for a given $z(t)$ just choose an arbitrary $g(u)$ satisfying (M.2), and then take

$$h(u) = \frac{z(su) - x\, g(u)}{y}$$

in order to obtain (M.1). Remark indeed that nothing forbids an explicit dependence of $g(u)$ and $h(u)$ on x and y in so far as the conditions (M.2) hold. As a consequence the expression (M.1) can be considered general enough for our purposes.

N. Cufaro Petroni, *Probability and Stochastic Processes for Physicists*, UNITEXT for Physics, https://doi.org/10.1007/978-3-030-48408-8

We will look then for a first order *ODE* such that the trajectory (M.1) will be its (unique) solution for the initial condition $z(0) = x$: it is straightforward to understand that the form of this equation, albeit independent from the initial condition x, will however explicitly depend on the final condition y aimed at by our trajectory. A first order *ODE* indeed allows a free choice for just one initial condition, while in general no independent final condition can be arbitrarily imposed if we want to find a possible solution. As a consequence the form of an *ODE* admitting both the extremal conditions (M.3) must depend on one of them: in other words there is no unique equation fitting both the conditions (M.3) for arbitrary values of x and y. In order to eliminate the initial condition $z(0) = x$ from our *ODE* let us remark that from (M.1) we get

$$\dot{z}(t) = \frac{x}{s}\,\dot{g}(^t/_s) + \frac{y}{s}\,\dot{h}(^t/_s) \tag{M.4}$$

so that from (M.1) and (M.4) we have

$$\frac{\dot{z}(t) - \frac{y}{s}\,\dot{h}(^t/_s)}{\frac{1}{s}\,\dot{g}(^t/_s)} = x = \frac{z(t) - y\,h(^t/_s)}{g(^t/_s)}$$

and rearranging the terms we find the linear *ODE*

$$\dot{z}(t) = \frac{\left[g(^t/_s)\,\dot{h}(^t/_s) - \dot{g}(^t/_s)\,h(^t/_s)\right]y + z(t)\dot{g}(^t/_s)}{s\,g(^t/_s)} \tag{M.5}$$

whose solutions (M.1) will connect every possible initial condition x to the same final value y inscribed into it. The simplest possible—albeit by no means the unique—interpolation adopts the following linear functions

$$g(u) = 1 - u \qquad h(u) = u \tag{M.6}$$

so that the corresponding connecting trajectories

$$z(t) = \left(1 - \frac{t}{s}\right)x + \frac{t}{s}\,y$$

are time-linear and satisfy the *ODE*

$$\dot{z}(t) = \frac{y - z(t)}{s - t}$$

A straightforward stochastic generalization of the *ODE* (M.5) is obtained just by the addition of a *Brownian noise* $W(t)$, namely a Wiener process with constant diffusion coefficient D

$$dZ(t) = \frac{\left[g(^t/_s)\,\dot{h}(^t/_s) - \dot{g}(^t/_s)\,h(^t/_s)\right]y + Z(t)\dot{g}(^t/_s)}{s\,g(^t/_s)}\,dt + dW(t) \qquad \text{(M.7)}$$

that apparently results in a *SDE* with space-linear, time-dependent coefficients of the form

$$dZ(t) = [a_0(t) + a_1(t)Z(t)]dt + [b_0(t) + b_1(t)Z(t)]dW(t) \qquad Z(0) = x,\ \boldsymbol{P}\text{-a.s.}$$

where it is understood that

$$a_0(t) = \frac{y}{s}\,\frac{g(^t/_s)\,\dot{h}(^t/_s) - \dot{g}(^t/_s)\,h(^t/_s)}{g(^t/_s)} = y\,g(^t/_s)\,\frac{d}{dt}\left[\frac{h(^t/_s)}{g(^t/_s)}\right]$$

$$a_1(t) = \frac{1}{s}\,\frac{\dot{g}(^t/_s)}{g(^t/_s)} = \frac{d}{dt}\left[\ln g(^t/_s)\right]$$

$$b_0(0) = 1$$

$$b_1(t) = 0$$

After some algebra, it is possible to show (see [20] p. 36–38) that the solution of (M.7) is

$$Z(t) = x\,g(^t/_s) + y\,h(^t/_s) + g(^t/_s)\int_0^t \frac{dW(u)}{g(^u/_s)} \qquad \text{(M.8)}$$

where the first two terms of the right-hand side exactly coincide with the non-random interpolating trajectories (M.1) which can therefore be recovered just by taking the expectation $\boldsymbol{E}\,[Z(t)]$ of (M.8). Since apparently we have $Z(0) = x$ and $Z(s) = y$, $\boldsymbol{P}$-a.s., the solution of (M.7) is also called a *Brownian bridge* $x \leftrightarrow y$. In particular when we take the linear functions (M.6) we get the solution

$$Z(t) = x\left(1 - \frac{t}{s}\right) + y\,\frac{t}{s} + (s - t)\int_0^t \frac{dW(u)}{s - u} \qquad \text{(M.9)}$$

namely the *time-linear Brownian bridge.*

These Brownian bridges are widely discussed in the literature (see [18] p. 358 and [19] vol. II, p. 172) where it is shown among others that the solution (M.8) is Gaussian with $\boldsymbol{P}$-a.s. continuous paths, with expectation

$$m(t) = \boldsymbol{E}\,[Z(t)] = x\,g(^t/_s) + y\,h(^t/_s)$$

and with autocovariance

$$C(t, u) = \boldsymbol{cov}\,[Z(t), Z(u)] = 2D\,g\left(^t/_s\right)g(^u/_s)\int_0^{t\wedge u} \frac{dv}{g^2(^v/_s)}$$

so that its laws can be deemed completely known. In particular the variance is

$$V[Z(t)] = C(t,t) = 2D\, g^2({}^t/_s) \int_0^t \frac{dv}{g^2({}^v/_s)}$$

In the case of the time-linear Brownian bridge (M.9) these formulas become

$$m(t) = x\left(1 - \frac{t}{s}\right) + y\,\frac{t}{s}$$

$$C(t,u) = D\,(s-t)(s-u)\int_0^{t\wedge u} \frac{dv}{(s-v)^2} = 2D\left((t\wedge u) - \frac{tu}{s}\right)$$

and its distributions coincide with that of a Wiener process conditioned at both the endpoints with $W(0) = x$ and $W(s) = y$: in fact, with the notation

$$\phi(x,t|x') = \frac{e^{-\frac{(x-x')^2}{4Dt}}}{\sqrt{4\pi Dt}}$$

it is possible to show (see [18] p. 358 and [19] vol. II, p. 172) that the finite dimensional distributions of our time-linear Brownian bridge (M.9) coincide with the following conditional *pdf*'s of a Wiener process for $0 < t_1 < \cdots < t_n < s$

$$\begin{aligned} f(x_1,t_1;\ldots;x_n,t_n\,|\,x,0;y,s) &= \frac{f(x_1,t_1;\ldots;x_n,t_n;y,s\,|\,x,0)}{f(y,s\,|\,x,0)} \\ &= \frac{f(y,s|x_n,t_n)\cdots f(x_2,t_2|x_1,t_1)f(x_1,t_1|x,0)}{f(y,s\,|\,x,0)} \\ &= \frac{\phi(y,s-t_n|x_n)\cdots\phi(x_2,t_2-t_1|x_1)\phi(x_1,t_1|x)}{\phi(y,s|x)} \end{aligned}$$

As a consequence we will call *time-linear Brownian bridge* every stochastic process with the previous finite dimensional distributions, and in particular it is possible to show that in this sense the process

$$Z(t) = x\left(1 - \frac{t}{s}\right) + y\,\frac{t}{s} + \left(W(t) - \frac{t}{s}W(s)\right)$$

always is a time-linear Brownian bridge $x \leftrightarrow y$.

Appendix N
Kinematics of Gaussian Diffusions (Sect. 10.3.1)

To support the notion that the mean acceleration (10.22) proposed in the Sect. 10.3.1 is a good candidate to appear in the Newton equation (10.23), we will provide a few explicit calculations for some processes we became acquainted with all along the previous chapters. To this end we will find expedient to take into account also the possibility—already suggested at the beginning of the Chap. 5—of processes $X(t)$ defined for every real $t \in \boldsymbol{R}$, namely of ***bilateral processes***. In this case however the knowledge of the transition *pdf* $f(x, t|y, s)$ *only in the retarded region* $t > s$ (plus an arbitrary initial condition) turns out to be no longer enough to find the global law of the process: for this purpose we would indeed be required first to extend $f(x, t|y, s)$ in the advanced region $t < s$, then to calculate the laws of the process for every time $t \in \boldsymbol{R}$ starting form the *pdf* at some time s (which anyway should no longer be considered as an *initial* instant) and finally to provide all the joint *pdf*'s through the chain-rule (7.14).

Without a command of the equations also in the advanced region, however, this procedure proves to be a dead end. We have seen indeed in the Sect. 10.2 that, in order to consistently extend the transition functions to the advanced $t < s$ region, we first need an *a priori* knowledge of the *pdf* $f(x, t)$ for every $t \in \boldsymbol{R}$. If however for instance we take an $f_0(x)$ in $t = 0$, the transition laws in the retarded region and the Eq. (7.16) apparently will give back only $f(x, t)$ for $t \geq 0$, while, given its typical diffusive character, it would be preposterous just to extend it backward in time beyond $t = 0$. Being on the other hand impossible to coherently choose $f(x, t)$ for every $t \in \boldsymbol{R}$ in an arbitrary way, this remark once more underline the need for equations holding for every $t \in \boldsymbol{R}$.

Under special conditions, however, such an extension to $t \in \boldsymbol{R}$ can be consistently achieved in a simple way: take for instance the case of a time homogeneous, Markovian, transition *pdf* given in the retarded region $t > s \geq 0$, and suppose that it admits an invariant measure, *be it a probability or not*. We will indeed keep open the possibility that our resulting stationary process $X(t)$ be endowed with a general σ-finite measure, rather than with a strictly finite, probabilistic one. In this case $X(t)$ will be called an ***improper process***, namely a process which is properly defined as

N. Cufaro Petroni, *Probability and Stochastic Processes for Physicists*, UNITEXT for Physics, https://doi.org/10.1007/978-3-030-48408-8

a measurable application from an underlying probabilizable space into a trajectory space, but which is endowed with a measure which is not finite. In particular we will find convenient the use of the σ-finite uniform Lebesgue measures on $\boldsymbol{R}$: think, as an analogy, to the well known case of the *plane waves* in quantum mechanics. Now, if we get such a Markov, time homogeneous process with its invariant measure, we will also be able to extend its transition laws into the advanced region $t < s$ in a natural way, and hence to get the bilateral, stationary process $X(t)$ we was looking for. To achieve this result we simply remark indeed that an invariant measure—being time independent—is trivially extended to $t \in \boldsymbol{R}$; and then that, by means of this invariant (either finite, if it exists, or σ-finite) measure, we will be able to consistently extend the transition *pdf*'s into the advanced region for every $s, t \in \boldsymbol{R}$ in an order whatsoever, so that in the end we will get a stationary, bilateral—but possibly improper—Markov process.

Non-stationary Wiener Process

For a Wiener process $W(t)$ with $t \geq 0$ and $W(0) = 0$, $\boldsymbol{P}$-a.s. we know that

$$f(x,t) = \frac{e^{-\frac{x^2}{2Dt}}}{\sqrt{2\pi Dt}} \sim \mathfrak{N}(0, Dt) \qquad 0 \leq t \tag{N.1}$$

$$f(x,t|y,s) = \frac{e^{-\frac{(x-y)^2}{2D(t-s)}}}{\sqrt{2\pi D(t-s)}} \sim \mathfrak{N}(y, D(t-s)) \qquad 0 \leq s \leq t \tag{N.2}$$

and hence from (1.17) the advanced transition *pdf* for $0 \leq t \leq s$ (namely $0 \leq {}^t/_s \leq 1$) is

$$f(x,t|y,s) = \frac{e^{-\frac{(y-x)^2}{2D(s-t)}}}{\sqrt{2\pi D(s-t)}} \frac{\frac{e^{-\frac{x^2}{2Dt}}}{\sqrt{2\pi Dt}}}{\frac{e^{-\frac{y^2}{2Dt}}}{\sqrt{2\pi Ds}}} = \frac{e^{-\frac{1}{2Ds}\left[\frac{(y-x)^2}{1-t/s}+\frac{x^2}{t/s}-y^2\right]}}{\sqrt{2\pi Ds\left(1-\frac{t}{s}\right)\frac{t}{s}}}$$

$$= \frac{e^{-\frac{\left(x-\frac{t}{s}y\right)^2}{2Ds\left(1-\frac{t}{s}\right)\frac{t}{s}}}}{\sqrt{2\pi Ds\left(1-\frac{t}{s}\right)\frac{t}{s}}} \sim \mathfrak{N}\left(\frac{t}{s}y\,,\; Ds\left(1-\frac{t}{s}\right)\frac{t}{s}\right) \tag{N.3}$$

so that, by summarizing, for every $t \geq 0$, we have

$$f(x,t|y,s) \sim \begin{cases} \mathfrak{N}\left(\frac{t}{s}y,\; Ds\left(1-\frac{t}{s}\right)\frac{t}{s}\right) & 0 \leq t \leq s \\ \mathfrak{N}(y, D(t-s)) & 0 \leq s \leq t \end{cases} \tag{N.4}$$

while for the conditional expectation and variance we get

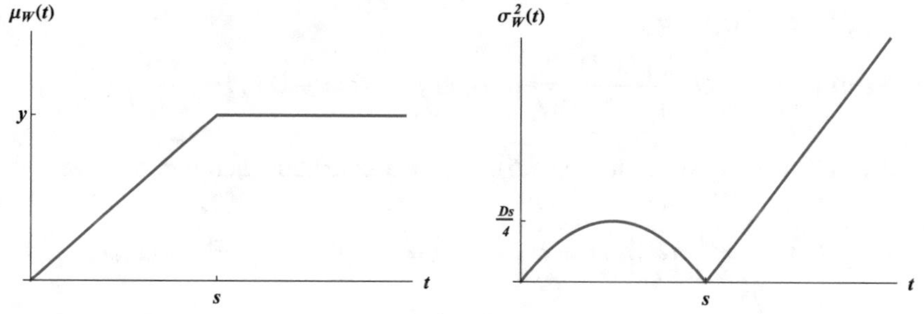

Fig. N.1 Conditional expectation and variance of a Wiener process, with $W(0) = 0$, in both the *advanced* $(0 \le t \le s)$ and the *retarded* $(0 \le s \le t)$ regions

$$\mu_W(t) = \boldsymbol{E}\left[W(t)|W(s) = y\right] = \begin{cases} \frac{t}{s}\, y & 0 \le t \le s \\ y & 0 \le s \le t \end{cases}$$

$$\sigma_W^2(t) = \boldsymbol{V}\left[W(t)|W(s) = y\right] = \begin{cases} Ds\left(1 - \frac{t}{s}\right)\frac{t}{s} & 0 \le t \le s \\ D\,(t - s) & 0 \le s \le t \end{cases}$$

which are plotted in the Fig. N.1.

Since for a Wiener process the coefficients of the Fokker–Planck equation (7.92) are $A(x,t) = 0,\ B(x,t) = D$, we find from (10.12) and (N.1) that

$$\widehat{A}(x,t) = -D\,\partial_x \ln f(x,t) = -D\,\partial_x\left(\frac{-x^2}{2Dt}\right) = \frac{x}{t}$$

and therefore the advanced Eq. (10.11) for $0 \le t \le s$ takes the form

$$\partial_t f(x,t|y,s) = -\partial_x\left[\frac{x}{t}\, f(x,t|y,s)\right] - \frac{D}{2}\,\partial_x^2 f(x,t|y,s)$$

A direct, but tiresome calculation would show that the advanced transition *pdf* (N.3) does in fact satisfy this equation, along with the two extremal conditions

$$f(x,0^+|y,s) = \delta(x) \qquad f(x,s^-|y,s) = \delta(x-y)$$

that apparently implement a Brownian bridge (see Appendix M) between 0 and s, as it is also visible from the behavior of the variance $\sigma_W^2(t)$ in the Fig. N.1.

Proposition N.1 *If, for* $t \ge 0$, $W(t)$ *is a Wiener process with* $W(0) = 0$, *then we have*

$$\overrightarrow{\partial} W(t) = 0 \qquad \overleftarrow{\partial} W(t) = \frac{W(t)}{t} \qquad \overleftrightarrow{\partial}^2 W(t) = -\frac{W(t)}{2t^2} \tag{N.5}$$

Proof Since $A(x,t) = 0$, from (10.27), (10.28) and (N.1) we immediately get

$$\vec{v}(x,t) = 0$$
$$\overleftarrow{v}(x,t) = -D\,\frac{\partial_x f(x,t)}{f(x,t)} = -D\,\partial_x \ln f(x,t) = -D\,\partial_x\left(-\frac{x^2}{2Dt}\right) = \frac{x}{t}$$

and hence the first two results in (N.5). For the acceleration therefore we have

$$\overleftrightarrow{\partial}^2 W(t) = \frac{1}{2}\vec{\partial}\overleftarrow{\partial} W(t) = \vec{\partial}\left[\frac{W(t)}{2t}\right] = \vec{\partial} h(W(t),t) = \vec{h}(W(x),t)$$

where $h(x,t) = x/2t$, and therefore from (10.39)

$$\overleftrightarrow{a}(x,t) = \vec{h}(x,t) = \partial_t h(x,t) + \frac{D}{2}\,\partial_x^2 h(x,t) = -\frac{x}{2t^2}$$

apparently leading to (N.5) ■

As a check we can also directly calculate the mean forward and backward velocities from their definitions (10.20) and (10.21): as for the forward velocity, since the *retarded* increment $W(t+\Delta t) - W(t)$ with $\Delta t > 0$ is $\mathfrak{N}(0, D\Delta t)$ independently from the condition $W(t) = x$, we easily have

$$\vec{v}(x,t) = \lim_{\Delta t\to 0^+} \boldsymbol{E}\left[\frac{W(t+\Delta t) - W(t)}{\Delta t}\,\middle|\, W(t) = x\right] = 0$$

For the backward velocity we must instead remember that, conditionally to the event $W(t) = x$, the advanced $W(t - \Delta t)$ with $\Delta t > 0$ is now distributed as (N.4) namely $\mathfrak{N}\left(\left(1 - \frac{\Delta t}{t}\right)x\,,\ Dt\left(1 - \frac{\Delta t}{t}\right)\frac{\Delta t}{t}\right)$. As a consequence we now have

$$\overleftarrow{v}(x,t) = \lim_{\Delta t\to 0^+} \boldsymbol{E}\left[\frac{W(t) - W(t-\Delta t)}{\Delta t}\,\middle|\, W(t) = x\right] = \lim_{\Delta t\to 0^+}\frac{x - \left(1 - \frac{\Delta t}{t}\right)x}{\Delta t} = \frac{x}{t}$$

As for the acceleration

$$\overleftrightarrow{\partial}^2 W(t) = \frac{1}{2}\vec{\partial}\left[\frac{W(t)}{t}\right]$$

to get the result in (N.5) it is enough to remark that

$$\begin{aligned}
&\boldsymbol{E}\left[\frac{1}{\Delta t}\left(\frac{W(t+\Delta t)}{t+\Delta t} - \frac{W(t)}{t}\right)\middle|\, W(t) = x\right] \\
&\qquad = \boldsymbol{E}\left[\frac{W(t+\Delta t) - W(t)}{\Delta t(t+\Delta t)} - \frac{W(t)}{t(t+\Delta t)}\,\middle|\, W(t) = x\right] \\
&\qquad = -\frac{x}{t(t+\Delta t)} \underset{\Delta t\to 0^+}{\longrightarrow} -\frac{x}{t^2}
\end{aligned}$$

It is immediate to see moreover that for our Wiener process we have

$$\boldsymbol{E}\left[\overleftarrow{\partial} W(t)\right] = \boldsymbol{E}\left[\overrightarrow{\partial} W(t)\right] = 0 \qquad \boldsymbol{V}\left[\overrightarrow{\partial} W(t)\right] = 0 \quad \boldsymbol{V}\left[\overleftarrow{\partial} W(t)\right] = \frac{D}{t}$$

$$\boldsymbol{E}\left[\overleftrightarrow{\partial}^2 W(t)\right] = 0 \quad \boldsymbol{V}\left[\overleftrightarrow{\partial}^2 W(t)\right] = \frac{D}{4t^3}$$

and therefore also

$$\lim_{t\to+\infty} \text{-}ms\ \overleftrightarrow{\partial}^2 W(t) = 0$$

namely, after a transient, the mean acceleration of a Wiener process asymptotically vanishes in *ms*.

Bilateral, Stationary, Improper Wiener Process

Strictly speaking, for $t \to +\infty$, a Wiener process $W(t)$ does not converge toward a process because the limit measure flattens to a uniform, not finite one. However we already pointed out that we will take into account also *improper* processes endowed with a suitable (not finite) measure: to this end we first show that the constant density of a uniform, Lebesgue measure

$$f_0(x) = g > 0$$

is indeed invariant for our Wiener transition *pdf* (N.2):

$$f(x,t) = \int_{\boldsymbol{R}} f(x,t|y,0)\, f_0(y)\, dy = g \int_{\boldsymbol{R}} \frac{e^{-\frac{(x-y)^2}{2Dt}}}{\sqrt{2\pi Dt}}\, dy = g = f_0(x) \tag{N.6}$$

This means first of all that a uniform Lebesgue measure represents an invariant (albeit not finite) measure for our improper process. Then we remark that the advanced transition *pdf* can now be found, and trivially extended to $-\infty < t \le s < +\infty$ as

$$f(x,t|y,s) = \frac{f(y,s|x,t)\, g}{g} = f(y,s|x,t) = \frac{e^{-\frac{(y-x)^2}{2D(s-t)}}}{\sqrt{2\pi D(s-t)}} \tag{N.7}$$

and hence by summarizing we have for every $t, s \in \boldsymbol{R}$

$$f(x,t|y,s) = \left\{ \begin{array}{ll} \frac{e^{-(x-y)^2/2D(t-s)}}{\sqrt{2\pi D(t-s)}} & -\infty < s \le t < +\infty \\ \frac{e^{-(y-x)^2/2D(s-t)}}{\sqrt{2\pi D(s-t)}} & -\infty < t \le s < +\infty \end{array} \right\} \sim \mathfrak{N}\,(y, D|t-s|)$$

which represents the complete transition law of our bilateral, stationary, improper Wiener process. Remark that, while expectation and variance of the improper process with the uniform Lebesgue measure are not defined, for the transition *pdf* we now have for every $s, t \in \boldsymbol{R}$

$$\boldsymbol{E}\,[W(t)|W(s) = y] = y \qquad \boldsymbol{V}\,[W(t)|W(s) = y] = D|t-s|$$

Since again in the Fokker–Planck equation (7.92) we have $A = 0$, from (10.12) and (N.6) we find

$$\widehat{A}(x,t) = -D\,\partial_x \ln g(x,t) = -D\,\partial_x \ln g = 0$$

so that now, in the advanced region $t \leq s$, the Fokker–Planck equation (10.11) of our bilateral, stationary, improper Wiener process takes the form

$$\partial_t f(x,t|y,s) = -\frac{D}{2}\,\partial_x^2 f(x,t|y,s)$$

and it is easy to check that the advanced transition *pdf* (N.7) satisfies this equation with the final condition

$$f(x,s^-|y,s) = \delta(x-y)$$

Proposition N.2 *If $W(t)$ is a bilateral, stationary, improper Wiener process, then we have*

$$\overrightarrow{\partial} W(t) = 0 \qquad \overleftarrow{\partial} W(t) = 0 \qquad \overleftrightarrow{\partial^2} W(t) = 0 \tag{N.8}$$

Proof Since in the Fokker–Planck equation of $W(t)$ it is $A = 0$, these results immediately follow from (10.27) (10.28) and (N.6). They can however also be obtained by direct calculation: while indeed the forward derivative (N.5) apparently coincides with that in (N.5), for the backward one we have that, conditionally to $W(t) = x$, the advanced $W(t - \Delta t)$ with $\Delta t > 0$ is distributed as $\mathfrak{N}\,(x,\ D\Delta t)$ so that

$$\overleftarrow{v}(x,t) = \lim_{\Delta t \to 0^+} \boldsymbol{E}\left[\frac{W(t) - W(t-\Delta t)}{\Delta t}\,\bigg|\, W(t) = x\right] = \lim_{\Delta t \to 0^+} \frac{x - x}{\Delta t} = 0$$

and consequently also the acceleration immediately vanish. ■

From these results it is then apparent that in the case of a stationary (improper) Wiener process the mean acceleration identically vanish, and therefore the process satisfies a Newton equation (10.23) with $F(t) = 0$: in other words the Wiener process describes the movement of a free particle, as it was expected. By comparing next the results (N.5) and (N.8) we see moreover that the mean velocities and acceleration of a non-stationary Wiener process converge to the corresponding quantities of the stationary process, and hence describe the movements of an asymptotically free particle.

Ornstein-Uhlenbeck Process

We preliminarily recall that the Fokker–Planck equation (7.95) of an Ornstein-Uhlenbeck process $X(t)$ is

$$\partial_t f(x,t) = \alpha\partial_x\,[x f(x,t)] + \alpha\beta^2\partial_x^2 f(x,t) \tag{N.9}$$

with the coefficients

$$A(x,t) = -\alpha x \qquad B(x,t) = D = 2\alpha\beta^2 \tag{N.10}$$

and that its solution for $(t \geq s)$ with the initial condition

$$f(x, s^+|y, s) = \delta(x-y) \tag{N.11}$$

is the transition *pdf*

$$f(x,t|y,s) = \frac{e^{-\frac{\left(x-ye^{-\alpha(t-s)}\right)^2}{2\beta^2\left(1-e^{-2\alpha(t-s)}\right)}}}{\sqrt{2\pi\beta^2\left(1-e^{-2\alpha(t-s)}\right)}} \sim \mathfrak{N}\left(ye^{-\alpha(t-s)}, \beta^2\left(1-e^{-2\alpha(t-s)}\right)\right) \tag{N.12}$$

in the retarded region $t \geq s \geq 0$.

Non-stationary Ornstein-Uhlenbeck Process
The non-stationary Ornstein-Uhlenbeck process $X(t)$, with $X(0) = 0$, has the *pdf*

$$f(x,t) = \frac{e^{-\frac{x^2}{2\beta^2\left(1-e^{-2\alpha t}\right)}}}{\sqrt{2\pi\beta^2\left(1-e^{-2\alpha t}\right)}} \sim \mathfrak{N}\left(0, \beta^2\left(1-e^{-2\alpha t}\right)\right) \tag{N.13}$$

solution of (N.9) for $t \geq 0$ with the initial condition $f(x, 0^+) = f_0(x) = \delta(x)$. From the *pdf*'s (N.12) and (N.13) we can therefore deduce the form of the transition *pdf* (1.17) in the advanced region $0 \leq t \leq s$

$$\begin{aligned} f(x,t|y,s) &= \frac{e^{-\frac{\left(y-xe^{-\alpha(s-t)}\right)^2}{2\beta^2\left(1-e^{-2\alpha(s-t)}\right)}}}{\sqrt{2\pi\beta^2\left(1-e^{-2\alpha(s-t)}\right)}} \frac{e^{-\frac{x^2}{2\beta^2\left(1-e^{-2\alpha t}\right)}}}{\sqrt{2\pi\beta^2\left(1-e^{-2\alpha t}\right)}} \frac{\sqrt{2\pi\beta^2\left(1-e^{-2\alpha s}\right)}}{e^{-\frac{y^2}{2\beta^2\left(1-e^{-2\alpha s}\right)}}} \\ &= \frac{e^{-\frac{\left(x-ye^{-\alpha(s-t)}\frac{1-e^{-2\alpha t}}{1-e^{-2\alpha s}}\right)^2}{2\beta^2\left(1-e^{-2\alpha(s-t)}\right)\frac{1-e^{-2\alpha t}}{1-e^{-2\alpha s}}}}}{\sqrt{2\pi\beta^2\left(1-e^{-2\alpha(s-t)}\right)\frac{1-e^{-2\alpha t}}{1-e^{-2\alpha s}}}} \\ &\sim \mathfrak{N}\left(ye^{-\alpha(s-t)}\frac{1-e^{-2\alpha t}}{1-e^{-2\alpha s}}, \beta^2\left(1-e^{-2\alpha(s-t)}\right)\frac{1-e^{-2\alpha t}}{1-e^{-2\alpha s}}\right) \end{aligned} \tag{N.14}$$

so that, by summarizing, we have for every $s, t \geq 0$

$$f(x,t|y,s) \sim \begin{cases} \mathfrak{N}\left(ye^{-\alpha(s-t)}\frac{1-e^{-2\alpha t}}{1-e^{-2\alpha s}}, \beta^2\left(1-e^{-2\alpha(s-t)}\right)\frac{1-e^{-2\alpha t}}{1-e^{-2\alpha s}}\right) & 0 \leq t \leq s \\ \mathfrak{N}\left(ye^{-\alpha(t-s)}, \beta^2\left(1-e^{-2\alpha(t-s)}\right)\right) & 0 \leq s \leq t \end{cases} \tag{N.15}$$

As a consequence the conditional expectation and variance are

$$\mu_X(t) = \boldsymbol{E}\left[X(t)|X(s)=y\right] = \begin{cases} ye^{-\alpha(s-t)}\frac{1-e^{-2\alpha t}}{1-e^{-2\alpha s}} & 0 \le t \le s \\ ye^{-\alpha(t-s)} & 0 \le s \le t \end{cases}$$

$$\sigma_X^2(t) = \boldsymbol{V}\left[X(t)|X(s)=y\right] = \begin{cases} \beta^2\left(1-e^{-2\alpha(s-t)}\right)\frac{1-e^{-2\alpha t}}{1-e^{-2\alpha s}} & 0 \le t \le s \\ \beta^2\left(1-e^{-2\alpha(t-s)}\right) & 0 \le s \le t \end{cases}$$

with their behavior displayed in the Fig. N.2. It is apparent therefore that also in this case the advanced *pdf* implements a Brownian bridge between the times 0 and s (see Appendix M). It is also easy to see from (10.12), (N.10) and (N.13) that now

$$\begin{aligned}\widehat{A}(x,t) &= -\alpha x - 2\alpha\beta^2\,\partial_x \ln f(x,t) = -\alpha x - \alpha\partial_x\left(\frac{-x^2}{1-e^{-2\alpha t}}\right)\\ &= -\alpha x + \frac{2\alpha x}{1-e^{-2\alpha t}} = \alpha x\,\frac{1+e^{-2\alpha t}}{1-e^{-2\alpha t}} = \frac{\alpha x}{\tanh\alpha t}\end{aligned}$$

so that the advanced Eq. (10.11) takes the form

$$\partial_t f(x,t|y,s) = -\partial_x\left[\frac{\alpha x}{\tanh\alpha t}f(x,t|y,s)\right] - \alpha\beta^2\partial_x^2 f(x,t|y,s)$$

and it would again be straightforward, but tiresome, to prove by direct calculation that in the advanced region $0 \le t \le s$ the transition *pdf* (N.14) is in fact its solution with the extremal conditions $f(x,0^+|y,s) = \delta(x)$ and $f(x,s^-|y,s) = \delta(x-y)$.

Proposition N.3 *If $X(t)$ is a non-stationary Ornstein-Uhlenbeck process with the condition $X(0) = 0$, then we have*

$$\overrightarrow{\partial}X(t) = -\alpha X(t) \qquad \overleftarrow{\partial}X(t) = \frac{\alpha X(t)}{\tanh\alpha t} \qquad \overleftrightarrow{\partial^2}X(t) = -\alpha^2 X(t)\frac{1+\sinh 2\alpha t}{2\sinh^2\alpha t} \tag{N.16}$$

Proof From (10.27) and (10.28), with (N.10) and (N.13), we immediately get

$$\begin{aligned}\overrightarrow{v}(x,t) &= -\alpha x\\ \overleftarrow{v}(x,t) &= -\alpha x - 2\alpha\beta^2\,\partial_x \ln f(x,t) = -\alpha x + \alpha\partial_x\left(\frac{x^2}{1-e^{-2\alpha t}}\right)\\ &= \alpha x\,\frac{1+e^{-2\alpha t}}{1-e^{-2\alpha t}} = \frac{\alpha x}{\tanh\alpha t}\end{aligned}$$

while for the acceleration we have from (10.39)

$$\overleftrightarrow{\partial^2}X(t) = \frac{\alpha}{2}\left[\overrightarrow{\partial}\left(\frac{X(t)}{\tanh\alpha t}\right) - \overleftarrow{\partial}X(t)\right] = \frac{\alpha}{2}\left[\overrightarrow{h}(X(t),t) - \frac{\alpha X(t)}{\tanh\alpha t}\right]$$

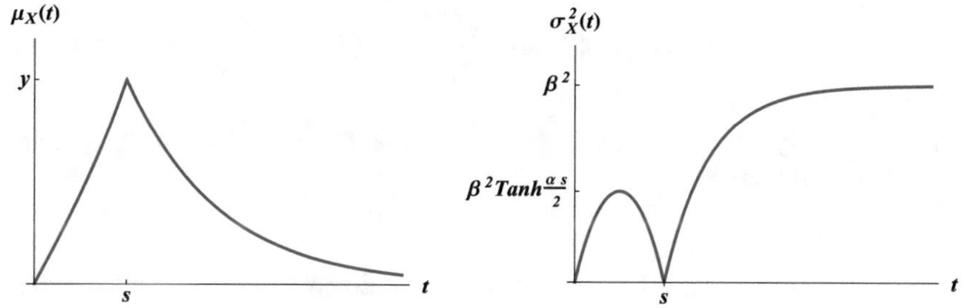

Fig. N.2 Expectation and variance of a non-stationary Ornstein-Uhlenbeck process in both the *advanced* ($0 \le t \le s$) and the *retarded* ($0 \le s \le t$) regions

with $h(x,t) = \frac{x}{\tanh \alpha t}$ and

$$\begin{aligned}\overrightarrow{h}(x,t) &= \partial_t h(x,t) - \alpha x\, \partial_x h(x,t) = \partial_t \left(\frac{x}{\tanh \alpha t}\right) - \frac{\alpha x}{\tanh \alpha t}\\ &= -\alpha x \left(\frac{1}{\sinh^2 \alpha t} + \frac{1}{\tanh \alpha t}\right)\end{aligned}$$

so that

$$\overleftrightarrow{\partial}^2 X(t) = -\alpha^2 X(t)\left(\frac{1}{\tanh \alpha t} + \frac{1}{2\sinh^2 \alpha t}\right) = -\alpha^2 X(t)\frac{1 + 2\cosh \alpha t \sinh \alpha t}{2\sinh^2 \alpha t}$$

leading finally to (N.16) ∎

Then for a non-stationary Ornstein-Uhlenbeck process we have

$$\boldsymbol{E}\left[\overleftarrow{\partial} X(t)\right] = \boldsymbol{E}\left[\overrightarrow{\partial} X(t)\right] = \boldsymbol{E}\left[\overleftrightarrow{\partial}^2 X(t)\right] = 0$$

$$\begin{aligned}\boldsymbol{V}\left[\overrightarrow{\partial} X(t)\right] &= \alpha^2\beta^2(1 - e^{-2\alpha t}) \underset{t\to+\infty}{\longrightarrow} \alpha^2\beta^2\\ \boldsymbol{V}\left[\overleftarrow{\partial} X(t)\right] &= \frac{\alpha^2\beta^2(1 - e^{-2\alpha t})}{\tanh^2 \alpha t} \underset{t\to+\infty}{\longrightarrow} \alpha^2\beta^2\\ \boldsymbol{V}\left[\overleftrightarrow{\partial}^2 X(t)\right] &= \alpha^4\left(\frac{1 + \sinh 2\alpha t}{2\sinh^2 \alpha t}\right)^2 \beta^2(1 - e^{-2\alpha t}) \underset{t\to+\infty}{\longrightarrow} \alpha^4\beta^2\end{aligned}$$

namely, after a transient, the mean velocities and acceleration of the asymptotic process continue to fluctuate around zero with constant variances.

Bilateral, Stationary Ornstein-Uhlenbeck Process

A bilateral, stationary Ornstein-Uhlenbeck process $X(t)$ has the invariant *pdf*

$$f(x,t) = f_0(x) = \frac{e^{-\frac{x^2}{2\beta^2}}}{\sqrt{2\pi\beta^2}} \sim \mathfrak{N}\left(0, \beta^2\right) \tag{N.17}$$

defined for every $t \in \boldsymbol{R}$ as a natural extension of the solution of (N.9) with $t \geq 0$ and the initial condition

$$f(x, 0^+) = f_0(x) \sim \mathfrak{N}\left(0, \beta^2\right)$$

while the transition *pdf* (N.12) still satisfies the Eq. (N.9) in the retarded region $s \leq t$ with the initial condition (N.11). From the retarded transition *pdf* (N.12) and the invariant *pdf* (N.17) we can now deduce the form of the advanced transition *pdf* (1.17) for $t \leq s$

$$\begin{aligned} f(x,t|y,s) &= \frac{e^{-\frac{\left(y-xe^{-\alpha(s-t)}\right)^2}{2\beta^2\left(1-e^{-2\alpha(s-t)}\right)}}}{\sqrt{2\pi\beta^2\left(1-e^{-2\alpha(s-t)}\right)}} \frac{e^{-\frac{x^2}{2\beta^2}}}{\sqrt{2\pi\beta^2}} \frac{\sqrt{2\pi\beta^2}}{e^{-\frac{y^2}{2\beta^2}}} \\ &= \frac{e^{-\frac{\left(y-xe^{-\alpha(s-t)}\right)^2+(x^2-y^2)\left(1-e^{-2\alpha(s-t)}\right)}{2\beta^2\left(1-e^{-2\alpha(s-t)}\right)}}}{\sqrt{2\pi\beta^2\left(1-e^{-2\alpha(s-t)}\right)}} = \frac{e^{-\frac{\left(x-ye^{-\alpha(s-t)}\right)^2}{2\beta^2\left(1-e^{-2\alpha(s-t)}\right)}}}{\sqrt{2\pi\beta^2\left(1-e^{-2\alpha(s-t)}\right)}} \\ &\sim \mathfrak{N}\left(ye^{-\alpha(s-t)},\, \beta^2\left(1-e^{-2\alpha(s-t)}\right)\right) \end{aligned} \tag{N.18}$$

so that, by summarizing, in both the advanced and the retarded regions we have

$$f(x,t|y,s) \sim \mathfrak{N}\left(ye^{-\alpha|t-s|},\, \beta^2\left(1-e^{-2\alpha|t-s|}\right)\right) \tag{N.19}$$

Remark that while for the stationary process with *pdf* (N.17) we apparently have $\boldsymbol{E}\left[X(t)\right] = 0$ and $\boldsymbol{V}\left[X(t)\right] = \beta^2$ for every $t \in \boldsymbol{R}$, for the transition *pdf* (N.19) we find for every $s, t \in \boldsymbol{R}$

$$\boldsymbol{E}\left[X(t)|X(s)=y\right] = ye^{-\alpha|t-s|} \qquad \boldsymbol{V}\left[X(t)|X(s)=y\right] = \beta^2\left(1-e^{-2\alpha|t-s|}\right)$$

It is easy to see moreover that from (10.12), (N.10) and (N.17) we now have

$$\widehat{A}(x,t) = -\alpha x - 2\alpha\beta^2\partial_x \ln f_0(x) = \alpha x$$

so that the Fokker–Planck equation of a stationary Ornstein-Uhlenbeck process in the advanced region $t \leq s$ becomes

$$\partial_t f(x,t|y,s) = -\alpha\partial_x\left[xf(x,t|y,s)\right] - \alpha\beta^2\partial_x^2 f(x,t|y,s)$$

We will neglect however to check by direct calculation that the transition *pdf* (N.18) is a solution of the said equation with the final condition

$$f(x, s^-|y, s) = \delta(x - y)$$

Proposition N.4 *If $X(t)$ is a bilateral, stationary Ornstein-Uhlenbeck process then we have*

$$\overrightarrow{\partial} X(t) = -\alpha X(t) \qquad \overleftarrow{\partial} X(t) = \alpha X(t) \qquad \overleftrightarrow{\partial}^2 X(t) = -\alpha^2 X(t) \tag{N.20}$$

Proof From (10.27) and (10.28), with (N.10) and (N.17), we immediately get

$$\begin{aligned}\overrightarrow{v}(x, t) &= -\alpha x \\ \overleftarrow{v}(x, t) &= -\alpha x - 2\alpha\beta^2\, \partial_x \ln f_0(x) = \alpha x\end{aligned}$$

while for the acceleration we have

$$\overleftrightarrow{\partial}^2 X(t) = \frac{\alpha}{2}\left[\overrightarrow{\partial} X(t) - \overleftarrow{\partial} X(t)\right] = -\alpha^2 X(t)$$

namely (N.20) ■

As a consequence for a bilateral, stationary Ornstein-Uhlenbeck process we also have

$$\boldsymbol{E}\left[\overleftarrow{\partial} X(t)\right] = \boldsymbol{E}\left[\overrightarrow{\partial} X(t)\right] = \boldsymbol{E}\left[\overleftrightarrow{\partial}^2 X(t)\right] = 0$$

$$\boldsymbol{V}\left[\overrightarrow{\partial} X(t)\right] = \boldsymbol{V}\left[\overleftarrow{\partial} X(t)\right] = \alpha^2\beta^2 \qquad \boldsymbol{V}\left[\overleftrightarrow{\partial}^2 X(t)\right] = \alpha^4\beta^2$$

Looking at the results (N.16) and (N.20) it is apparent once more that the mean velocities and acceleration of the non-stationary process exponentially converge in *ms* to the corresponding quantities of the stationary process. At variance with the Wiener case, however, these asymptotic velocities and acceleration do not vanish, but rather fluctuate around zero with constant variances. The behavior of the stationary process, on the other hand, hints at the presence of an elastic external binding force as apparently shown by (N.20): the stationary process will satisfy indeed a Newton equation (10.23) with $F(t) = -\alpha^2 X(t)$. Since we already know from the Example 9.6 that the stationary Ornstein-Uhlenbeck process also is the Smoluchowski approximation of an harmonically bound Brownian motion, this supports once again the choice of (10.22) as a mean acceleration.

Appendix O
Substantial Operators (Sect. 10.3.3)

The forward and backward substantial operators adopted in the Proposition 10.4 are generalizations of the so-called *substantial derivatives* of the fluid mechanics (see [21] p. 8 and [22] p. 83): if $A(x,t)$ is a velocity field, we first define the *streamlines* $x(t)$ of a classical dynamical system as the solutions of the *ODE*

$$\dot{x}(t) = A(x(t), t) \tag{O.1}$$

and then a function $y(t) = h(x(t), t)$ that represents the evolution of a quantity $h(x,t)$ along a streamline $x(t)$. Its total derivative then is

$$\dot{y}(t) = \partial_t h(x(t), t) + \dot{x}(t)\,\partial_x h(x(t), t) = h_t(x(t), t) + A(x(t), t)\,h_x(x(t), t) = \mathcal{D}h(x(t), t)$$

where we have introduced the shorthand notation

$$\mathcal{D}h(x,t) = h_t(x,t) + A(x,t)h_x(x,t) \qquad\qquad \mathcal{D} \equiv \partial_t + A\partial_x \tag{O.2}$$

The new symbol $\mathcal{D}$ is usually called *substantial derivative* and must not be mistaken for a plain (partial or total) derivative: the derivative $\mathcal{D}h(x,t)$ can indeed be properly defined only when we also take into account the velocity field $A(x,t)$. In other words a substantial derivative is the particular form taken by the total derivative $\dot{y}(t)$ of $y(t) = h(x(t), t)$ when we throw in the information that $x(t)$ is a streamline of the dynamical system (O.1).

It is possible now to generalize these ideas to the case of a diffusion process whose trajectories $X(t)$ will play a role analog to that of the streamlines $x(t)$ when the *SDE* (8.26) replaces the *ODE* (O.1): remember that according to the Proposition 8.19 the coefficient $a(x,t)$ of the *SDE* coincides with the $A(x,t)$ of the Fokker–Planck equation. We must take into account however that here we will have to do with two (forward and backward) mean derivatives of the compound process $Y(t) = h(X(t), t)$, and therefore it is not surprising to find the results of the Proposition 10.4

N. Cufaro Petroni, *Probability and Stochastic Processes for Physicists*, UNITEXT for Physics, https://doi.org/10.1007/978-3-030-48408-8

of which we will give here a simple proof. For the mean *forward stream* we have indeed

$$
\begin{aligned}
\overrightarrow{h}(x,t) &= \lim_{\Delta t\to 0^+} \boldsymbol{E}\left[\frac{h\left(X(t+\Delta t),t+\Delta t\right)-h\left(X(t),t\right)}{\Delta t}\,\middle|\, X(t)=x\right] \\
&= \lim_{\Delta t\to 0^+} \frac{\boldsymbol{E}\left[h\left(X(t+\Delta t),t+\Delta t\right)\mid X(t)=x\right]-h(x,t)}{\Delta t} \\
&= \lim_{\Delta t\to 0^+} \frac{1}{\Delta t}\left[\int_{\boldsymbol{R}} h(y,t+\Delta t)f(y,t+\Delta t|x,t)\,dy - h(x,t)\right] \\
&= \lim_{\Delta t\to 0^+} \frac{1}{\Delta t}\left\{\int_{\boldsymbol{R}} [h(y,t)+\partial_t h(y,t)\Delta t + o(\Delta t)]\, f(y,t+\Delta t|x,t)\,dy \right. \\
&\qquad\qquad \left. - \int_{\boldsymbol{R}} h(y,t)\delta(y-x)\,dy\right\} \\
&= \int_{\boldsymbol{R}} \partial_t h(y,t)\delta(y-x)\,dy \\
&\qquad + \lim_{\Delta t\to 0^+}\int_{\boldsymbol{R}} h(y,t)\frac{f(y,t+\Delta t|x,t)-f(y,t^+|x,t)}{\Delta t}\,dy \\
&= \partial_t h(x,t) + \int_{\boldsymbol{R}} h(y,t)[\partial_s f(y,s|x,t)]_{s=t^+}dy
\end{aligned}
$$

and then the result (10.39) stems from the same line of reasoning leading to (10.27) of Proposition 10.3 when we replace x with $h(x,t)$. In the same vein, for the mean *backward stream* we have

$$
\begin{aligned}
\overleftarrow{h}(x,t) &= \lim_{\Delta t\to 0^+} \boldsymbol{E}\left[\frac{h\left(X(t),t\right)-h\left(X(t-\Delta t),t-\Delta t\right)}{\Delta t}\,\middle|\, X(t)=x\right] \\
&= \lim_{\Delta t\to 0^+} \frac{h(x,t)-\boldsymbol{E}\left[h\left(X(t-\Delta t),t-\Delta t\right)\mid X(t)=x\right]}{\Delta t} \\
&= \lim_{\Delta t\to 0^+} \frac{1}{\Delta t}\left[h(x,t)-\int_{\boldsymbol{R}} h(y,t-\Delta t)f(y,t-\Delta t|x,t)\,dy\right] \\
&= \lim_{\Delta t\to 0^+} \frac{1}{\Delta t}\left\{\int_{\boldsymbol{R}} h(y,t)\delta(y-x)\,dy \right. \\
&\qquad \left. - \int_{\boldsymbol{R}} [h(y,t)-\partial_t h(y,t)\Delta t + o(\Delta t)]\, f(y,t-\Delta t|x,t)\,dy\right\} \\
&= \int_{\boldsymbol{R}} \partial_t h(y,t)\delta(y-x)\,dy \\
&\qquad + \lim_{\Delta t\to 0^+}\int_{\boldsymbol{R}} h(y,t)\frac{f(y,t^-|x,t)-f(y,t-\Delta t|x,t)}{\Delta t}\,dy \\
&= \partial_t h(x,t) + \int_{\boldsymbol{R}} h(y,t)[\partial_s f(y,s|x,t)]_{s=t^-}dy
\end{aligned}
$$

and the result (10.40) follows again along the same line of reasoning leading to (10.28) of Proposition 10.3, when we replace x with $h(x,t)$.

Appendix P
Constant Diffusion Coefficients (Sect. 10.4)

In order to show that we can always suppose to have a Fokker–Planck equation of the form (10.48) with a constant diffusion coefficient, it will be expedient to take advantage of the change of variable Itō formula (8.28) in the Corollary 8.17. We will first remember to this purpose that according to the Propositon 8.19 the *pdf* $f(x,t)$ of a diffusion processes $X(t)$, solution of the following Itō *SDE* with a given Wiener process $W(t)$ of diffusion coefficient D

$$dX(t) = a(X(t),t)\,dt + b(X(t),t)\,dW(t) \tag{P.1}$$

in turn satisfies the forward Fokker–Planck equation (10.9) that we reproduce here

$$\partial_t f(x,t) = -\partial_x\left[A(x,t)f(x,t)\right] + \frac{1}{2}\,\partial_x^2\left[B(x,t)f(x,t)\right]$$

where we also have

$$A(x,t) = a(x,t) \qquad\qquad B(x,t) = D\,b^2(x,t)$$

Is is indeed possible to prove (see [20] Chap. 3 p. 70) even a reciprocal form of the Proposition 8.19, namely that if $X(t)$ is a diffusion process defined for given $A(x,t)$ and $B(x,t) \geq 0$ through the Fokker–Planck equation (10.9), then—under rather mild conditions on A and B—there exists a Wiener process $W(t)$ such that $X(t)$ satisfies the *SDE* (P.1) with $a(x,t) = A(x,t)$ and $b(x,t) = \sqrt{D^{-1}B(x,t)} \geq 0$. As a consequence of this reciprocity we will be able to discuss a change of variable of a Fokker–Planck equation through the corresponding transformation of the associated *SDE* provided by the Itō formula.

According to the Corollary 8.17, a transformation $y = g(x,t)$, admitting the inverse $x = h(y,t)$, will define a new process $Y(t) = g\big(X(t),t\big)$ solution of the *SDE*

$$dY(t) = \widetilde{a}\big(Y(t),t\big)\,dt + \widetilde{b}\big(Y(t),t\big)\,dW(t)$$

N. Cufaro Petroni, *Probability and Stochastic Processes for Physicists*, UNITEXT for Physics, https://doi.org/10.1007/978-3-030-48408-8

where the transformed coefficients are

$$\widetilde{a}(y,t) = \Big[g_t(x,t) + a(x,t)g_x(x,t) + \frac{D}{2}b^2(x,t)g_{xx}(x,t\Big]_{x=h(y,t)}$$

$$\widetilde{b}(y,t) = \big[b(x,t)g_x(x,t)\big]_{x=h(y,t)}$$

When $b(x,t) > 0$ we can in particular implement the following monotone and invertible transformation

$$y = g(x,t) = \int \frac{dx}{b(x,t)}$$

with

$$g_t(x,t) = -\int \frac{b_t(x,t)}{b^2(x,t)}\,dx \qquad g_x(x,t) = \frac{1}{b(x,t)} \qquad g_{xx}(x,t) = -\frac{b_x(x,t)}{b^2(x,t)}$$

As a consequence we immediately get for the diffusive term $\widetilde{b}(y,t) = 1$, while the new drift coefficient is

$$\widetilde{a}\,(y,t) = -\Big[\int \frac{b_t(x,t)}{b^2(x,t)}\,dx - \frac{a(x,t)}{b(x,t)} + \frac{D}{2}\,b_x(x,t)\Big]_{x=g(y,t)}$$

so that the transformed *SDE* boils down to

$$dY(t) = \widetilde{a}\big(Y(t),t\big)\,dt + dW(t)$$

It is apparent then that the *pdf* $f_Y(y,t)$ of the new process $Y(t)$ will be now a solution of a new Fokker–Planck equation with a constant diffusion coefficient

$$\partial_t f_Y(y,t) = -\partial_y\big[\widetilde{A}(y,t) f_Y(y,t)\big] + \frac{D}{2}\,\partial_y^2 f_Y(y,t)$$

where $\widetilde{A}(y,t) = \widetilde{a}(y,t)$, while $\widetilde{B}(y,t) = D\,\widetilde{b}^2(y,t) = D$. This conclusively shows that in most cases—at the cost of some possible extra complication in the drift term—we will be able to transform the Fokker–Planck equation of our process into an equation with constant diffusion coefficient by means of a suitable variable transformation, so that the restriction imposed to our systems by the adoption of the Eq. (10.48) should not be deemed too severe.

Appendices References

1. Accardi, L., Fedullo, A.: On the statistical meaning of complex numbers in quantum mechanics. Lett. N. Cim. **34**, 161 (1982)
2. Bell, J.S.: On the Einstein Podolsky Rosen Paradox. Physics **1**, 195 (1964)
3. Shiryaev, A.N.: Probability. Springer, New York (1996)
4. Bertrand, J.: Calcul des Probabilités. Gauthier-Villars, Paris (1889)
5. Gardiner, C.W.: Handbook of Stochastic Methods. Springer, Berlin (1997)
6. Bernoulli, J.: Ars Coniectandi. Thurneysen, Basel (1713)
7. de Moivre, A.: The Doctrine of Chances. Woodfall, London (1738)
8. de Laplace, P.S.: Théorie analytique des probabilités. Courcier, Paris (1812)
9. Papoulis, A.: Probability. Random Variables and Stochastic Processes. McGraw Hill, Boston (2002)
10. van Kampen, N.G.: Stochastic Processes in Physics and Chemistry. North-Holland, Amsterdam (1992)
11. Lévy, P.: Exemples de processus pseudo-markoviens. C. R. Acad. Sci. Paris **228**, 2204 (1949)
12. Feller, W.: Non-Markovian processes with the semigroup property. Ann. Math. Stat. **30**, 1252 (1959)
13. Parzen, E.: Stochastic Processes. Holden-Day, San Francisco (1962)
14. Mandelbrot, B.B., van Ness, J.W.: SIAM Review **10**, 422 (1968)
15. Samorodnitsky, G., Taqqu, M.S.: Stable non-Gaussian Random Processes. Chapman&Hall/CRC, Boca Raton (2000)
16. Dieker, A.: http://www.columbia.edu/~ad3217/fbm.html
17. Neckel, T., Rupp, F.: Random Differential Equations in Scientific Computing. Versita, London (2013)
18. Karatzas, I., Shreve, S.E.: Brownian Motion and Stochastic Calculus. Springer, Berlin (1991)
19. Shreve, S.E.: Stochastic Calculus for Finance, vol. I-II. Springer, New York (2004)
20. Gihman, I.I., Skorohod, A.V.: Stochastic Differential Equations. Springer, Berlin (1972)
21. Landau, L.D., Lifshitz, E.M.: Fluid Mechanics. Pergamon, Oxford (1987)
22. Bird, R.B., Stewart, W.E., Lightfoot, E.N.: Transport Phenomena. Wiley, New York (2002)

N. Cufaro Petroni, *Probability and Stochastic Processes for Physicists*, UNITEXT for Physics, https://doi.org/10.1007/978-3-030-48408-8

Index

N. Cufaro Petroni, *Probability and Stochastic Processes for Physicists*, UNITEXT for Physics, https://doi.org/10.1007/978-3-030-48408-8

Printed in the United States
by Baker & Taylor Publisher Services